Lee G. Anderson

Return Please

Bruce,
as you requested
you may borrow
this as an
example

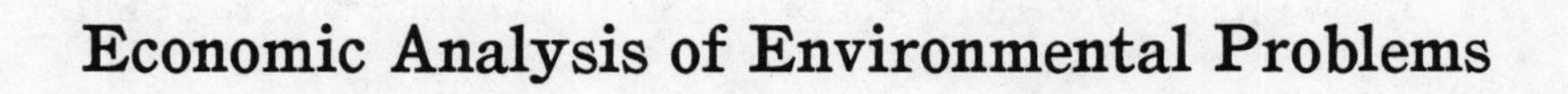

Economic Analysis of Environmental Problems

NATIONAL BUREAU OF ECONOMIC RESEARCH

UNIVERSITIES–NATIONAL BUREAU CONFERENCE SERIES

1. Problems in the Study of Economic Growth (in mimeograph)
2. Conference on Business Cycles
3. Conference on Research in Business Finance
4. Regularization of Business Investment
5. Business Concentration and Price Policy
6. Capital Formation and Economic Growth
7. Policies to Combat Depression
8. The Measurement and Behavior of Unemployment
9. Problems in International Economics
10. The Quality and Economic Significance of Anticipations Data
11. Demographic and Economic Change in Developed Countries
12. Public Finances: Needs, Sources, and Utilization
13. The Rate and Direction of Inventive Activity: Economic and Social Factors
14. Aspects of Labor Economics
15. Investment in Human Beings
16. The State of Monetary Economics
17. Transportation Economics
18. Determinants of Investment Behavior
19. National Economic Planning
20. Issues in Defense Economics
21. The Role of Agriculture in Economic Development
22. The Technology Factor in International Trade
23. The Analysis of Public Output
24. International Mobility and Movement of Capital
25. Conference on Secular Inflation
26. Economic Analysis of Environmental Problems

Economic Analysis of Environmental Problems

Edited by
Edwin S. Mills

A Conference of the Universities–National Bureau
Committee for Economic Research
and Resources for the Future, Inc.

National Bureau of Economic Research
New York 1975

Distributed by
Columbia University Press
New York and London

Printed in the United States of America.

Library of Congress Cataloging in Publication Data
Main entry under title:

Economic analysis of environmental problems.

(Universities—National Bureau conference series; 26)
Papers and comments from the conference held at the University of Chicago, Nov. 10–11, 1972.
Includes index.
1. Environmental policy—United States—Congresses.
2. Environmental policy—Mathematical models—Congresses.
3. Pollution—Congresses. I. Mills, Edwin S.
II. Universities—National Bureau Committee for Economic Research. III. Resources for the Future.
IV. Series.
HC110.E5E245 301.31'0973 74–82378

ISBN 0–87014–267–4

Relation of the National Bureau Directors to Publications Reporting Conference Proceedings

Since the present volume is a record of conference proceedings, it has been exempted from the rules governing submission of manuscripts to, and critical review by, the Board of Directors of the National Bureau.

(*Resolution adopted July 6, 1948, as revised November 21, 1949, and April 20, 1968*)

Funds for the economic research conference program of the National Bureau of Economic Research are supplied by the National Science Foundation

Contents

Introduction — *Edwin S. Mills* — 1

Theoretical Analysis

Optimum Investment in Social Overhead Capital — *Hirofumi Uzawa* — 9
Comment: *Robert Dorfman* — 22

Macroeconomic Aspects of Environmental Policy — *Karl Göran-Mäler* — 27
Comment: *Colin Wright* — 56

The Technology of Public Goods, Externalities, and the Exclusion Principle — *Charles Plott and Robert Meyer* — 65
Comment: *Robert Haveman* — 90

The Instruments for Environmental Policy — *Wallace Oates and William Baumol* — 95
Comment: *Charles Upton* — 128

The Resource Allocation Effects of Environmental Policies — *George Tolley* — 133
Comment: *Gardner Brown, Jr.* — 163

Empirical Analysis of Domestic Problems

Operational Problems in Large Scale Residuals Management Models — *Walter Spofford, Clifford Russell and Robert Kelly* — 171
Comment: *J. Hayden Boyd* — 234

Input-Output Analysis and Air Pollution Control — *Robert Kohn* — 239
Comment: *Frederick Peterson* — 272

Studies of Residuals Management in Industry — *Blair Bower* — 275
Comment: *Paul MacAvoy* — 321

Acute Relationships Among Daily Mortality, Air Pollution, and Climate — *Lester Lave and Eugene Seskin* — 325
Comment: *Thomas Hodgson* — 347

"People or Ducks?" Who Decides? — *John Jackson* — 351
Comment: *Richard Quandt* — 389

International and Comparative Analysis

On the Economics of Transnational Environmental Externalities — *Ralph C. d'Arge* — 397

Appendix—A Note on Transfrontier Externalities Transactions Costs, and Long-Run International Adjustment Firms and Factors of Production: *Ralph C. d'Arge and William Schulze* — 417
Comment: *Larry Westphal* — 427

Management of Environmental Quality: Observations on Recent Experience in the United States and the United Kingdom — *Maynard Hufschmidt* — 435
Comment: *A. Myrick Freeman* — 453

Index — 461

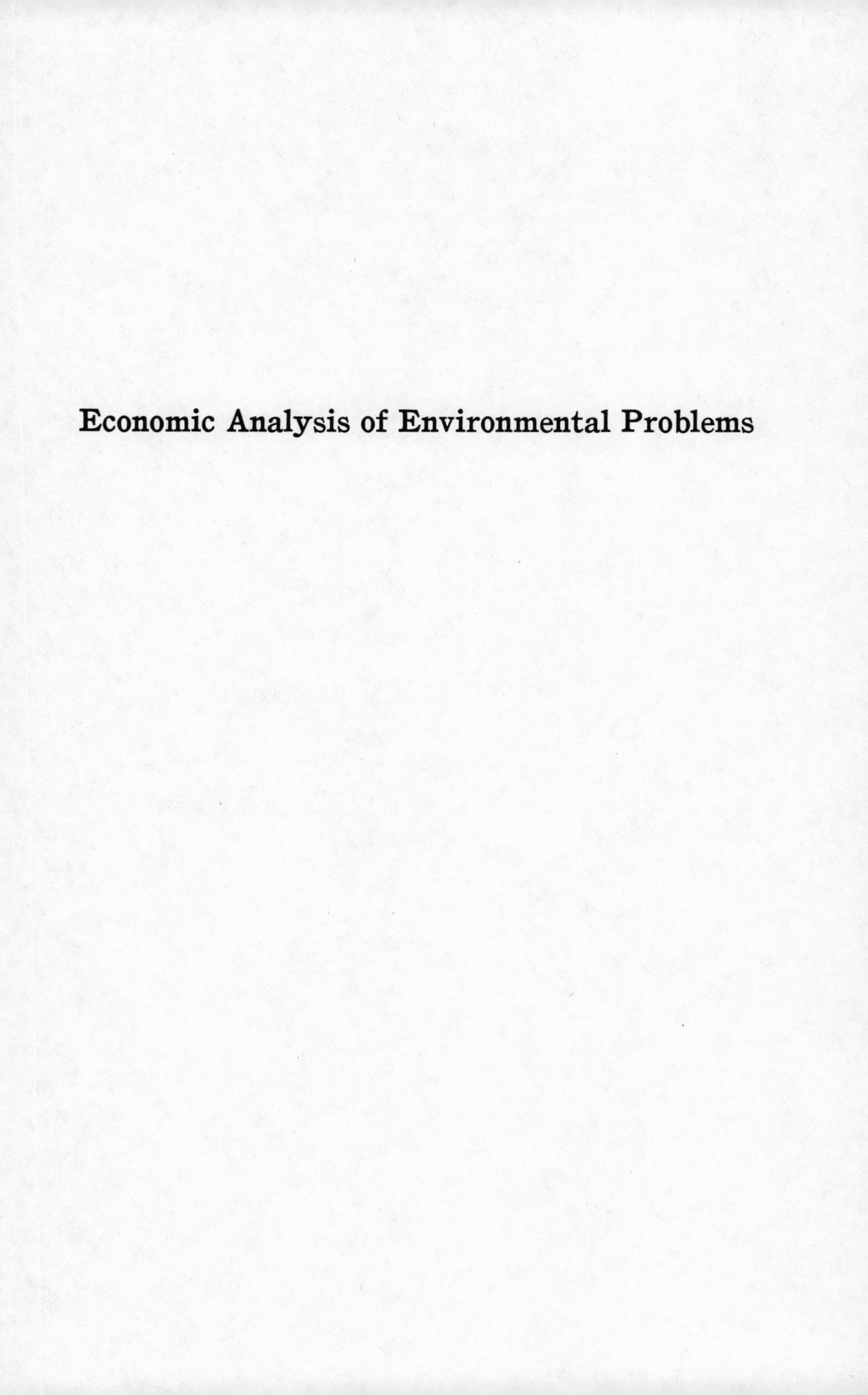

Economic Analysis of Environmental Problems

Introduction

Edwin S. Mills, Princeton University

The papers and comments printed in this volume were delivered at a conference held at the University of Chicago, November 10 and 11, 1972. The conference and conference volume were jointly sponsored by the Universities-National Bureau Committee for Economic Research and Resources for the Future, and were financed partly from a grant to the National Bureau by the National Science Foundation. Both the exploratory committee and the planning committee consisted of Otto Davis, Allen Kneese and the Editor. The Editor is indebted to his colleagues for their work on these committees, and to all the participants in the conference.

Despite the outpouring of printed material on environmental problems in recent years, careful economic research on the subject is still a scarce commodity. It is therefore hoped that the research brought together in this volume will be particularly useful to scholars, students and public and private officials interested in the subject.

The papers and comments in this volume speak for themselves and need not be summarized here. Instead, this introduction will present a brief guided tour through economic issues related to the environment, pointing out niches into which the papers fit and some niches that are still empty or nearly empty.

The essence of economic activity is the removal of materials from the environment, their transformation by production and consumption, and their eventual return to the environment. With relatively unimportant exceptions, the mass of materials removed must equal the mass of materials returned during any substantial period of time. Resource economics is concerned with the kinds, amounts, and forms of materials removed, places from which materials are removed, and the institutions, markets and laws by which removal is governed. Environmental economics is concerned with similar issues related to the return of materials to the environment.

It is natural to classify environmental studies in the ways that other economic studies are classified: micro, macro; theoretical, empirical; domestic, international; etc. The papers in this volume fall naturally into the three parts listed in the contents.

Of the five theoretical papers in part one, the first two, by Uzawa and Mäler, are macroeconomic. In the long run, the most important environmental issue is the relationship between environmental quality and economic growth in a finite world. Economic growth normally entails increases in the amounts of materials removed from and returned to the environment. In the absence of efforts to avoid it, environmental deterioration must result when materials returns exceed the absorptive capacity of the environment. This statement applies to a local or regional environment, such as a flowing stream, but it also applies to the global environment, especially to the earth's air mantle. Resources can be devoted to environmental protection in two ways: materials can be transformed by production, consumption and waste treatment in ways that make their return innocuous; and materials can be reused or recycled rather than returned to the environment. Failure to realize that resources can be devoted to environmental protection as well as to direct production of goods is a fatal defect of much popular environmental literature and technical environmental literature by noneconomists. This failure is the source of anti-growth bias in some of that literature. Both Uzawa and Mäler, by incorporating resource use for environmental protection in growth models, contribute to our understanding of their relationship. An important direction for future work on growth theory ought to be the incorporation of both resource and environmental considerations in long-run growth models.

Mäler's paper also discusses short run macro stabilization policy in a model with environmental considerations built in. He shows that effluent or discharge taxes may have quite different stabilization effects from income or other taxes. This must be one of the first attempts to include environmental variables in a short run macro stabilization model. Clearly, more such research is needed.

The last three theoretical papers are microeconomic. The most important microeconomic environmental issue is to understand the precise nature of the market failure that pollution entails. Economists have studied this issue since Pigou, and the recent research by, and inspired by, Coase has reopened the subject. The paper by Plott and Meyer explores fundamental aspects of the concept of market failure. The authors consider in what sense and to what extent it is useful to characterize the reasons for market failure by the currently fashionable term "transaction costs." To the extent that it is useful, they discuss what can be said about the transaction costs of various government policies relative to the costs of private transactions.

The next theoretical issue considered is the measurement of costs and benefits of environmental protection. If it is accepted that there is market failure regarding the environment, how are the costs and benefits of measures to improve the environment to be appraised? Measuring costs of environmental protection raises few new theoretical issues, but calculating the benefit side involves important unanswered questions. Environmental quality differs from the usual economic goods in several ways and has several quite different kinds of benefits. Neither this volume, nor the profession has much to contribute on benefit measurement to date.

A third theoretical issue is the choice of public policy instruments for environmental protection. Most research in recent years has focused on the choice between direct regulation of materials returns or effluents, and the use of taxes or charges on effluents. In fact, a much broader range of policies is available to protect the environment. The papers by Oates and Baumol and by Tolley contribute to the evaluation of alternative public policies. It can be said that no environmental issue has been so well studied: The great American pollution paradox is that no serious economic study has concluded that discharge fees should not be a major tool of public policy, whereas no serious environmental measure has included them. The paradox is deep in a society in which almost everything else is taxed, including those areas opposed by powerful interest groups.

The empirical studies reported in part two are all microeconomic. One important kind of empirical environmental study can be termed industry studies. These include studies of the kinds, magnitudes and speeds of response to various public policies to protect the environment. The term industry here must be interpreted to include public institutions, such as municipal waste treatment plants, as well as private firms. Several earlier studies investigated responses of industry to effluent charges on waterborne wastes in particular river basins. The first two empirical studies in this volume are much more comprehensive industry studies. The paper by Spofford, Russell and Kelly reports a large-scale systems study of all discharges in the Delaware Basin. It includes air, water, and solid wastes and a large number of alternative ways of changing or reducing discharges. The paper by Kohn is an input-output study of air pollution control in the St. Louis metropolitan area. Both are optimization studies in that they compute least-cost combinations of ways to achieve given environmental standards. The third empirical paper is a different kind of industry study. In it, Bower reports on several studies undertaken at Resources for the Future of the process, product, and waste treatment changes that can be employed in several industries to reduce the amounts or harmfulness of waste discharges.

The paper by Lave and Seskin is the only paper in the volume on the benefits of pollution abatement. In it, the authors study health effects of episodes of severe air pollution in metropolitan areas. It is curious that we now have better and more comprehensive studies of benefits of air pollution abatement than of water pollution abatement. Many studies, some of the best of them by Lave and Seskin, have estimated effects of air pollutants on human health, and some have estimated property damage. There are only a few studies of benefits of water pollution abatement, mostly restricted to recreational benefits, despite the fact that basic physical, chemical and biological processes are better understood for water pollution than for air pollution. High priority should be attached to further studies of water pollution abatement benefits.

The final paper in part two is a study of the political process by which environmental issues are settled. In it, Jackson studies in detail the controversy surrounding the decision whether to build a second jetport in the Minneapolis-St. Paul metropolitan area. More such studies are needed, especially of the passage of major national environmental legislation.

Part three consists of two international environmental studies. Several kinds of international environmental studies might be undertaken. First and most obvious is environmental effects that spill over national boundaries. Examples most often quoted are airborne discharges that pollute the air over neighboring countries. Other possible examples pertain to the global environment, e.g., heating of the atmosphere or pollution of the oceans. However, no economic, e.g., benefit-cost, studies appear to have been published on global problems, probably because underlying technical relationships are not well understood. A second kind of international study is of the kinds of arrangements, institutions and policies that nations might adopt to deal with environmental effects that spill over national boundaries. The paper by d'Arge explores basic aspects of this problem. A third kind of international study would be effects of domestic environmental protection policies on a country's international trade and balance of payments. Environmental protection measures tend to make domestic goods expensive relative to foreign goods, hence increasing imports and decreasing exports. Unfortunately, we have no paper on the subject in this volume. The fourth kind of international study is a cross-country comparison of environmental protection policies. Wise men tell us we do not learn the lessons of history. A generalization of the proposition is that we fail to learn from both time series and cross-sectional data. Hufschmidt's paper is a contribution to refuting the generalization. In it, he compares recent air and water pollution control programs in the United Kingdom and the United States.

THEORETICAL ANALYSIS

Optimum Investment in Social Overhead Capital

Hirofumi Uzawa, University of Tokyo

Introduction

In most industrialized countries, the "environment" has come to play a significant role in recent years both in the process of resource allocation and in the determination of real income distribution. This is primarily due to the fact that, in these countries, the environment has become scarce relative to those resources which may be privately appropriated and efficiently allocated through the market mechanism. There is no inherent mechanism in a decentralized market economy whereby the scarcity of social resources may be effectively restored. In decentralized economies, most research has been concerned with regulating the use of the environment; few researchers have analyzed the effects the accumulation of the environment may have upon the pattern of resource allocation and income distribution in general.

In order to analyze the role played by the environment in the processes of resource allocation and income distribution, it may be convenient to introduce a broader concept of "social overhead capital," of which the environment may be regarded as an important component. Social overhead capital is defined as those resources which are not privately appropriated, either for technological reasons or for social and institutional reasons, and which may be used by the members of the society free of charge or with nominal charges. Social overhead capital may be collectively produced by the society, as in the case of social capital such as roads, bridges, sewage systems, and ports, or it may be simply endowed in the society as in the case of natural capital such as air, water, etc.

Thus all scarce resources may be classified into two categories; private means of production and social overhead capital. In a decentralized market economy, private means of production are allocated through the market mechanism, resulting in an efficient allocation of given amounts of scarce resources, but the management of social overhead capital has to be delegated to a certain social institution.

Such social institutions in charge of social overhead capital are concerned with two functions. First, they have to devise regulatory measures in order to see to it that the given stock of social overhead capital may be efficiently allocated among the members of the society and effectively utilized by each member. Second, they are concerned with the construction of social overhead capital by using private and social means of production in such a way that the resulting pattern of resource allocation and income distribution is optimum from the social point of view. In this paper, I should first like to discuss the criteria by which social institutions in charge of social overhead capital allocate scarce resources in order to attain a dynamic pattern of resource allocation which is optimum from the social point of view. Next I should like to analyze the problem of what sort of criteria one has to impose upon the behavior of social institutions which are in charge of the management and construction of such social overhead capital. It will in particular be shown that, if the effect of social overhead capital is neutral in the sense precisely defined below, then the dynamically optimum pattern of accumulation of social overhead capital may result when the use of the services from such a social overhead capital is priced according to its marginal social cost and an interest subsidy is given to the extent to which the market (real) rate of interest differs from the social rate of discount.

Social Overhead Capital

Before I proceed with the main discussion, I should like to present a general framework in which some of the more crucial aspects of social overhead capital may be detailed. The services provided by social overhead capital are usually analyzed in terms of Samuelsonian public goods, as introduced in Samuelson's now classical papers [2, 3]. However, most of the familiar examples of social overhead capital such as highways, air, etc., do not satisfy the definition of pure public goods. There are two aspects of the services provided by social overhead capital with which I am particularly concerned. The first aspect is related to the choice made by each member of the society of the level at which he uses the services; the Samuelsonian concept of pure public goods excludes such a possibility. For most services provided by social overhead capital, the members of the society have, within a significant range, freedom in determining the amounts of the services they use.

The second aspect which has been neglected in the Samuelsonian analysis is related to the phenomenon of congestion. For most social overhead capital, the capacity of overhead capital is generally limited and the effectiveness of a certain amount of the services to be derived from the given stock of overhead capital is affected by the amounts of the same services other members of the society are using.

The concept of social overhead which has been introduced in Uzawa [6] may take care of these two aspects which are characteristic of most services provided by the government as public goods. The basic approach in Uzawa [6] may be briefly outlined as follows.

All the means of production are classified into two categories; private means of production and social overhead capital. Private means of production, simply referred to as private capital, may be privately appropriated and each member of the society may dispose of private capital which he possesses in such a way that his utility or profit is maximized. On the other hand, social overhead capital, simply referred to as social capital or overhead capital, may not be privately appropriated and the use of the services derived from overhead capital may be regulated either by the government or by a social institution to which the management of overhead capital is delegated.

In the following discussion, it is assumed that both private capital and overhead capital are respectively composed of homogeneous quantities, and that the output produced by various production units is identical. The main propositions obtained in this paper may be extended, without much difficulty, to the general situation where there may exist several kinds of overhead capital as well as private capital. Is is also assumed that private capital is a variable factor of production in the sense that it costs less to shift its usage from one line to another. Again, the analysis may, with slight modifications, be extended to the general situation where some private capital goods are fixed factors of production, involving significant adjustment costs either in the process of accumulation or in the shift in their allocation.

The effects of social overhead capital upon the productive process are formulated in terms of the short-run production function which summarizes production processes of each producing unit. Namely, the output Q_β produced by a typical production unit β is assumed to depend upon the amount of the services of social overhead capital as well as that of private capital. Let K_β and X_β be respectively the amount of the services of private and social capital used by producing unit β, and let X and V

be respectively the total amount of the services of social capital being utilized and the stock of social capital existing at each moment of time. If α and β stand for the typical consuming unit and producing unit, respectively, then

$$X = \sum_{\alpha} X_{\alpha} + \sum_{\beta} X_{\beta}. \tag{1}$$

The production function for producing unit β may be written as:

$$Q_{\beta} = F^{\beta}(K_{\beta}, X_{\beta}, X, V). \tag{2}$$

The dependence of the production function upon X and V indicates that the phenomenon of congestion occurs with respect to the use of the services of social overhead capital. The effect of congestion may be brought out by the assumptions:

$$F_X{}^{\beta} < 0, \quad F_V{}^{\beta} > 0. \tag{3}$$

The Samuelsonian case of pure public goods may be regarded as the limiting case where the production function F^{β} is independent of X_{β} and X.

The standard properties concerning the production function are assumed for the present case. In particular, it will be assumed that the marginal rates of substitution between various factors are diminishing and that the production processes are subject to constant returns to scale. Namely, the production function F^{β} given by (2) is concave with respect to K_{β}, X_{β}, X, and V, and it is linear homogeneous with respect to all of its variables. Furthermore, it is assumed that private capital and social capital are complementary in the sense that the increase in the use of the services of social capital shifts the schedule of the marginal product of private capital upward; namely,

$$F_{K_{\beta}{}^{\beta}X_{\beta}} > 0. \tag{4}$$

Social overhead capital may be defined *neutral,* when the negative effect due to the increase in the aggregate level of the use of the services of overhead capital is precisely counterbalanced by the corresponding increase in the stock of social overhead capital. Such would be the case if, for each producing unit β, the production function is homogeneous of order zero with respect to the aggregate level of the use of social capital X and the stock of social capital V. In this case, the production F may be written in the following form:

$$Q_{\beta} = F^{\beta}(K_{\beta}, X_{\beta}, X/V). \tag{5}$$

(Using the convention that functional symbols are used to indicate the nature of dependency in general.)

If social overhead capital is homogeneous of order zero with respect to X and V, then the following condition is satisfied:

$$F_X{}^\beta X + F_V{}^\beta V = 0. \tag{6}$$

In this paper, social overhead capital will be defined neutral with respect to production, if the following condition is satisfied:

$$\sum_\beta F_X{}^\beta X + \sum_\beta F_V{}^\beta V = 0. \tag{7}$$

Thus, if, for each production unit β, the production function is homogeneous of order zero with respect to X and V, then social overhead capital in question is neutral with respect to production.

The neutrality condition may be stated in terms of various elasticities. Let the elasticities η_X and η_V be defined as follows:

$$\eta_X = \frac{-\sum_\beta F_X{}^\beta X}{\sum_\beta F^\beta}, \quad \eta_V = \frac{-\sum_\beta F_V{}^\beta V}{\sum_\beta F^\beta}. \tag{8}$$

Thus, social overhead capital is neutral with respect to the production if and only if the elasticity is equal to the elasticity η_X;

$$\eta_V = \eta_X. \tag{9}$$

The effects exerted by social overhead capital upon the processes of consumption may be similarly formulated. For consuming unit α, the level of utility U_α depends upon the amount of services of social overhead capital, X_α, as well as the amount of private consumption C_α. The phenomenon of congestion may occur with respect to the processes of consumption, so that the utility function for each consuming unit α may be in general written as

$$U_\alpha = U^\alpha(C_\alpha, X_\alpha, X, V). \tag{10}$$

The phenomenon of congestion may be again formulated by the following conditions:

$$U_X{}^\alpha < 0, \quad U_V{}^\alpha > 0. \tag{11}$$

It is assumed that the utility function of U_α exhibits the feature of diminishing marginal utility and that the variables appearing in (10) exhaust all the variables which are limitational in the processes of consumption; namely, it may be assumed that, for each consuming unit α, the utility function U^α is concave and linear homogeneous with respect to C_α, X_α, X, V.

Social overhead capital is defined *neutral with respect to consumption,* if the following conditions are satisfied:

$$\sum_\alpha \frac{U_X{}^\alpha X}{U_{C_\alpha}{}^\alpha} + \sum_\alpha \frac{U_V{}^\alpha V}{U_{C_\alpha}{}^\alpha} = 0. \tag{12}$$

Finally, it is assumed that the cost in real terms in providing the services of social overhead capital depends upon the stock of overhead capital as well as upon the amount of the services provided; namely, the current cost in real terms W is a function of X and V:

$$W = W(X, V). \tag{13}$$

It may be in general assumed that the larger the amount of the services of overhead capital provided, the higher is the current cost, but the larger the stock of overhead capital, the lower is the current cost. In symbols,

$$W_X > 0, \quad W_V < 0. \tag{14}$$

Furthermore, it will be assumed that the cost function $W(X, V)$ is linear homogeneous with respect to X and V, so that it may be written as:

$$W = w(x)V, \quad x = X/V. \tag{15}$$

It is now necessary to introduce two concepts which will play a central role in the analysis of social overhead capital. They are the marginal social cost associated with the use of social overhead capital and the marginal social product of overhead capital.

The marginal social cost associated with the use of overhead capital, θ, is defined as the aggregate of marginal losses incurred by all the economic units in the economy due to the marginal increase in the use of the services of social overhead capital. Thus the marginal social cost may be represented by the following formula:

$$\theta = \sum_\alpha \frac{-U_X{}^\alpha}{U_{C_\alpha}{}^\alpha} + \sum_\beta - F_X{}^\beta + W_X. \tag{16}$$

On the other hand, the *marginal social product of overhead capital, r,* is defined as the aggregate of the marginal increase due to the marginal increase in the stock of overhead capital. In symbols, the marginal social product r may be given by:

$$r = \sum_{\alpha} \frac{U_{V}{}^{\alpha}}{U_{C_\alpha}{}^{\alpha}} + \sum_{\beta} F_{V}{}^{\beta} + W_{V}. \tag{17}$$

Social overhead capital now may be defined as *neutral* if the following condition is satisfied:

$$rV = \theta X. \tag{18}$$

It may be easily seen that, if all the production functions and utility functions are homogeneous of order zero with respect to X and V, then social overhead capital is neutral in the sense defined here.

Social overhead capital may be classified as *socially biased* if rV exceeds θX. The implication of such a classification will be discussed later.

The marginal social cost θ and the marginal social product r both depend upon the relative magnitude of the endowments of private capital and social overhead capital, as well as upon the allocation of these resources between individual economic units. In general, the higher the ratio of the private capital over the stock of overhead capital, the higher is the marginal social cost and also the higher is the marginal social product of overhead capital. The marginal social cost may be regarded as an index to measure the scarcity of social overhead capital relative to the stock of private capital or relative to the level of economic activities.

Efficient Pricing for the Use of Social Overhead Capital

Let us first discuss the problem of how to make efficient use of the services derived from a given stock of social overhead capital. Suppose that the stock of overhead capital V and the stock of private capital K are given and remain constant. The allocation of private capital between producing units, the distribution of output between consuming units, and the distribution of the services of social overhead capital are termed efficient if they result in the situation where the level U of social utility, to be defined as the aggregate of individual utility levels, is maximized among all the feasible patterns of resource allocation. Mathematically, the problem is stated as follows. To find the pattern of resource allocation $(C_\alpha, X_\alpha, K_\beta, X_\beta)$ which maximizes the level of social utility

$$U = \sum_\alpha U_\alpha, \tag{19}$$

subject to the constraints:

$$\sum_\alpha C_\alpha + W = \sum_\beta Q_\beta, \tag{20}$$

$$\sum_\beta K_\beta = K, \tag{21}$$

$$\sum_\alpha X_\alpha + \sum_\beta X_\beta = X, \tag{22}$$

where

$$U_\alpha = U^\alpha(C_\alpha, X_\alpha, X, V), \tag{23}$$

$$Q_\beta = F^\beta(K_\beta, X_\beta, X, V). \tag{24}$$

As explained in Uzawa [6], such a problem may be easily solved, by applying the method of Lagrange multipliers. The solution then may be shown to be identical with the one which is obtained by the principle of the marginal social cost pricing. Namely, the efficient allocation may be obtained by the following mechanism.

Private capital is allocated through a perfectly competitive market, while the services of social overhead capital may be distributed among consuming and producing units in such a way that each individual unit is charged the marginal social cost for the use of the services rendered by social overhead capital. It may be noted that either the administrative costs associated with such a pricing scheme are assumed to be neglible or it is possible to find an alternative pricing scheme, without involving significant administrative costs, which results in the identical distribution of the services of social overhead capital.

As indicated in the previous section, when social overhead capital becomes scarce relative to the endowment of private capital, then the marginal social cost becomes higher and individual economic units are charged a higher price for the use of social overhead capital. However, it is generally the case that some of the scarce resources are classified as social overhead capital because of the impact they may have upon the equalization of the income distribution. Hence, if a high price were charged for the use of such social overhead capital, it would become difficult to justify the decision in classifying such resources as social overhead capital. The implications of such a phenomenon may be more explicitly brought out if the problem of accumulation of social overhead capital is discussed. In the next section, I should like to discuss the problem of optimum investment in social overhead capital.

Optimum Investment in Social Overhead Capital

In this section, I should like to extend the previous analysis to the situation where one is concerned with the process of capital accumulation for both private and social capital, and to examine the pattern of resource allocation over time which is optimum from a dynamic point of view. It will be shown that the principle of the marginal social costs may be extended to this dynamic case and the criteria for optimum allocation of investment between private and social capital will be obtained within the framework of the Ramsey theory of optimum growth.

In order to simplify the exposition, it will be assumed that the rate by which consumers discount their future levels of utility is constant and identical for all consumers in the society. Let δ be the rate of discount. The level of social utility U may now be expressed by

$$U = \int_0^\infty U(t)e^{-\delta t}\,dt, \tag{25}$$

where the utility level $U(t)$ at a point of time t may be given y

$$U(t) = \sum_\alpha U_\alpha(t), \tag{26}$$

with

$$U_\alpha(t) = U^\alpha[C_\alpha(t), X_\alpha(t), X(t), V(t)]. \tag{27}$$

Let V_0 be the stock of social overhead capital existing at the initial point of time 0. I am concerned with the problem of finding a path of private consumption for each consumer, of allocation of private and social capital between various economic units, and of capital accumulation for both private and social capital over time such that the resulting level of social utility (25) is maximized over all feasible paths. In order to discuss this optimum problem, I should like to pay particular attention to the difference between private and social capital with regard to the extent to which investment is used to increase the stock of capital (to be measured in the efficiency unit). In general, social overhead capital is difficult to reproduce in the sense that a significant amount of scarce resources have to be used in order to increase the stock of capital. For private capital however, investment may without much difficulty be converted into the accumulation of capital. It may be possible to formulate the relationships between the amount of investment and the resulting increase in the stock of capital in terms of a certain functional relationship. I have elsewhere discussed this problem for the case of private capital and a similar conceptual framework may be applied to the case involving social overhead capital (Penrose [1] and Uzawa [4, 5]).

Let I_V be the amount of real investment devoted to the accumulation of social overhead capital. If social overhead capital V is measured in a specified efficiency unit, the amount of real investment I_V may not necessarily result in the increase in the stock of capital by the same amount. Instead, there exists a certain relationship between the amount of real investment I_V and the corresponding increase $\dot{V}$ in the stock of social overhead capital on one hand, and the current stock of social overhead capital V on the other:

$$I_V = \phi_V(\dot{V}, V). \tag{28}$$

The relationship (28) may be interpreted as follows: in order to increase the stock of social overhead capital V by the amount $\dot{V}$, real investment I_V given by (28) has to be spent on the accumulative activities for social overhead capital. In what follows, it will be assumed that the function ϕ exhibits a feature of constant returns to scale with respect to $\dot{V}$ and V, thus one may write (28) as

$$I_V/V = \phi_V(\dot{V}/V). \tag{29}$$

Since it may be assumed that the marginal costs of investment are increasing as the level of investment is increased, the function ϕ_V satisfies the following conditions:

$$\phi_V'(.) > 0, \quad \phi_V''(.) > 0. \tag{30}$$

Similar relationships may be postulated for the accumulation of private capital for each producing unit; namely, for each producer β, the amount of real investment I_β required to increase the stock of capital K_β by the amount $\dot{K}_\beta$ may be determined by the following Penrose function:

$$I_\beta/K_\beta = \phi_\beta(\dot{K}_\beta/K_\beta), \tag{31}$$

where the Penrose function ϕ_β again satisfies the conditions:

$$\phi_\beta'(.) > 0, \quad \phi_\beta''(.) > 0. \tag{32}$$

For the current cost W, we have,

$$W = w\left(\frac{X}{V}\right)V, \quad w'(.) > 0, \quad w''(.) > 0. \tag{33}$$

The optimum problem may now be more precisely stated as follows: A path of resource allocation over time,

$$[C_\alpha(t), I_\beta(t), I_V(t), X_\alpha(t), X_\beta(t), K_\beta(t), V(t)], \tag{34}$$

is defined as a feasible path if it satisfies the following conditions:

$$Q(t) = \sum_\alpha C_\alpha(t) + \sum_\beta I_\beta(t) + I_V(t) + W(t), \tag{35}$$

$$Q(t) = \sum_\beta F^\beta[K_\beta(t), X_\beta(t), X(t), V(t)], \tag{36}$$

$$X(t) = \sum_\alpha X_\alpha(t) + \sum_\beta X_\beta(t), \tag{37}$$

$$\frac{I_\beta(t)}{K_\beta(t)} = \phi_\beta[z_\beta(t)], \quad \frac{\dot{K}_\beta(t)}{K_\beta(t)} = z_\beta(t), \tag{38}$$

$$\frac{I_V(t)}{V(t)} = \phi V[z_V(t)], \quad \frac{\dot{V}(t)}{V(t)} = z_V(t), \tag{39}$$

$$W(t) = w[X(t)]V(t), \quad x(t) = X(t)/V(t), \tag{40}$$

$$K_\beta(0) = K_\beta{}^0, \quad V(0) = V^0 \text{ given.} \tag{41}$$

I am then interested in finding a feasible path of resource allocation over time which maximizes the social utility (25). This optimum problem is difficult to solve, and I shall instead be concerned with finding a path of resource allocation which reasonably approximates the optimum path. Among such approximate paths, the one with the simplest structure will be obtained by examining the conditions which the imputed prices of private and social capital have to satisfy.

Let $p_\beta(t)$ and $p_V(t)$ be respectively the imputed prices, at time t, of private capital K_β and social overhead capital V, and let $p(t)$ and $\theta(t)$ be the imputed prices of output Q and the use of social overhead capital X.

These imputed prices correspond to the Lagrange multipliers associated with the constraints for the optimum problem. The Euler-Lagrange conditions which the optimum path has to satisfy may be rearranged to yield the following conditions:

$$U^{\alpha}{}_{C_{\alpha}} = p, \quad U^{\alpha}{}_{X_{\alpha}}/U^{\alpha}{}_{C_{\alpha}} = \theta, \tag{42}$$

$$F^{\beta}{}_{X_{\beta}} = \theta, \tag{43}$$

$$\theta = \sum_{\alpha} \frac{(-U^{\alpha}{}_{X})}{U_{C_{\alpha}}{}^{\alpha}} + \sum_{\beta}(-F^{\beta}{}_{X}) + W'\left(\frac{X}{V}\right), \tag{44}$$

$$\frac{\dot{p}_{\beta}}{p_{\beta}} = \delta - z_{\beta} - \frac{r_{\beta} - \phi_{\beta}(z_{\beta})}{\phi'_{\beta}{}'(z_{\beta})}, \tag{45}$$

where

$$\phi_{\beta}'(z_{\beta}) = \frac{p_{\beta}}{p}, \quad r_{\beta} = F_{K_{\beta}}{}^{\beta}, \tag{46}$$

$$\frac{\dot{p}_{V}}{p_{V}} = \delta - z_{v} - \frac{r_{V} - \phi_{V}(z_{V})}{\phi_{V}'(z_{V})}, \tag{47}$$

where

$$\phi_{V}'(z_{V}) = \frac{p_{V}}{p}, \quad r_{V} = \sum_{\alpha} \frac{U_{V}{}^{\alpha}}{U_{C_{\alpha}}{}^{\alpha}} + \sum_{\beta} F_{V}{}^{\beta}. \tag{48}$$

I have omitted the time suffix t to avoid ambiguity.

The quantity on the right-hand side of equation (44) corresponds to the concept of the marginal social costs associated with the use of social overhead capital in the context of the dynamic optimization. It may be noted that the marginal costs associated with the depreciation of social overhead capital are evaluated in terms of its imputed price p_V/p measured in real terms. The quantity r_{β} defined in (46) is nothing but the marginal product of private capital, while the r_V defined in (48) is the marginal social product of social overhead capital measured in real terms. Namely, the r_V represents the marginal gain to the society measured in real terms due to the marginal increase in the stock of social overhead capital V.

The conditions (44–46) suggest that, in order to attain an optimum allocation of scarce resources in the short run, one has to impose charges equal to the marginal social costs for the use of social overhead capital, with the marginal social costs being defined in the modified sense (46). On the other hand, the pattern of accumulation of private and social capital may be described by the conditions (45–48) describing the rules by which the imputed prices change over time. In order to approximate the structure of the optimum path of capital accumulation, I should like to consider the case where the imputed prices are assumed constant at each point in time. Namely, the rates of accumulation of private and social capital are obtained by assuming that equations (45) and (47) are equated to zero. It can be shown that the path of capital accumulation obtained by such a procedure reasonably approximates the optimum path, although the sense in which reasonable approximation is used needs a more complicated formalization.

If the imputed prices are assumed constant over time, the rates of capital accumulation z_β and z_V may be obtained by solving the following conditions:

$$\frac{r_\beta - \phi_\beta(z_\beta)}{\delta - z_\beta} = \phi_\beta'(z_\beta), \tag{49}$$

$$\frac{r_V - \phi_V(z_V)}{\delta - z_V} = \phi_V'(z_V). \tag{50}$$

The determination of the rates of accumulation, z_β and z_V, is illustrated by the following diagrams.

It is easily seen, from the diagrams in Figures 1 and 2, that the rates of accumulation of private and social capital are uniquely determined, that the higher the marginal product of private capital, the higher is the

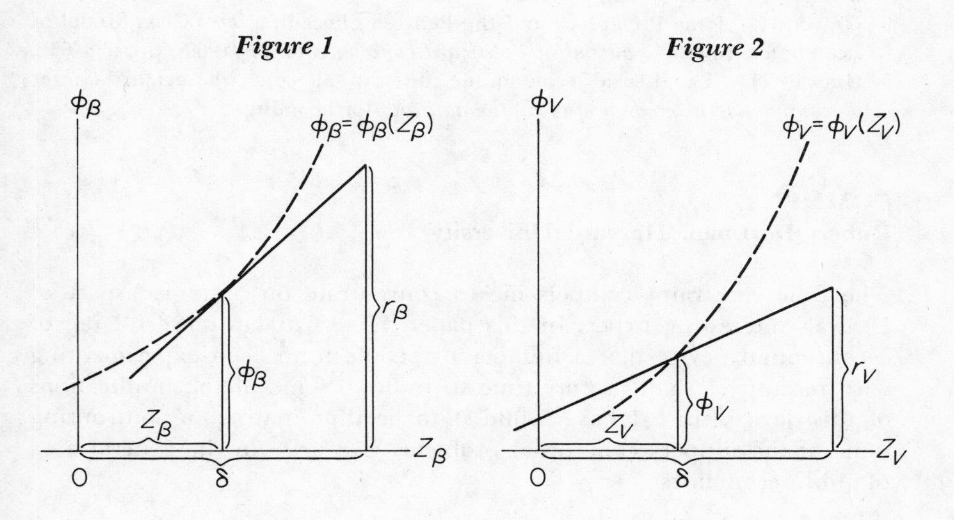

corresponding rate of accumulation for private capital, and that the higher the marginal social product of social capital, the higher is the rate of accumulation. On the other hand, an increase in the social rate of discount δ will lower the rate of accumulation both for private and social capital.

Thus, the (approximate) optimum rates of accumulation for private and social capital will be determined once the marginal private or social product of these capital are known. However, the marginal products of both private and social capital depend upon the extent to which social overhead capital is used by the member of the society. The amount of the services of social overhead capital used is in turn related to the imputed price p_V/p of social overhead capital, as is seen from the definition of the marginal social costs.

References

1. Penrose, E. T. *The Theory of the Growth of the Firm.* Oxford: Blackwell's, 1959.
2. Samuelson, P. A. "The Pure Theory of Public Expenditures," *Review of Economics and Statistics,* Vol. XXXVI (1954), No. 4, pp. 387–389.
3. Samuelson, P. A. "Diagrammatic Exposition of a Pure Theory of Public Expenditures," *Review of Economics and Statistics,* Vol. XXXVII (1955), pp. 350–356.
4. Uzawa, H. "The Penrose Effect and Optimum Growth," *Economic Studies Quarterly,* Vol. 19 (1968), pp. 1–14.
5. Uzawa, H. "Time Preference and the Penrose Effect in a Two-Class Model of Economic Growth," *Journal of Political Economy,* Vol. 77 (1969), pp. 628–652.
6. Uzawa, H. "La theorie economique du capital collectif social," *Cahier d'econometrie et economique mathematique,* forthcoming.

COMMENT

Robert Dorfman, Harvard University

The time constraint compels me to concentrate on a single aspect of Uzawa's interesting paper. In this paper, Uzawa makes powerful use of a conceptual device that a number of people have been experimenting with recently. I shall use my time to indicate some of the implications of this device, since I have found it to be illuminating in confronting some of the slippery conceptual problems that arise in the general area of public economics.

First, I would like to sketch the context in which the use of this device arises. In the traditional theory of consumption we imagine a typical consumer, Mr. α, who chooses the amounts he purchases of different commodities, x_i^α, so as to maximize the value of an indicator of his utility, $u^\alpha(x_1^\alpha, x_2^\alpha, \ldots, x_n^\alpha)$, subject of course, to a budget constraint. This formulation is adequate for studying Mr. α's personal consumption decisions, but when we turn attention to the economy-wide allocation of resources we have to bring some additional considerations into the picture. In particular we have to take public goods and externalities into account. In 1954 Samuelson demonstrated how to introduce public goods into the analysis. The essential idea is to include the level of provision of public goods in the individual's utility indicator by writing the typical indicator as $u^\alpha(x_1^\alpha, x_2^\alpha, \ldots, x_n^\alpha, Z_1, \ldots, Z_m)$. In this notation the Z_j indicate the levels of provision of public goods or collective goods. Notice, they do not have any superscripts.

Too much has already been written about the definition of these public goods and I shall not further burden the literature. The essential operational characteristics that distinguish them from private goods are that (1) each enters with the very same magnitude into the utility indicators of several or all consumers, and (2) the individual consumer does not select this magnitude, nor does it enter explicitly into his budget constraint. The levels of the public goods are regarded as being chosen by some economic entity, normally a governmental body or a nonprofit institution, that does not have a utility function of its own, but is supposed to be concerned with the effect of its choice on the utilities and productivities of the other decision units in the economy. The well-known upshot of the analysis in this form is a set of criteria for the optimal levels of these public goods.

But externalities are still absent from the formulation. Here is where the conceptual device that I have mentioned enters. Let us write the typical utility indicator as $u^\alpha(x_1^\alpha, x_2^\alpha, \ldots, x_n^\alpha, X_1, X_2, \ldots, X_n, Z_1, \ldots, Z_m)$. In this notation the X_i represent the levels of what I shall call externality-conveying goods. Like the public goods variables, they carry no superscripts and their magnitudes are common to the utility indicators of several or all consumers. They differ from the public goods in the locus of the decisions made about them. Indeed, no explicit decisions are made about them at all, but their levels are by-products of decisions made about private goods. We may formalize this notion by writing $X_i = f_i(x_i^\alpha, x_i^\beta, \ldots, x_i^\omega)$ so that the level of each externality-conveying good is some function of the consumers' choices concerning their private consumptions of that very same good. Uzawa concentrates on the special case where $X_i = \sum x_i^\alpha$ and so shall I for the most part, since it is an important special case and is sufficient to bring out most of the significant issues. In particular, this special case arises where the externality is some form of congestion. For example, if x_i^α represents the number of visits that the α family makes to a public park then u^α is likely to be positively affected by an increase in x_i^α, but negatively affected by an increase in $X_i = \sum x_i^\alpha$. In this formulation the number of acres in a public park would also be relevant, and would be indicated by one of the components Z_j since it is a public good. This simple formulation thus captures all three aspects of the enjoyment of a good used in common: the direct private decision, the interaction of private decisions, and the level of provision of the common facility.

In addition to congestion externalities, the simple summation formulation is appropriate for some types of environmental pollution. Street litter is a plausible instance, as is atmospheric pollution when effluents are considered evenly dispersed over a wide area. There are other forms of externality for which more complicated formulas are needed. The pollution of public waters is an important instance, but for this case a weighted sum is likely to be adequate in many applications.

The general formula suggested above does appear to cover these and a great many other kinds of externalities, but can be made more general. One obvious generalization is to permit a particular externality-conveying good, X_i, to be a function of the private consumptions of a number of different private goods. Actually this generalization is not very useful since the effects of different forms of consumption on a particular externality can be regarded as additive to a linear approximation. A more interesting generalization is to recognize that the functional relationship of an externality to private levels of consumption may be different for different consumers. This is the case for some kinds of congestion, and also for concentrations of atmospheric pollutants which are not uniform over a wide area.

In spite of these possibilities for elaboration, the simple formulations are perfectly adequate for conveying the main insights, which are likely to be obscured by attempting to achieve generality. Besides, in practical applications the data requirements increase rapidly with the complexity of the expression. It seems wisest to rest with simple expressions of the relationship between externalities and the private decisions that give rise to them.

To this point I have mentioned only consumers, but producers are also affected by public goods and externalities, and precisely the same considerations can be used to introduce the levels of public goods and externality-conveying goods into their production functions or technology sets. From this point on I shall presume that this has been done.

I have now sketched a tripartite vision of economic goods: private goods, externality-conveying goods, and public goods. The essential distinction among them is the locus of decisions concerning them and the considerations taken into account in making these decisions. More familiar formulations rest on such difficult concepts as "appropriability," "excludibility," "rivalness," and so forth, and I shall not recount the complications that those concepts lead into when we try to apply them. All those interrelationships are comprehended in the simple formulation $u^{\alpha}(x_1^{\alpha}, x_2^{\alpha}, \ldots, x_n^{\alpha}, X_1, X_2, \ldots, X_n, Z_1, \ldots, Z_m)$.

The virtue of this formulation is that it enables us to think about private goods, externalities, and public goods within a single unified framework. The conditions for Pareto optimality can be deduced from it, and there are no surprises. In the case of a pure private good, one for which $X_i =$ constant, both the conventional marginal inequalities for consumer choice and the standard requirement that each consumer's marginal rate of substitution between any two such goods must be equal to producers' marginal rate of transformation between them emerges. With respect to pure public goods, this formulation leads to the familiar Bowen-Samuelson requirement that the marginal rate of transformation between a pure private good and a pure public good should be equal to the sum of consumers' marginal rates of substitution between that private good and that public good.

The condition for the optimal provision of externality-producing goods is a bit more complicated, but still much as expected. I do not see any way to state these conditions precisely without doing the underlying algebra, from which I forbear. The essential idea is that a wedge has to be driven between the private consumer's perception of the value of such private goods (expressed by his marginal rates of substitution) and the social resource cost of providing the goods (expressed by marginal rates of transformation). This wedge has to be equal to the sum of the externalities conveyed by the use of the good, to all consumers and producers. (These externalities are also marginal rates of substitution, exemplified by the number of units of some private good that a typical consumer would be willing to relinquish for a one-unit diminution in some X_i.) The formulas are complicated by the presence of externalities that bridge the consumption and production spheres: externalities that consumers impose on producers, producers on consumers, consumers on each other and producers on each other. One interesting consequence is that the social production function is not well defined when external effects are present, because consumption decisions change the range of choice for production decisions.

This point of view provides, as I said before, some significant clarifications in thinking about public economics. It emphasizes the importance of institutional arrangements in characterizing commodities and their social consequences, and subordinates technological characteristics. It explains why a public park is a public good while a commercial ski lift or resort is not, in spite of the virtual identity of their technological properties. It explains why a highway is a public good although the consumption of its services may be fiercely competitive. It circumvents the truly baffling problems of defining such concepts as excludability and appropriability, which may be important legally or technically, but are irrelevant to the analysis of resource allocation. It sorts out the private, external, and public aspects of such intangibles as education and law and order.

For all these reasons, I commend it to you strongly as the most satisfactory conceptual framework for dealing with public goods, spillover effects, and related phenomena. I predict that Professor Uzawa's paper is only the first in a great number that will invoke this point of view in the analysis of public goods and externalities.

Macroeconomic Aspects of Environmental Policy

Karl Göran-Mäler, University of Stockholm

Introduction

This paper is divided into two parts. The first part deals with long-term planning of the environment. The discussion focuses on a neoclassical growth model of the Ramsey type, in which environmental considerations have been introduced. The objective of the planning authority or the government is represented by an intertemporal welfare function which is equal to the present value of future utilities. The instantaneous utility function has as arguments consumption per capita and environmental quality. The environmental quality depends on the discharge of waste products generated in firms. The firms have two possibilities to reduce their waste discharges. They can decrease their rate of production or they can allocate more capital and labor to waste treatment. It is shown that if some natural assumptions are satisfied, there exists an optimal path, and that this optimal path will tend to a steady state in the long run.

The second part of the paper is devoted to a discussion of short-run macro problems when the economy switches from one steady state to another. These problems arise because the labor market is not perfect: the wage rate is fixed from outside in case there is an excess supply of labor. Only when there is an excess demand for labor will the wage rate rise. It is shown that if effluent charges are used as an instrument in environmental policy, these charges (or taxes) cannot be regarded as ordinary taxes, because an increase in the charge may have expansive effects on the economy in contrast to income taxes. In order to analyse these questions, it is necessary to have an explicit monetary mechanism by which the general price level is determined. In a pure barter economy, where the price level is arbitrary, the effluent charge will function as a numeraire, and a change in the charge will not affect the allocation of resources in the economy. Finally, a similar but very brief analysis is made for the situation in which effluent standards are used as main instruments in environmental policy.

Long-term Planning

The production structure

We will assume the existence of a single homogeneous good that can be used for private and public consumption, and for capital accumulation. This good is produced according to the production function

$$Q = H(K_1, L_1), \tag{1}$$

where Q is output (gross national product), K_1 is the capital stock allocated to production, and L_1 is the labor input in production. It is assumed that H is twice continuously differentiable and strictly concave, and that it has positive partial derivatives. Moreover, it is assumed that H satisfies the Inada conditions: $H_1(0, L_1) = H_2(K_1, 0) = +\infty$, $H_1(\infty, L_1) = H_2(K_1, \infty) = 0$. In the course of production wastes are generated. It is assumed that it is technically possible to reduce the amount of waste by treatment. Therefore, the need for ultimate waste disposal in the environment is given by the treatment function:

$$z = M(K_1, L_1, K_2, L_2), \tag{2}$$

where z is the amount of waste discharged into the environment. K_1 and L_1 appear as arguments in the M function because it is assumed that the amount of primary waste is determined not only by the amount of output, but also by the way this output is produced. K_2 and L_2 are the amounts of capital and labor input allocated to waste treatment.

M is assumed to be twice continuously differentiable and convex. Moreover, we assume that

$$\frac{\partial M}{\partial K_1} > 0, \quad \frac{\partial M}{\partial L_1} > 0, \quad \frac{\partial M}{\partial K_2} < 0, \quad \frac{\partial M}{\partial L_2} < 0.$$

The convexity implies that

$$\frac{\partial^2 M}{\partial K_1^2} \geq 0, \quad \frac{\partial^2 M}{\partial L_1^2} \geq 0, \quad \frac{\partial^2 M}{\partial K_2^2} \geq 0, \quad \frac{\partial^2 M}{\partial L_2^2} \geq 0,$$

which seems natural to assume. We also assume that

$$\frac{\partial^2 M}{\partial K_1 \partial K_2} > 0, \quad \frac{\partial^2 M}{\partial K_1 \partial L_2} > 0, \quad \frac{\partial^2 M}{\partial L_1 \partial K_2} > 0, \quad \frac{\partial^2 M}{\partial L_2 \partial K_2} > 0.$$

In order to simplify the notation, let us denote derivatives by subscripts:

$$\frac{\partial M}{\partial K_1} = M_1, \text{ etc.}$$

We assume that both factors are completely malleable. This seems to be a legitimate assumption in long-run modeling, and Arrow and Kurz [1] have shown that the removal of the malleability assumption does not change the asymptotic properties of optimal growth models. Only the transition from the initial situation to the ultimate equilibrium is changed. When we come to discuss short-run problems, in section 3, capital will no longer be assumed to be malleable.

We will assume that wastes are generated only in connection with production, and not in connection with consumption. This is obviously a far from realistic assumption. Moreover, we will not consider raw materials input, although the extraction of raw materials and their subsequent uses is very important in connection with environmental questions. For a model that takes both these factors into consideration see [9, Chap. 3].

It is now possible to aggregate the production function and the waste treatment function into a single production function by assuming a maximizing behavior. Consider the problem

$$\max H(K_1, L_1),$$

subject to

$$\begin{aligned} &K_1 + K_2 \leqq K \\ &L_1 + L_2 \leqq L \\ &M(K_1, L_1, K_2, L_2) \leqq z \\ &K_i \geqq 0, \quad L_i \geqq 0, \quad z \geqq 0; \quad i = 1, 2. \end{aligned}$$

It is easily seen that the constraints satisfy the constraint qualification; therefore, the Kuhn-Tucker theorem can be applied. This theorem yields nonnegative multipliers r, w, and q such that the following conditions are satisfied:

$$\begin{aligned} H_1 - qM_1 - r &\leqq 0 \\ H_2 - qM_2 - w &\leqq 0 \\ - qM_3 - r &\leqq 0 \\ - qM_4 - w &\leqq 0. \end{aligned} \tag{3}$$

As usual, the multipliers can be given an economic interpretation: r can be interpreted as the rental for capital, w, the wage rate, and q, the effluent charge. This maximization problem defines maximal output as a function of available factor inputs, K and L, and of the discharge of wastes, z:

$$Q = F(K, L, z). \tag{4}$$

It is proved in the theory of nonlinear programming [11] that F is a concave function of all its arguments. Since H is strictly concave, so is F. Moreover, it can be proved that $F_1 = r$, $F_2 = w$, and $F_3 = q$. It is however not possible to determine the sign of the second-order cross derivatives. The assumption that the Inada conditions are satisfied for H implies that they are also satisfied for F. For the rest of this section, the aggregate production function will be used.

Let C and G stand for private and public consumption respectively. Assume that due to wear and tear the capital stock depreciates at exponential rate μ. The allocation of output must then satisfy the conditions

$$\dot{K} + \mu K + C + G \leqq F(K, L, z), \tag{5}$$

where $\dot{K}$ is the rate of change of the capital stock or the net investment.

The environmental interaction function

Due to the production processes, wastes are generated in the amount z. These wastes must be disposed of in some way, and we assume that the only way is to discharge them into the environment (all other ways are thought to be included in the waste treatment function). This discharge will, however, affect the quality of the environment. We assume that it is possible to measure the quality of the environment in one variable Y.

This is a very restrictive assumption because it prevents a discussion of reversible and nonreversible changes in the environment simultaneously.

The scale we measure Y in is chosen so that $Y = 1$ when no discharge of wastes has ever been made. $Y = 1$ corresponds thus to a virgin environment. The discharge of wastes will cause Y to decrease, but the environment is assumed to have an assimilative capacity which starts a self-purification process. However, the strength of this purification process depends on the quality of the environment, and we assume that the closer Y is to 1, the slower the purification process works. We can summarize these assumptions in the following equation:

$$\dot{Y} = \lambda(1 - Y) - \gamma z, \tag{6}$$

in which λ and γ are constants. In the absence of any discharges of wastes, the solution to this equation will converge to 1. Formally this equation is identical to the celebrated Streeter-Phelps equation which governs the dissolved oxygen in a river into which organic wastes are discharged.

To each waste load z, constant over time, there is a stationary state to which the quality of the environment will approach. This stationary state is given by

$$Y = 1 - \frac{\gamma}{\lambda} z.$$

The objective function

We assume that the preferences of the government can be represented by a utility functional:

$$W = \int_0^T U(c, g, Y)\, dt; \tag{7}$$

where c and g are per capita private and public consumption respectively, U an instantaneous utility function, and T the horizon. We assume that U is twice continuously differentiable and strictly concave, and has positive partial derivatives. Moreover, we assume that $U_1(0, g, Y) = +\infty$, $U_2(c, 0, Y) = +\infty$. The horizon T is predetermined in this model, but we assume that T is very large (one could replace the assumption of a finite horizon with an assumption of infinity without substantially changing the results). Furthermore, it is assumed that the capital stock at the horizon must not be smaller than a predetermined size K_T, i.e.,

$$K(T) \geqq K_T, \tag{8}$$

and that environmental quality at the horizon must not be smaller than a predetermined size Y_T, i.e.,

$$Y(T) \geqq Y_T. \tag{9}$$

Moreover the initial size of the capital stock and environmental quality is given by the past,

$$\begin{aligned} K(0) &= K_0 \\ Y(0) &= Y_0. \end{aligned} \tag{10}$$

We assume that K_T and Y_T are such that there exists at least one path from K_0, Y_0 to K_T, Y_T satisfying the constraints.

Optimal growth

For the discussion of optimal growth, only one final assumption is required, namely, that the labor supply is inelastic, constant over time, and is a constant fraction of the population. The problem of optimal growth can now be formulated:

Find K, Y, c, g, and z as functions of time, satisfying (5), (6), (8), (9), (10), (11) and the obvious nonnegativity constraints, so that the utility functional (7) is maximized. As L is constant, we can without loss of generality set $L = 1$, and $c = C$, $g = G$. This is a problem in calculus of variations. We state the necessary and sufficient conditions for an optimal growth path in terms of Pontryagin's maximum principle (in our application these conditions are the classical Euler-Lagrange equations; see [1] and [4]). According to this principle, if $K(t)$, $Y(t)$ is an optimal path, there exist two functions of time $p(t)$ and $\delta(t)$, such that the optimal values of the controls $C(t)$, $G(t)$ and $z(t)$ maximize the Hamiltonian

$$H = U(C, G, Y) + p[F(K, L, z) - \mu K - C - G] + \delta[\lambda(1 - Y) - \gamma z]. \tag{11}$$

Furthermore, the two auxiliary variables satisfy the following differential equations

$$\dot{p} = -\frac{\partial H}{\partial K} = -p(F_1 - \mu), \tag{12}$$

$$\dot{\delta} = -\frac{\partial H}{\partial Y} = -U_3 + \lambda\delta, \tag{13}$$

and satisfy the transversality conditions

$$p(T)[K(T) - K_T] = 0; \tag{14}$$

$$\delta(T)[Y(T) - Y_T] = 0. \tag{15}$$

Necessary conditions that the controls maximize the Hamiltonian are

$$\frac{\partial H}{\partial C} = U_1 - p \leqq 0; \tag{16}$$

$$\frac{\partial H}{\partial G} = U_2 - p \leqq 0; \tag{17}$$

$$\frac{\partial H}{\partial z} = pF_3 - \gamma\delta \leqq 0; \tag{18}$$

with equality when the corresponding variable is positive. That these conditions are necessary follows directly from Pontryagin's maximum principle. Our assumptions about concavity guarantee that they also are sufficient. This follows from a theorem proved by Mangasarian [6], but it is also easy to establish the sufficiency directly:

Let K^*, Y^*, C^*, G^*, z^* be a path that satisfies these necessary conditions, and let K, Y, C, G, z be another feasible path, i.e., a path that satisfies (5), (6), (8), (9), and the nonnegativity constraints. We can now make the following estimates. Since U is concave:

$$\int_0^T U(C, G, Y)\, dt - \int_0^T U(C^*, G^*, Y^*)\, dt \tag{i}$$

$$\leqq \int_0^T [U_1(C - C^*) + U_2(G - G^*) + U_3(Y - Y^*)]\, dt. \tag{ii}$$

By (16), (17), and (13)—if $U_1 - p < 0$, then $C^* = 0$, and $U_1(C - C^*) \leq pC$ because $C > 0$, etc.—

expression (ii)

$$\leqq \int_0^T [p(C - C^*) + p(G - G^*) + \lambda\delta(Y - Y^*) - \dot{\delta}(Y - Y^*)]\, dt. \tag{iii}$$

By equation (5) and partial integration of (iii),

expression (iii)

$$= -\delta(T)(Y)(T) - Y^*(T) + \delta(0)Y(0) - Y^*(0) \qquad \text{(iv)}$$
$$+ \int_0^T \{p[F(K, L, z) - F(K^*, L^*, z^*) - (\dot{K} - \dot{K}^*) - \mu(K - K^*)]$$
$$+ \lambda\delta(Y - Y^*) + \delta(\dot{Y} - \dot{Y}^*)\}\, dt.$$

By equations (9), (11), and (15) and concavity of F and equation (6), expression (iv)

$$\leqq \int_0^T [p(F_1 - \mu)(\dot{K} - \dot{K}^*) - p(K - K^*) + pF_3(z - z^*)$$
$$+ \lambda\delta(Y - Y^*) - \lambda\delta(Y - Y^*) - \gamma\delta(z - z^*)]\, dt. \qquad \text{(v)}$$

By equation (12) and partial integration of (v), expression (v)

$$= -p(T)[K(T) - K^*(T)] + p(0)[K(0) - K^*(0)]$$
$$+ \int_0^T [-\dot{p}(K - K^*) + \dot{p}(K - K^*) + (pF_3 - \gamma\delta)(z - z^*)]\, dt.$$

since (8), (10), (15), and (18) $\leqq 0$, therefore

$$\int_0^T U(C, G, Y)\, dt \leqq \int_0^T U(C^*, G^*, Y^*)\, dt,$$

which proves the sufficiency.

If we assume that K_T and Y_T are chosen so that there exists at least one choice of controls $C(t)$, $G(t)$, $z(t)$ such that the corresponding state variables K and Y satisfy $K(T) \geqq K_T$, $Y(T) \geqq Y_T$; then there exists also an optimal path. This follows from existence theorems in control theory (see [12]) or from existence theorems in the theory of ordinary differential equations (see [2]). It is also easy to see that the optimal path must be unique, because if K, Y, C, G, z differed from K^*, Y^*, C^*, G^*, z^* then we would have a strict inequality in the first estimate in the sufficiency proof above, because U is strictly concave. This shows that the optimal path must be unique.

It is obvious that p and δ can be interpreted as shadow prices: p is the imputed demand price on capital while δ is the imputed demand price for environmental quality. With this interpretation, (12) and (13) express the assumption of perfect certainty about future prices, and (14) and (15) express the usual demand and supply relation at the horizon, if capital is in excess supply, then the demand price is zero and similarly for environmental quality. It is possible to interpret $q = \gamma\delta$ as a demand price for waste disposal services, or as an effluent charge. The effluent charge is thus equal to the value of the deterioration of environmental quality that is caused by one more unit of waste discharge. From (18) it is seen that along an optimal path, firms will discharge wastes in such amounts that the marginal productivity of waste discharges will equal the effluent charge. We can also introduce the wage rate from the following condition:

$$pF_2 - w \leqq 0. \tag{19}$$

With these interpretations, the profit (π) of the firms can be written as

$$\pi = pF(K, L, z) - rpK - \mu pK - wL - qz, \tag{20}$$

where r is the interest rate. The interest rate is equal to

$$\frac{\dot{p}}{p} = -F_1 + \mu, \tag{21}$$

and the own rate of interest is

$$\frac{\dot{p}}{p} + F_1 - \mu = 0. \tag{22}$$

The own rate of interest is zero because there is no growth in labor supply or in technical knowledge in this model. If profits are maximized, it is seen that conditions (12), (18), and (19) are satisfied.

It is interesting to note that if we make the usual neoclassical assumption about the production function H, namely that H is linear homogeneous, and if we make the assumption that the waste treatment function M is homogeneous of degree zero (that is, doubling of production input and doubling of waste treatment input do not change the need for waste discharges), F will be linear homogeneous in K and L and profits will be negative. However, if M is assumed to be homogeneous of degree zero. we must drop the assumption that M is convex. Along an optimal growth path, the firms must therefore be subsidized. This is of course due to the increasing returns to scale that are embodied in this kind of technology. By increasing the use of labor and capital in the same proportion it is possible to increase output in the same proportion without increasing the use of the third factor, waste disposal services. If, however, effluent standards are used instead of effluent charges, the profits will be zero along an optimal path. In this case profit is equal to $pF - rpK - wL$, and profit is maximized subject to $z \leqq \bar{z}$. The first-order conditions are the same as before:

$$\begin{aligned} pF_1 - rp &= 0 \\ pF_2 - w &= 0 \\ pF_3 = q &= 0 \end{aligned}$$

and profit becomes

$$pF_1K + pF_2L - rpK - wL = 0.$$

The explanation is of course that the firms need the scarcity rents that are created from environmental control to cover the deficits. In order to avoid subsidies, effluent standards may therefore be desirable. Effluent charges have other merits, however, which cannot be analysed in an aggregate model like this (see [7]). It is hardly realistic, however, to assume that M is homogeneous of degree zero, and we will not use that assumption.

Steady state

The steady state is defined as the singular solution to (5), (6), (12), and (13), or as the solution to

$$\begin{aligned} C + G + \mu K - F(K, L, z) &= 0 \\ \lambda(1 - Y) - \gamma z &= 0 \\ F_1 - \mu &= 0 \\ -U_3 + \lambda\delta &= 0 \\ U_1 - p &\leqq 0 \\ U_2 - p &\leqq 0 \\ pF_3 - \gamma\delta &\leqq 0 \end{aligned} \tag{23}$$

It is easy to show that there exists a unique solution to this system, as follows:

Consider the maximization problem

$$\max U(C, G, Y)$$

subject to

$$C + G + \mu K - F(K, L, z) \leqq 0$$
$$\lambda(1 - Y) - \gamma z = 0$$
$$C \geqq 0, \quad G \geqq 0, \quad K \geqq 0, \quad z \geqq 0.$$

The constraints are consistent; so there exist values of C, G, K, Y, and z which satisfy the constraints. Moreover, the constraints define a closed subset of R^5. Y is bounded above by 1, and $C + G$ is bounded above by $\max F - \mu K$. The continuous function U is thus bounded above on a closed set and therefore attains its maximum on this set. A maximum $(C^*, G^*, K^*, Y^*, z^*)$ thus exists. As U is strictly concave, this maximum is unique. It can be easily shown that the Kuhn-Tucker theorem may be applied to this problem. This step yields nonnegative multipliers p and δ, such that the Lagrangean function

$$L = U - p(C + G + \mu K - F) - \delta[-\lambda(1 - Y) + \gamma z]$$

has a saddlepoint at the maximum. A necessary and sufficient condition for L to have a saddlepoint at C^*, G^*, K^*, Y^*, z^* is that at this point the following conditions be satisfied:

$$\begin{array}{ll} U_1 - p \leqq 0 & C^*(U_1 - p) = 0 \\ U_2 - p \leqq 0 & G^*(U_2 - p) = 0 \\ -p(\mu - F_1) \leqq 0 & K^* p(\mu - F_1) = 0 \\ U_3 - \lambda\delta = 0 & \\ pF_3 - \gamma\delta \leqq 0 & z^*(pF_3 - \gamma\delta) = 0. \end{array}$$

If $K^* = 0$, then $C^* = G^* = 0$, and the conditions are not satisfied. Thus $K^* > 0$, and $-p(\mu - F_1) = 0$. Obviously $p > 0$, and thus $\mu = F_1$. The system is now identical to system (23), which therefore must have a unique solution.

In this steady state or stationary equilibrium, the interest rate is zero, the demand price for environmental quality is equal to (except for a scale factor) the marginal utility of the environment.

Asymptotic behavior

It is well known that the Euler-Hamilton equations for autonomous problems in the calculus of variation have certain special properties that characterize the extremes. One such property is that the linear approximations to the systems have characteristic roots that appear pairwise symmetric around the origin. Let us derive this property.

In order to simplify the derivation, let us derive the property for an abstract system:

$$\dot{y} = \Delta_p H(y, p, x); \tag{24}$$

$$\dot{p} = -\Delta_y H(y, p, x); \tag{25}$$

$$0 = \Delta_x H(y, p, x); \tag{26}$$

where y is an n-vector corresponding to the vector (K, Y) in our application; p is an n-vector corresponding to (p, δ); and x is an m-vector corresponding to (C, G, z). The symbol Δ_p denotes the gradient of H regarded as a function of p only. The symbols Δ_y and Δ_x are used in an exactly analogous way.

Denote the singular solution by (y^*, p^*, x^*), i.e.,

$$\begin{aligned} \Delta_p H(y^*, x^*) &= 0 \\ \Delta_y H(y^*, p^*, x^*) &= 0 \\ \Delta_x H(y^*, p^*, x^*) &= 0. \end{aligned}$$

We assume that there exists a unique singular solution. Since differentiability is assumed, it is possible to expand (24), (25), and (26) in a Taylor series in the following way:

$$\begin{aligned} \dot{y} = H_{yp}(y - y^*) + H_{pp}(p - p^*) + H_{xp}(x - x^*) \\ + \zeta_y(y - y^*, p - p^*, x - x^*); \end{aligned} \tag{27}$$

$$\begin{aligned} \dot{p} = -H_{yy}(y - y^*) - H_{py}(p - p^*) - H_{xy}(x - x^*) \\ + \zeta_p(y - y^*, p - p^*, x - x^*); \end{aligned} \tag{28}$$

$$\begin{aligned} 0 = H_{yx}(y - y^*) + H_{px}(p - p^*) + H_{xx}(x - x^*) \\ + \zeta_x(y - y^*, p - p^*, x - x^*); \end{aligned} \tag{29}$$

where H_{yp} is a matrix of second-order derivatives. $\partial^2 H/\partial y_i \partial p_j$ and the other H's are other matrices of second-order derivatives. The ζ-functions are such that:

$$\lim_{(y,p,x)\to(y,p,x)} \frac{||\zeta(y - y^*, p - p^*, x - x^*)||}{||y - y^*|| + ||p - p^*|| + ||x - x^*||} = 0. \tag{30}$$

We assume that H_{xx} is nonsingular (which is the case in our application).

We can then solve for $x - x^*$, obtaining

$$x - x^* = -H_{xx}^{-1}H_{yx}(y - y^*) - H_{xx}^{-1}H_{px}(p - p^*) - H_{xx}^{-1}\zeta_x. \quad (31)$$

By substituting this expression for $x - x^*$ in (27) and (28) we obtain

$$\begin{aligned}\dot{y} = (H_{yp} - H_{xp}H_{xx}^{-1}H_{yx})(y - y^*) \\ + (H_{pp} - H_{xp}H_{xx}^{-1}H_{px})(p - p^*) + \epsilon_y;\end{aligned} \quad (32)$$

$$\begin{aligned}\dot{p} = -(H_{yy} - H_{xy}H_{xx}^{-1}H_{yx})(y - y^*) \\ + (H_{py} - H_{xy}H_{xx}^{-1}H_{px})(p - p^*) + \epsilon_p;\end{aligned} \quad (33)$$

where the ϵ functions satisfy a condition similar to (30).

Let us now put aside the ϵ-functions and consider only the linear system. It has the characteristic equation

$$\begin{bmatrix} H_{yp} - H_{xp}H_{xx}^{-1}H_{yx} - \tau & H_{pp} - H_{xp}H_{xx}^{-1}H_{px} \\ -(H_{yy} - H_{xy}H_{xx}^{-1}H_{yx}) & -(H_{py} - H_{xy}H_{xx}^{-1}H_{px}) - \tau \end{bmatrix} = 0. \quad (34)$$

This equation can also be written

$$\begin{bmatrix} H_{yy} - H_{xy}H_{xx}^{-1}H_{yx} & H_{py} - H_{xy}H_{xx}^{-1}H_{px} + \tau \\ H_{yp} - H_{xp}H_{xx}^{-1}H_{yx} - \tau & H_{pp} - H_{xp}H_{xx}^{-1}H_{px} \end{bmatrix} = 0. \quad (35)$$

But (35) implies that if τ_1 is a characteristic root, then so is $-\tau_1$, because apart from τ the determinant is symmetric. The characteristic roots are thus pairwise symmetric around the origin.

I now show that in this system, zero cannot be a characteristic root and that no root is purely imaginary. Note that the characteristic equation can also be written

$$\begin{bmatrix} H_{xx} & H_{xy} & H_{xp} \\ H_{yx} & H_{yy} & H_{yp} - \tau \\ H_{px} & H_{py} + \tau & H_{pp} \end{bmatrix} = 0. \quad (36)$$

Note also that H is linear in p, so that $H_{pp} = 0$, and that in our application H is strictly concave in x and y. This implies that the quadratic form

$$[dx, dy]\begin{bmatrix} H_{xx} & H_{xy} \\ H_{yx} & H_{yy} \end{bmatrix}\begin{bmatrix} dx \\ dy \end{bmatrix},$$

must be negative definite, in particular for dx, dy satisfying $H_{px}dx + H_{py}dy = 0$. (Here dx and dy are m- and n-dimensional vectors, respectively.) It is well known that this implies that the determinant

$$D = \begin{bmatrix} H_{xx} & H_{xy} & H_{xp} \\ H_{yx} & H_{yy} & H_{yp} \\ H_{px} & H_{py} & 0 \end{bmatrix},$$

has the same sign as $(-1)^{n+m}$. As H is nonconcave in x, the determinant

$$[H_{xx}],$$

has the same sign as $(-1)^m$. If we expand the characteristic determinant, we find that the constant term in the characteristic equation is equal to

$$\frac{1}{[H_{xx}]} D.$$

The constant term is accordingly positive, because in our application $n = 2$, $m = 3$. This shows that zero cannot be a characteristic root. In our application, the characteristic equation is fourth degree, and because of the symmetric distribution of the roots, the coefficients for τ^3 and τ are zero. A straightforward but complex computation shows that the coefficient for τ^2 is negative.

Assume now that all roots are purely imaginary. In that case the characteristic equation takes the form

$$(\tau - di)(\tau + di)(\tau - bi)(\tau + bi) = \tau^4 + (d^2 + b^2)\tau^2 + b^2d^2 = 0.$$

This is impossible, however, because the coefficient for τ^2 must be negative.

Assume now that two roots di and $-di$, are purely imaginary, and let the two other roots be $a + bi$, $a - bi$. Due to the symmetric distribution of the roots, $-a - bi$ and $-a + bi$ must also be roots, which is possible only if $b = 0$. The characteristic equation then takes the form

$$(\tau - di)(\tau + di)(\tau - a)(\tau + a) = \tau^4 + (d^2 - a^2)\tau^2 - a^2d^2 = 0,$$

which is impossible because the constant term is positive. We can therefore conclude that no root is purely imaginary.

We have now proved that the linear approximation of (5), (6), (12), (13), (16), (17), and (18) has four characteristic roots of which no one has a zero real part, and such that if τ_1 is a root, then so is $-\tau_1$. For a system of differential equations that possesses these properties, there exists an n-dimensional manifold, such that solutions with initial points on this manifold will converge to the singular solution, and all other solutions will ultimately leave any neighborhood of the singular solution. (See [2], Chap. 13, theorem 4.1.) This means in particular that for an infinite horizon, if the transversality conditions are satisfied (these conditions are no longer necessary with an infinite horizon), the prices must converge to a finite limit and the initial point must be on the stable manifold. (This presupposes that there exists an optimal path when the horizon is infinite; the existence theorems referred to earlier do not hold for an infinite horizon.)

For a finite horizon, the following theorem may be proved (see [8], Chap. 3): Given $m > 0$ and $t_1 > 0$, there exist positive numbers ϵ and N, such that if

$$|K_0 - K| + |Y_0 - Y| + |K_T - K| + |Y_T - Y| < \epsilon,$$

and if $T > N$, then it is true that

$$|K(t) - K^*| + |Y(t) - Y^*| < m\epsilon,$$

for $t_1 < t < T - t_1$.

The interpretation of this theorem is obvious. If the initial point (K_0, Y_0) and the terminal point (K_T, Y_T) are not too far away from the steady state, and if the horizon is long enough, then the optimal path will enter a small neighborhood of the steady state and stay there for most of the time. We have now characterized a very important property of optimal growth paths in this model. In particular, we can use this property as an argument to defend the procedure we will follow in the discussion of short-run macro problems. In this discussion we will assume that the economy is on a steady state. We know that this is approximately true if the economy has followed an optimal path for some time.

Some extensions

The model described above can be extended in several directions:

The Introduction of Time Preferences. We have so far assumed that the government does not discount future utilities, but regards them as equivalent to present utilities. We know however that future streams of goods and services are discounted to present values when actual decisions are made, and it seems therefore probable that the governments do have a positive time preference. Assuming that this time preference can be represented by a constant discount rate σ, the objective function should be written

$$\int_0^T U(C, g, Y)e^{-\sigma t}\, dt.$$

The necessary conditions now take the form

$$\begin{aligned} \dot{p} &= -p(F_1 - \mu - \sigma) \\ \dot{\delta} &= -U_3 + (\lambda + \sigma)\delta \end{aligned}$$

as well as (16), (17), and (18). The transversality conditions become

$$\begin{aligned} e^{-\sigma t}p(T)[K(T) - K_T] &= 0 \\ e^{-\sigma t}(T)[Y(T) - Y_T] &= 0. \end{aligned}$$

The sufficiency proof is carried out almost exactly as before, and the existence theorems referred to above are equally applicable in this case. The proof of the existence of a singular solution is not applicable here, but it is possible to give another proof of the existence.

The big difference is in the distribution of the characteristic roots. It is no longer true that the characteristic roots to the linear approximation are distributed symmetrically around the origin. It is, however, possible to prove that if τ_1 is a characteristic root, then so is $\sigma - \tau_1$. If the discount rate is not too large, then two of the roots will have positive real parts and two will have negative real parts. But the theorems referred to at the end of the last section require only that two roots have negative real parts and two have positive real parts, and they are therefore still applicable. If the discount rate is not too large the same turnpike property will therefore hold in this case, too.

The interest rate is now $r = (\dot{p}/p) + \sigma$; and the own rate of interest, σ. Even in the steady state, the interest rate will be different from zero.

Exogenous Growth of Labor Supply. If labor supply is growing exponentially at a rate n, a similar analysis may be performed only if the production function is linear homogeneous and the waste treatment function is homogeneous of degree zero. But if that is the case, we already know that the production units will be operated at a loss in a market economy (as long as the effluent charge is positive), and must be subsidized. Let $k = K/L$, i.e., the capital labor ratio. The constraints can now be written

$$\dot{k} = F(k, 1, z) - c - g - (n + \mu)k;$$
$$\dot{Y} = \lambda(1 - Y) - \gamma z.$$

The necessary conditions become

$$\dot{p} = -p(F_1 - n - \mu - \sigma)$$
$$\dot{\delta} = -U_3 + (\lambda + \sigma)\delta$$

as well as (16), (17), and (18). The transversality conditions require only the formal substitution of k for K.

The sufficiency and existence proofs are exactly as before. If $\sigma = 0$ the proof of existence of a singular solution may be repeated. Finally, the characteristic roots still have the property that if τ_1 is a root, then so is $-\tau_1$. The turnpike property is therefore still valid for the model with these changes. The interest rate now becomes $(\dot{p}/p) + \sigma + n$.

Autonomous Technical Change. It is possible to introduce Harrod-neutral technical change in the model without changing the basic qualitative properties of the optimal paths if the instantaneous utility function is such that the marginal utilities of private and public consumption are homogeneous of degree -1 as functions of c and g. We will not explore this further, however, because Harrod neutrality in the aggregate production function requires the same kind of technical progress in the function H as in the function M.

Short-run Stabilization Problems

Behavioral assumptions for firms

In the last section the optimal growth path was defined and some of its qualitative properties were derived. However, the possibilities of achieving this growth path by the use of some collection of fiscal and monetary policy instruments were not discussed. If the economy is completely centralized and if the government has perfect information on the production possibilities (which, in any case, is necessary for the definition of the optimal path), then it is of course possible to implement any feasible growth path, and in particular the optimal growth path. If, however, the economy is decentralized and if the government has only a limited collection of policy instruments (such as the income tax rate, the money supply, the effluent charge, etc.) then it is no longer certain that the optimal path can be implemented. Here, I do not go into this new and exciting field for economic analysis (but see [1] for one approach). My aim is much more modest. I start with the assumption that the economy is already at an optimal path and discuss the short-run adjustment problems (unemployment, inflation) that may appear due to various imperfections that are not accounted for in the model discussed above. Here, the discussion has as its starting point an employment model extended so as to include waste generation and environmental quality.

Let us recall the definition of profits from the second section. The present value of future profits is

$$\int_0^T (pF - p\dot{K} - \mu pK - wL - qz)e^{-rt}\,dt + p(T)K(T)e^{-rt}; \tag{1}$$

where r is the interest rate. Maximizing profits yields the following necessary conditions (which also are sufficient due to our concavity assumptions):

$$\frac{\dot{p}}{p} = -(F_1 - \mu - r); \tag{2}$$

$$w = pF_2; \tag{3}$$

$$q = pF_3. \tag{4}$$

At any given time, the demand for capital, labor, and waste disposal services are functions of the relative prices q/p, w/p, and the real interest rate $r - (\dot{p}/p)$. This implies that the demand for net investment is a function of the time derivatives of these real variables. The idea of a schedule of the marginal efficiency of investment, giving a functional relation between the rate of interest and the rate of investment, is thus not a meaningful concept in this kind of model. This point has been particularly stressed by Haavelmo [3] and by Arrow and Kurz [1]. We will however modify our assumptions somewhat in order to be able to derive an investment function which includes the real interest rate.

In the last section it was shown that an optimal path will for the most part stay in a neighborhood of the stationary equilibrium, provided that the time horizon is long enough and that the initial endowments of capital and environmental quality are not too different from the values in the stationary equilibrium. We can therefore assume that the economy is so close to the steady state that the actual economy can be sufficiently approximated by the steady state. From here on I will therefore characterize the economy by the stationary equilibrium:

$$F(K, L, z) - \mu K - C - G = 0$$
$$Y = 1 - \frac{\gamma}{\lambda} z$$
$$\begin{aligned} F_1 - \mu - r = 0 \quad & pF_2 = w \\ U_3 - \lambda\delta = 0 \quad & pF_3 = q \\ U_1 = U_2 = p \quad & \gamma\delta = q \end{aligned} \tag{5}$$

r is the interest rate and is equal to the utility discount rate. In this stationary equilibrium, net investment is by definition zero and gross investment equals μK.

As before, the profits of firms are

$$\int_0^T (pF - p\dot{K} - \mu pK - wL - qz)e^{-rt}\, dt + p(T)K(T)e^{-rt}.$$

In stationary equilibrium, the necessary conditions for maximum profits become

$$pF_1 = \mu + r; \tag{6}$$

$$pF_2 = w; \tag{7}$$

$$pF_3 = q. \tag{8}$$

We will study transitions between different stationary equilibria of firms. If any of the prices change, so will the optimal factor input. We will assume that firms can immediately adjust their input of labor and their waste generation to changing prices, but that adjustment of the capital stock takes time. Assume that the transition time between two different stationary equilibria is $\bar{t}$. The difference in the capital stock between the two times constitutes net investment demand during this time period. This net investment demand is determined from (6), (7), and (8), and is obviously a function of $r, w, q,$ and p.

The demand for labor and for waste disposal services is not determined from this set of equations, however, because (6) and (7) describe perfect adjustment of all factors, while the capital stock is not adjusted immediately. Neither is the demand for labor and waste disposal capacity determined by (7) and (8) (keeping the capital stock constant), because the F-function is defined as the maximum output when capital and labor can be perfectly adjusted. In order to derive the short-run demand for labor and waste disposal services, we have to consider short-run profits:

$$\max pH(K_1, L_1) - w(L_1 + L_2) - qM(K_1, L_1, K_2, L_2); \tag{9}$$

where the maximum is taken over L_1 and L_2. This yields the following necessary conditions that determine the short-run demand for labor and short-run supply of goods and residuals:

$$pH_2 - qM_2 - w = 0; \tag{10}$$

$$- qM_4 - w = 0; \tag{11}$$

$$z - M = 0. \tag{12}$$

We assume that capital stock is constant in the transition period and then suddenly changes to the new stationary equilibrium. In consequence we also assume that short-run demand is constant in the transition period, and then at the end of this period suddenly changes to the new equilibrium demand. These assumptions are in contrast to the assumption, in the preceding section, of a gradually changing capital stock. This change in assumptions is not fundamental, however, because it is possible to reach the conclusions to be derived below by maintaining the idea of a continuous changing capital stock, but at the cost of some technical complications. In effect, what we are doing is switching from a model with continuous time to a model with discrete time.

We can without loss of generality assume that $\bar{t} = 1$. This implies that we can identify the total short-run demand during the transition period from the solution to (10), (11), and (12). Let us summarize the implications of these assumptions: The demand for capital is given by equations (6), (7), and (8); and the demand for labor and waste disposal services, by equations (10), (11), and (12). We thus obtain the following behavioral functions for firms:

$$I = I\left(\frac{w}{p}, \frac{q}{p}, r\right); \tag{13}$$

this gives the demand for new capital goods as a function of the real wage rate, the real effluent charge, and the interest rate;

$$L^D = L\left(\frac{w}{p}, \frac{q}{p}\right); \tag{14}$$

this gives the demand for labor during the transition period. Note that the interest rate does not appear as an argument in this demand function because the capital stock is kept constant.

$$z = z\left(\frac{w}{p}, \frac{q}{p}\right); \tag{15}$$

this gives the demand for waste disposal services:

$$S = S\left(\frac{w}{p}, \frac{q}{p}\right); \tag{16}$$

this gives the supply of goods produced and services. This function is defined by

$$S\left(\frac{w}{p}, \frac{q}{p}\right) = H\left[K_1, L_1^D\left(\frac{w}{p}, \frac{q}{p}\right)\right],$$

where $L_1{}^D$ is the demand for labor input for production of goods.

Using the assumptions we already have made for the production function and the waste treatment function, it is possible to show that

$$\frac{\partial S}{\partial p} > 0, \quad \frac{\partial S}{\partial w} < 0, \quad \frac{\partial S}{\partial q} < 0$$

$$\frac{\partial I}{\partial p} > 0 \qquad \frac{\partial I}{\partial r} < 0.$$

$$\frac{\partial L^D}{\partial w} < 0$$

$$\frac{\partial z}{\partial p} > 0 \qquad \frac{\partial z}{\partial q} < 0.$$

If the following additional assumptions are made for the waste treatment function, that

$$M_{11} - M_{12} > 0 \tag{i}$$

$$M_{22} - M_{12} > 0 \tag{ii}$$

$$M_{11} - 2M_{12} + M_{22} > 0 \tag{iii}$$

$$M_2M_{11} - M_{12}(M_1 + M_2) + M_1M_{22} < 0, \tag{iv}$$

it is possible to show that

$$\frac{\partial L^D}{\partial q} = \frac{\partial z}{\partial w} > 0$$

$$\frac{\partial L^D}{\partial p} > 0.$$

We also assume that $(\partial I/\partial w) < 0$ and $(\partial I/\partial q) > 0$.

This last assumption can be justified in several ways. The argument used here is the following. An increase in q means that the cost for waste disposal increases. If, as seems to be the case, waste treatment is a capital-intensive activity, more capital will be allocated to waste treatment than the corresponding fall in the optimal capital stock in production. The net result is thus an increase in the demand for new capital goods.

These assumptions are to a very large extent quite arbitrary, but the point is that they are not completely unreasonable. The conclusions drawn will therefore have some validity, and they will show the necessity for great care in discussing the macroeconomic short-run effects of changes in the effluent charge.

We have now completely characterized the behavioral functions for firms. We have however not introduced any financial factors. The analysis has been carried out in real terms. I will shortly show that it is necessary to introduce some monetary theory into the model in order to study the effects of changes in the effluent charge.

The necessity of a monetary theory

Before discussing the role of the government and the behavior of consumers in any detail, let us note the following important implications of a general equilibrium model for a barter economy in which an effluent charge is imposed on the waste dischargers. It is well known that price homogeneity implies that only relative prices are determined. In the absence of credits and money the government's budget must be balanced, and Walras's law implies that it is sufficient to study only one market, say the labor market (recall that in our model there are only two markets, one for labor and one for produced goods; the effluent charge is a tax). We have already derived the demand for labor as a function of the real wage rate and the real effluent charge:

$$L^D\left(\frac{w}{p}, \frac{q}{p}\right).$$

If the supply of labor is completely inelastic, which we assume it to be, and equal to L, we have as the condition for equilibrium in the economy

$$L^D\left(\frac{w}{p}, \frac{q}{p}\right) = L,$$

where q is determined by the government. But this means that given q, all relative prices are determined. But as long as the price p is not determined in the model, a change in q will not have any effect on the real variables, because the price level will change in the same proportion. It is therefore not possible to achieve a decrease in the flow of wastes by increasing the effluent charge, because the generation of wastes is itself a function of relative prices only.

The simple explanation for this result is that q is not an equilibrium price determined on a market. If we had added a supply function for waste disposal services to the model, this result would not have been obtained. But—and this is the important point—a supply function implies a kind of effluent standard, which is not implied by an exogenously determined effluent charge. If there is a market for waste disposal services, then the waste dischargers must not discharge more residuals than the supply of such services and they have to pay the equilibrium price on this market. On the other hand, effluent charges imply that the discharge of residuals is in no way limited by the supply, and the charge cannot therefore be interpreted as an equilibrium price.[1]

It seems necessary, therefore, to introduce some mechanism by which the price level is determined, in order to analyze in a fruitful way the effects of effluent charges. In a later section, I will specify a Patinkin-like monetary model [10], in which the price level is determined endogenously, but for the moment assume that the price level is in some way fixed and that $p = 1$. The equilibrium condition can now be written as $L^D(w, q) = L$, and it is possible to investigate the effects of a change in q. We know that

$$\frac{dw}{dq} = -\frac{L_2^D}{L_1^D},$$

which can be positive or negative depending on the sign of L_2^D. If the demand for labor increases with an increase in the effluent charge, the wage rate will increase. This gives us a hint that for short-run stabilization policies, an increase in the effluent charge may imply an increase in the activity in the economy. This possibility will now be studied more closely by using an extension of the simple model set out above. In doing this, I will follow [5] closely.

1. If instead of using a charge as a tool in environmental policy, the authorities construct property rights which are negotiable on a market, this difficulty would not have shown up.

Government and Consumers

We have already derived behavioral functions for the firm, giving the demand for capital goods, labor, and waste disposal services and the supply of goods as functions of the real wage rate, the real effluent charge, and the interest rate. These functions were derived, however, for the case of a barter economy. We now assume that the firms do not want to hold money balances and that investments are financed by borrowed funds. This means that we can mantain the behavioral functions already derived. We have to add one more behavioral function, however, giving the net demand for credit instruments.

We assume that bonds are the only type of credit instrument in the economy. The bonds are perpetuities and yield \$1 per period. The price of a bond is therefore equal to $1/r$. Let B^f be the number of bonds the firms want to hold. The real value of these bonds is B^f/pr, and the net demand for holding bonds is given by

$$\frac{B^f}{pr} = -I\left(\frac{w}{p}, \frac{q}{p}, r\right) - K. \tag{17}$$

The aggregate profit becomes (we will ignore capital depreciation for simplicity):

$$\pi = pS - wL^D - qz + B_0{}^f; \tag{18}$$

where $B_0{}^f$ is the interest to be paid on the initial holding of bonds. This definition of aggregate profit is consistent with our earlier discussion of the firm.

Let us now turn to the government. The government demands goods and services in the quantity G. It finances its expenditure by the proceeds from the effluent charge (qz), by raising the income tax T, and from interest on government bonds B^g. Any deficit is financed either by an expansion in the money supply or by selling bonds. Let us assume that initially the government has no debt. The government budget can then be written:

$$pG - T - qz = \Delta M + \frac{B^g}{r}. \tag{19}$$

B^g will in general be negative, ΔM is the expansion (or contraction if negative) in the money supply. G, T, q, and B^g/pr are assumed to be determined by the government. z is determined from (15), and ΔM is determined as a residual.

Let us now turn to the consumers. Let M_0 be the cash balance held by households at the beginning of the transition period, and let $B_0{}^h$ be the number of bonds held by households at the same point of time. The gross income or wealth of consumers is then

$$E = wL^s + B_0{}^h + \pi + M_0 + \frac{B_0{}^h}{r}; \tag{20}$$

that is, the sum of wage income, interest on bondholdings, profits, initial cash balance, and value of initial bondholding. The income is spent on the purchase of consumer goods C, on demand for real cash balances, on demand for real bondholding, and on income taxes. The budget constraint can therefore be written:

$$E - T = pC + M + \frac{B^h}{r}.$$

The demand functions can now be written (we maintain the assumption that the supply of labor is completely inelastic and equal to L):

$$C = C\left(\frac{E - T}{p}\right) \tag{21}$$

$$\frac{M}{p} = m\left(\frac{E - T}{p}\right) \tag{22}$$

$$\frac{B^h}{pr} = b^h\left(\frac{E - T}{p}\right). \tag{23}$$

The Complete Model

We have now specified the behavioral functions and are now in a position to give a complete presentation of the short-run macro model.

Equilibrium in the market for commodities:

$$S\left(\frac{w}{p}, \frac{q}{p}\right) = G + C\left(\frac{E - T}{p}\right) + I\left(\frac{w}{p}, \frac{q}{p}, r\right). \tag{24}$$

Equilibrium in the labor market:

$$L^D\left(\frac{w}{p}, \frac{q}{p}\right) = L. \tag{25}$$

Equilibrium in the bond market:

$$-I\left(\frac{w}{p}, \frac{q}{p}, r\right) - K + b^g\left(\frac{E-T}{p}\right) + B^g = 0. \tag{26}$$

Equilibrium in the money market:

$$m^h\left(\frac{E-T}{p}\right) = \frac{M_0}{p} + \Delta M. \tag{27}$$

If we substitute the expression for total profit given in (18) into (20) we get the definition of income:

$$E = pS\left(\frac{w}{p}, \frac{q}{p}\right) - qz\left(\frac{w}{p}, \frac{q}{p}\right) - prK + B_0{}^h + \frac{B_0{}^h}{r} + M_0. \tag{28}$$

Finally we have the change in money supply:

$$\Delta M = T + qz\left(\frac{w}{p}, \frac{q}{p}\right) - pG - \frac{B^g}{r}. \tag{29}$$

We have thus six equations in the five unknowns, w, p, r, E, and ΔM. According to Walras's law one equation is redundant, however, and we can drop (27). At the same time we can also drop equation (29), and we are left with four equations in the unknowns w, p, r, and E. The variables K, $B_0{}^h$, M_0 are historically given; and G, T, b^g, and q are determined by the government. We assume that the system has a unique, economically meaningful solution.

It is now possible to examine the effects of changes in the governmentally controlled variables. The system is, however, too complicated to admit a straightforward determination of these effects. We will introduce one imperfection in the model, namely, that the wage level is fixed from outside at a higher level than the equilibrium level. This means that there will be unemployment. If, through an expansion in aggregate demand the demand for labor increases the full employment level, we assume that the wage rate will adjust upward. Our assumption then is simply that the wage rate will not fall when there is excess supply of labor, but will increase to a new equilibrium level if there is excess demand for labor.

We will also assume that a goal of the government is to maintain the interest rate at the level associated with the new steady state. The interest rate will therefore be considered as a constant (and equal to $\bar{r}$). The government's demand for goods and services is determined by the long-term goal discussed in the previous section. The effluent charge is determined by environmental considerations.

These assumptions imply that we can drop equation (27) and substitute $\bar{r}$ for r in the rest of the equations, and that we can drop equation (25) and substitute $\bar{w}$ (the exogenously given wage rate) for w. The system is now reduced to two equations, (24) and (28), in the two unknowns p and E. If we differentiate these two equations totally and solve for dp/dq, we obtain

$$\frac{dp}{dq} = \frac{p\,\dfrac{\partial S}{\partial q} - p\,\dfrac{\partial I}{\partial q} - C'\left[S + p\,\dfrac{\partial S}{\partial p} - z(1 + s_q{}^z)\right]}{-p\,\dfrac{\partial S}{\partial q} + p\,\dfrac{\partial I}{\partial q} + C'\left(S + p\,\dfrac{\partial S}{\partial p} - q\,\dfrac{z}{p}\right) - C'\,\dfrac{E - T}{p}}; \qquad (30)$$

where $\epsilon_q{}^z$ is the elasticity of z with respect to q.

If we assume that the market for goods and services is stable, an application of Samuelson's correspondence principle shows that the denominator is negative. If the waste generation function z is such that an increase in the effluent charge increases the revenue from the charge (i.e., if the function is inelastic as a function of q), the expression inside the brackets in the numerator is negative. It is, however, impossible to say anything about the signs of the first two terms in the numerator. The effect on the price level of a change in the effluent charge is therefore ambiguous. The first two terms represent the direct effects on supply and demand for goods and services, while the third term represents the effect on demand for goods and services from a change in disposable income. If this last term can be neglected, then dp/dq is positive. Disregard of the third term can be justified in the following way: If we had divided the consumers into wage earners and capitalists, then the effect on disposable income would presumably have fallen mainly on the capitalists, and as their marginal propensity to consume presumably is lower, the third term will be small. An increase in the effluent charge can therefore be assumed to cause an increase in the price level.

Let us now investigate the effect on the demand for labor. Differentiating (14), we have:

$$\frac{dL^D}{dq} = \frac{\partial L^D}{\partial_q} + \frac{\partial L^D}{\partial p}\frac{dp}{dq}.$$

All terms are positive, and an increase in the effluent charge will therefore increase the demand for labor. This shows that the effluent charge, considered as a tax, has quite different effects than, for example, the income tax. While an increase in the income tax will have contracting effects on the economy, an increase in the effluent charge may have expanding effects. In particular, this implies that if the government wants to increase its demand for goods and services, it cannot finance this increase by an increase in the effluent charge if the economy is close to full employment, because both these measures will presumably increase the total demand to such an extent that inflation will result.

This analysis is based on several almost arbitrary assumptions, and the only way to "prove" the results we have obtained is of course by careful empirical research. The analysis is, however, not void due to this, because it points out the possibility that the effects on total demand of effluent charges may be quite different from the effects of other kinds of taxes. It may therefore not be true that an effluent charge can be regarded as an ideal tax, at least not in the short run with imperfections in the labor market.

Effluent Standards

We can very easily adapt the previous model to the situation in which effluent standards instead of effluent charges are used as instruments in environmental policy. The demand and supply functions of firms will now be functions of the real wage rate, the effluent standard $\bar{z}$, and the interest rate.

$$S = S\left(\frac{w}{p}, \bar{z}\right);$$

$$L^D = L^D\left(\frac{w}{p}, \bar{z}\right);$$

$$I = I\left(\frac{w}{p}, \bar{z}, r\right).$$

The government's budget will now look like

$$M - \frac{b^g}{r} = G - T;$$

and the disposable income of households is

$$E - \dot{T} = pS - rpK + B_0{}^h + \frac{B_0{}^{hh}}{r} + M_0.$$

If we make the same assumptions concerning imperfections in labor market and goals of government, we can compute the effect on the price level of a change in the effluent standard:

$$\frac{dp}{dz} = p\,\frac{-\dfrac{\partial S}{\partial \bar{z}} + \dfrac{\partial I}{\partial \bar{z}} + C'\,\dfrac{\partial S}{\partial \bar{z}}}{p\,\dfrac{\partial S}{\partial p} - p\,\dfrac{\partial I}{\partial p} + C'\left(\dfrac{E-T}{p} - S - p\,\dfrac{\partial S}{\partial p} + \bar{r}K\right)}.$$

Now the sign of the derivative can be determined unambiguously, and $dp/d\bar{z}$ is negative. Stricter standards will therefore mean an increase in the price level, a fall in the supply, and may imply an increase in the demand for labor.

References

1. Arrow, K., and M. Kurz. *Public Investment, the Rate of Return, and Optimal Fiscal Policy*. Baltimore: Johns Hopkins Press, 1970.
2. Coddington, E., and N. Levinson. *Theory of Ordinary Differential Equations*. New York: Wiley, 1955.
3. Haavelmo, T. *A Study in the Theory of Investment*. Chicago: University of Chicago Press, 1961.
4. Hestenes, M. *Calculus of Variations and Optimal Control Theory*. New York: Wiley, 1966.
5. Lindbeck, A. *Monetary-Fiscal Analysis and General Equilibrium*. Helsinki: 1967.
6. Mangasarian, O. "Sufficient conditions for the optimal control of nonlinear systems," *SIAM Journal of Control* 4:139:52.
7. Mäler, K. G. "Effluent Charges versus Effluent Standards." Paper prepared for Conference on Urbanization and the Environment of the International Economic Association. Copenhagen, 1972. Mimeographed.
8. ———. "Studier i Intertemporal Allokering," Stockholm School of Economics, 1969. Mimeographed.

9. ———. *A Study in Environmental Economics.* Forthcoming.
10. Patinkin, D. *Money, Interest, and Prices.* London: Harper and Row, 1962.
11. Rockefeller, T. *Convex Analysis.* Princeton: Princeton University Press, 1970.
12. Strauss, A. *An Introduction to Optimal Control Theory.* Heidelberg: Springer-Verlag, 1969.

COMMENT

Colin Wright, Claremont Graduate School

Two problems that seem to plague individuals who accept invitations to discuss conference papers are those of complete agreement or complete disagreement. When under the spell of the former, the reviewer might experience difficulty in writing a respectable quota of nonsycophantic prose while with the latter the difficulty lies in exercising judicious constraint in not submitting an article in the guise of review comments. Though I do not suffer from either of these maladies, I am afflicted with a problem that might prove more embarrassing than the role of submitting unctuous praise or unremitting criticism—I am not sure that I fully understood the contents of this paper. So in the sense that the paper under review was a "preliminary paper" so shall this review be a "preliminary critique."

This is a most difficult paper to review because the author attempts too much. Though I would not impose the requirement that papers at conferences such as this be polished pieces of scholarship, neither would I wish that an author subject his readers to *several* papers under the guise of one paper. Nor would I refer him at crucial points in the development of the paper to as yet unpublished or generally unavailable papers. A further difficulty that I experienced and, I would think, many others would experience, is related to our specialization of research interest and research techniques. My bag of tricks includes rudimentary aspects of Pontryagin's maximum principle, but falls short of manipulating n-dimensional stationary manifolds with fourth degree characteristic equations having symmetric roots! Because of these gaps in my understanding I accepted unquestioningly the author's pronouncements of certain steady state conditions and have allowed him to chauffeur me along the turnpike.

Outline of the model

Mäler has introduced pollution into an aggregate growth model and a short run macroeconomic model. Pollution (z) is generated by the production of a single good (Q) produced in an idealized economy. Thus we have a utility or felicity function that depends not only upon consumption (divided into private and public in this model and noted as c and g respectively) but also depends negatively upon an index of environmental quality (y). The index of environmental quality is related to the output of waste by the differential equation

$$\dot{y} = \lambda(1 - y) - \gamma z. \tag{1}$$

It should be noted for future reference that the current waste disposal rate (i.e., emission rate) does enter the utility functional though indirectly through the environmental quality index (y). At a stationary equilibrium we have

$$y = 1 - \frac{\gamma}{\lambda} z. \tag{2}$$

A production function for the single good is noted as

$$Q = H(K_1, L_1) \tag{3}$$

and what might be called a pollution production function is written as

$$z = M(K_1, L_1, K_2, L_2) \tag{4}$$

with the presence of K_1 and L_1 to signify that not only the amount of Q but the way in which Q is produced may affect the production of waste.

The production function for Q and the pollution production function are aggregated to obtain

$$Q = H(K, L, z). \tag{5}$$

This is accomplished by engaging in "technical optimization." It is then noted that F is a concave function of all of its arguments. (It is also claimed that if H is strictly concave, so is F. With a "little help from my friends" I have been informed that this is not technically correct, and that in the theory of technological optimization the "strictly concave' argument holds only if the constraints and conditions in the optimization problem hold with equality. With inequality conditions only concavity can be proved. This is only a minor point, however. (See Daniel C. Vandermeulen. *Linear Economic Theory*. Englewood Cliffs, N.J.: Prentice-Hall, 1971, Chapter 8.)

After all of these preliminaries, the author invokes Pontryagin's maximum principle to determine the characteristics of the optimum path for the controls c, g, and z. Sufficient conditions are derived and certain characteristics of the steady state are outlined—specifically the characteristic roots of a linear approximation to the system are shown to appear pairwise and symmetric around the origin. A few extensions to the basic model are considered briefly—time preference is introduced, exogenous growth in the labor supply is introduced, and autonomous technical change is considered.

This dynamic model is then set aside for a consideration of short run stabilization problems. The way in which the short run problems in this section are handled is rationalized by the discovery that the optimal path will stay most of the time in a neighborhood of the stationary equilibrium provided that the time horizon is long enough and that the initial endowment of capital and environmental quality are not too different from the values in the stationary equilibrium. The last assumption is, in my view, an assumption of heroic proportions if some semblance of relevance and or reality lurks behind his model and does, I think, take the author beyond the province of those necessary assumptions involved to make the mathematics "manageable."

Taxes and standards

The development of Mäler's dynamic model appears to be technically flawless and some of his conclusions appear to be intuitively obvious. What I should like to consider is some of his pronouncements relative to his steady state conditions. I will do so, however, on my own terms which, I feel, are more enlightening and yet do not fall into the category of responding unduly to the second of the reviewer's curses.

To make these comments I first introduce the Lagrangean

$$\begin{aligned} u^I[c, g, y(z)] &+ \sum_{i=1}^{I-1} w^i[u^i(\cdot) - u^{-i}] + \phi_1[H(K_1, L_1) - c - g] \\ &+ \phi_2[z - M(K_1, L_1, K_2, L_2)] + \phi_3[K - K_1 - K_2] + \phi_4[L - L_1 - L_2]. \end{aligned} \tag{6}$$

Note that I use the same utility function as Mäler, but adjusted only by defining (y) as a function of (z), as in (2), and thereby eliminating the need for a constraint in the form of (z). I do not engage in technical optimization. The w^i are Lagrangean multipliers, which may be interpreted as welfare weights, all normalized on w^I.[1] Differentiating this system with respect to c, g, z, K_1, K_2, L_1, and L_2, rearranging, and noting

$$-w^i u_y{}^i \frac{\partial y}{\partial z} \text{ as } \lambda,$$

which we shall note as λ, we obtain

$$\begin{aligned}
&\text{a. } w^i u_c{}^i - \phi_1 = 0 && i = 1, \ldots, I \\
&\text{b. } w^i u_g{}^i - \phi_1 = 0 && i = 1, \ldots, I. \\
&\text{c. } \phi_1 H_1 + \lambda M_1 - \phi_3 = 0 && \\
&\text{d. } \qquad\; + \lambda M_3 - \phi_3 = 0 && \\
&\text{e. } \phi_1 H_2 + \lambda M_2 - \phi_4 = 0 && \\
&\text{f. } \qquad\; + \lambda M_4 - \phi_4 = 0 &&
\end{aligned} \tag{7}$$

These equations define Pareto optimum conditions—the choice of the w^i singling out a specific optimum. What is required is a mechanism by which utility maximizing individuals and profit maximizing firms effect such an optimum.

Utility maximizing individuals confronted with given prices will equate their marginal rates of substitution to the price ratio. Interpreting ϕ_2 as the shadow price of c and g (which are the same in a model such as this and such as Mäler's) assures us that conditions (7:a, b) are satisfied. The w^i, which are the reciprocals of the individual's marginal utility of income, are manipulated by the choice of head taxes h^i. It is in this manner that a specific Pareto optimum is determined.

We now introduce the following profit function

$$\pi = pH(K_1, L_1) - \rho(K_1 + K_2) - \omega(L_1 + L_2) - T[M(K_1, L_1, K_2, L_2)]. \tag{8}$$

This differs slightly in form from Mäler's but not in content. The change I have introduced is that of a tax function. Such a function is directly related to the amount of pollution (z). However, since z is a function of the distribution of capital and labor, we can invoke the function of a function rule and express the tax function as $T[M(K_1, L_1, K_2, L_2)]$. Differentiating this function with respect to those variables under the control of the firm, the K_i and the L_i, we obtain

1. See Robert Strotz, "Urban Transportation Parables," *The Public Economy of Urban Communities*, J. Margolis, editor (Baltimore: Johns Hopkins Press, 1965).

$$\begin{aligned} &\text{a. } pH_1 - \rho - T'M_1 = 0 \\ &\text{b. } \quad\;\; - \rho - T'M_3 = 0 \\ &\text{c. } pH_2 - \omega - T'M_2 = 0 \\ &\text{d. } \quad\;\; - \omega - T'M_4 = 0. \end{aligned} \tag{9}$$

Noting that ϕ_3 and ϕ_4 are the shadow prices of capital and labor respectively we assert their equivalence to ρ and ω. Defining T' as the change in the tax bill due to a change in (z) and $T'M_i$ as the marginal tax rate (or subsidy since M_3 and M_4 are negative) due to the incremental use of the *i*th factor as it enters the pollution production function, we note that $\Sigma w^i u_y^i(\partial y/\partial z)$, the welfare weighted sum of marginal disutilities due to the deterioration of environmental quality, is equal to T'. Thus the tax imposed upon a firm is related to the use of the various inputs and the effects such inputs have upon the production or reduction of pollution and the effect such changes have upon the level of utility. Thus the firm will be taxed in its use of K_1 and L_1 and subsidized in its use of K_2 and L_2.

Mäler asserts that linear homogeneity in H and zero homogeneity in M require that a firm be subsidized in the face of a pollution charge, i.e., the tax in (8). We should first note that the assumption of linear homogeneity in models having many firms implies that firm size and the number of firms is indeterminate and marginal products, and hence factor prices, are determined as though there existed a single producer who equates price and marginal cost. This is essentially what Mäler is or should be doing with his single firm model.

We next note that from (9:a) the price of the product (p) in the absence of an appropriately chosen corrective tax-subsidy function is $p = \rho/H_1 = \omega/H_2$, which can be interpreted as price equal to (private) marginal cost. In the light of an optimum correction to the existence of pollution we have $p = \rho + T'M_1/H_1 = \omega + T'M_2/H_2$. Since L_1 and K_1 contribute to pollution and since as z increases the tax would increase, we have $T'M_1$ and $T'M_2$ positive. Thus the cost to the firm of its inputs has increased though not necessarily by the same proportion. As costs increase and price remains momentarily constant, losses will be experienced; but by a change in the industry supply function caused by a decrease in the size and or number of firms brought about by the change in the firms' cost functions, price will increase to the level noted above and output will diminish. Simultaneous with the tax on K_1 and L_1 the firm is encouraged to use L_2 and K_2 through their subsidization (i.e., $T'M_3$ and $T'M_4$ being negative). Surely profits for the remaining firms will be normal, not negative; total profit will decrease though not necessarily its rate.

Either Mäler's confusion, or my confusion of Mäler, comes from his assertion that the characteristics of the H and M functions result in technology exhibiting increasing returns. As he says "by increasing the use of labor and capital (presumably L_1 and K_1 and L_2 and K_2) in the same proportion it is possible to increase output without increasing the use of the third factor, waste disposal." But such a system does not require firm subsidization in the sense I believe Mäler intends it—analogous to the standard case of increasing returns to scale firms operating at price equal to marginal cost. In his model the use of factors in production must be taxed and the use of factors in waste treatment subsidized—the size of the tax and subsidy being determined by the equilibrium value of the emission rate z.

Whether the tax bill is smaller or larger than the subsidy receipts is irrelevant for the analysis at this level since what is required is that factors be paid the value of the (social) marginal product. It can be shown that if the tax/subsidy function $T[M]$ is homogeneous of zero degree, then tax payments would equal subsidy receipts though such equality will not be interpreted as a zero tax by the firm. This follows, of course, because of the effect of the $T'M_i$ upon the relative prices of the factors in different uses.

Should a net subsidy be the case, as Mäler suggests, the question arises as to where funds to pay the subsidy are obtained. In the model presented in this review the receipts from the effluent tax may be used parametrically to pay the subsidy, the balance being paid out of the forced surplus of the head taxes and subsidies that were chosen to achieve a particular Pareto optimum. The "forced surplus" thus narrowing the range of Pareto optima that can be chosen. Mäler's model provides no direct answer to this query.

To investigate certain aspects of effluent standards, which I take to imply that some direct control is imposed upon the rate at which z can be emitted, I wish to return to equations (7). Solving these equations would result in an emission rate of say z^*. If z^m denotes the emission rate in the absence of the corrective tax, then the quantity $(z^m - z^*)$ denotes the level of abatement that the firm has engaged in. This abatement imposes costs upon the firm, but such costs are smaller than the tax saving $[T(z^m) - T(z^*)]$. Emitting pollution at the rate z^* does, in the model using effluent fees, require that the firm not only expend money to hire factors of production for abatement activities but also that it pay a tax of $T(z^*)$. These tax receipts in models such as this could be paid to those suffering the pollution in the form of a lump sum transfer or returned to the polluting firm in the form of a lump sum payment, regarded as being parametric to the firm. Such maneuvers produce income effects only and do not alter the formal marginal conditions.

If effluent standards are imposed and are imposed in such a manner that $\bar{z} = z^*$, the difference between the tax solution and the effluent standards solution is the presence in the former of a tax payment of $T(z^*)$. Such a tax, if viewed as parametric, has only income effects, and does not alter the marginal conditions. The tax formulation and the direct control methods differ in producing the appropriate marginal conditions when $z^* \neq \bar{z}$.

Firms do not operate at zero (normal) profits, as Mäler states, for adjustments in price and output counteract the use of unproductive resources (i.e., K_2 and L_2) by the firm.

Macroeconomic implications

In the second major part of his paper Mäler again causes me some puzzlement. His definition of the demand for waste disposal services as a function of the real wage and the real effluent charge seems at variance with previous sections. Previously such a function was noted as $z = M(L_1, K_1, L_2, K_2)$ which would seem to require that the interest rate be an argument in the new version of this function in the macro-model. In addition, we see that whereas the characteristics of the effluent charge (i.e., the $T'M_i$) were determined by the M function now the new waste disposal function has the effluents as a variable. These confusions arise, I think, because Mäler seems to vacillate between calling and using the M function as a *treatment* function and a *disposal* function. These are conceivably two different genres. If z^m is the amount of pollution emitted in the absence of control and z the amount after control, then $(z^m - z)$ is the amount treated and z the amount disposed of. Such a discrepancy is not sloughed off by letting z^m be constant—especially in a dynamic model—though I have not sorted out all of the ramifications of this difference.

In his discussion of the effect of an increase of the effluent charge on the emission rate Mäler makes the following points: (1) given the effluent charge relative prices are determined, (2) as long as the price level is not determined in the model an increase in the effluent charge will not decrease the emission rate because (3) all prices will change in the same proportion, therefore (4) "it seems necessary to introduce some mechanism by which the price level is determined" and (5) point (3) occurs because a supply function for waste disposal services was not given.

Are not (3) and (4) contradictory? The former implies a certain monetary mechanism—albeit a faulty one. Also, if the effluent charge changes relative prices (as is implied by (1) and which certainly is a characteristic of the model in this review) will not this be sufficient to change the emission rate? That is to imply that an increase in the effluent charge should change the relative prices of labor and capital used in production relative to abatement with a resulting increase in the relative price of output and a resulting decrease in its consumption. Perhaps, however, a one commodity model gets the theorist into trouble on such matters!

What does it mean to have a supply of waste disposal services? It would seem to mean a schedule of alternative quantities supplied at alternative prices. Is not this what the effluent charge *function* does? This function is not exogenously determined as Mäler claims, but it is determined by the marginal products of the treatment function and the (presumed) increasing marginal disutility of degradations to environmental quality. This was reflected in

$$T'M_i = \sum_i w^i u_y{}^i \frac{\partial y}{\partial z} M_i.$$

I personally find the attempt to determine the effect an increase in the effluent has upon the price level uninteresting and again puzzling. On the one hand Mäler assumes an inelastic supply of labor and on the other he assumes that the wage rate may only adjust upwards. The latter allows him to play around with a dichotomous economy of capitalists and workers by which he increases the probability of finding a positive relationship between the effluent charge and the price level. Yet such an assumption raises in my mind the question of how the firm responds to the effluent charge, particularly when it is noted that in the earlier part of the paper such a charge should logically fall upon all factors of production and treatment. Perhaps this is the way by which he is able to assert that the firm needs to be subsidized along its optimal path.

In summary let me add that I found this to be a stimulating paper with certain rough edges and certain problems in moving from Pontryagin's world to Patinkin's world. I am not convinced that the move was made—at least not along my turnpike.

The Technology of Public Goods, Externalities, and the Exclusion Principle

Charles R. Plott, California Institute of Technology and Robert A. Meyer, University of California, Berkeley

Presented here[1] are some notions which we hope will help researchers in their attempts to model various aspects of the complex situations which, currently, come under the heading of "externality problems." Roughly, the major theoretical idea is to exploit the advantages of separating into different structural models the consumer-based activities of consumption and acquisition and the producer-based activities of production, marketing, and revenue collection. The links between these activities can then be used to characterize types or classes of externality problems.

Before continuing, we offer the reader two serious disclaimers. First, we offer no hard results (theorems). In fact, we do not even argue to a specific set of conclusions. Of course, implicit, as well as explicit, in our discussion is our belief that the currently accepted views of how one might go about modeling externality problems are too narrow in scope. Second, we offer no detailed example which would indicate that our classification really works. We simply address some points of view which we find help us in our attempts to lay things out in a coherent manner. We have attempted to streamline the ideas to fit within standard mathematical economics constructions, so that it is possible to use standard tools in applications. The ideas allow attention to be focused on special aspects of externality modeling, thereby allowing, to some degree, specialization of research efforts.

Production is the only basic activity in the standard characterization of a public good [5]. Consumption is not a separate activity since it occurs automatically with production. If an irritant consumed by a person reaches him by means of some chain of events, e.g. irritant consumed depends upon the amount of smoke reaching him and thus upon the amount of smoke emission from the source, $c_i = c_i[s_i(x)]$, then the utility function is simply relabeled. The intermediate consumption variables are eliminated and utility is expressed as a function of the amount of smoke produced at the source, e.g. $\tilde{U}_i\{c_i[s_i(x)]\} \equiv \tilde{U}_i(x)$. Rather than the several variables "irritant 1 consumes, irritant 2 consumes, . . . , smoke at the source," we only have the latter, "smoke at the source."

The fact that there is only one variable involved in the standard model has an immediate and important consequence. Within models that have this attribute, a researcher is not free to interpret actions taken by a receptor as an expression of preference. Any given receptor must realize that the effects on total supply and, thus on his utility, of any action taken by him, depends critically upon the action taken by other receptors. Thus, decisions made by him may reflect both his preference *and* his expectations about the decisions of others, rather than his preference alone. Such expressions cannot be expected to properly guide output decisions. The current efforts at finding a solution such as outlined in [3], [7], and [9], involve some type of pseudo-price mechanism (Lindahl prices) which, by clever means, extracts the true preference from the consumer. This preference then serves as the basis for output decisions. These procedures are most complicated and in view of the complexities may never have applied manifestations. Of course, our speculations along these lines are based on current theoretical discoveries.[2]

This conception of a public good is so broad it comes dangerously close to being empty in content. Almost any problem can be called a public goods problem. One may then inquire about changes in institutional arrangements which assure that the choices of magnitudes made by the choosing agents are coordinated in a manner such that the system equilibria are Pareto optimal. The lack of coordination resulting in the non-Pareto optimal behavior of the system can always be identified as the lack of institutions within the system which would assure Pareto optimal equilibria. The failure of the necessary institution to be voluntarily forthcoming can then be called a "public goods" problem in cases where the system has the proper "convexity" properties. This last sentence is a major point made by Arrow [1].

Let us carry this argument a little further. A monopoly equilibrium is not Pareto optimal (under reasonable conditions). In fact, it can be shown that under a wide class of conditions, consumers could collectively bribe the monopolist to price according to marginal cost. The resulting contract could make both the monopolist and the consumers better off.

Why do such contracts fail to arise naturally thus eliminating the monopoly (or replacing it by a cooperative)? Because the contract is a public good. All consumers benefit from the contract but it is to their strategic advantage to withhold, in hopes that the other consumers will finance the project or at least take on a bigger share.

The point is that the monopoly problem can actually be viewed as a public goods problem. It would appear, in fact, that *any* equilibrium which is not Pareto optimal can be viewed as a public goods problem by definition. The problem is that the public good, the institution that would yield Pareto optimal change, is not forthcoming. While this view may be useful in many cases, it would appear that some effort should be expended in developing concepts, or characteristics of models, which, independent of optimality properties, can be identified as involving externalities. It is upon this "technology" of externality and public goods relationships that we wish to focus.

Two Examples

Before continuing, we will present two examples that motivate almost everything we have to say about the exclusion principle. The examples indicate the delicate interaction between the outcome of a market-type process and the legal-institutional setting. They also serve as examples, within our framework, of how and why traditional solutions to externality problems, which come under the heading "internalization of costs," seem to work. The theme is that property rights can sometimes be used to repair social institutions (or lack of institutions) which are "inefficient."

Demsetz [6] claims that Canadian Indians learned that beavers' hutches could be farmed. If the kill on a pond was too great, the stock of beavers would be exhausted; this, as it turns out, was actually beginning to happen due to an increased demand for their pelts. Consequently, ponds were marked according to the tribe that claimed farming rights. One could say that a title to a pond was established according to who discovered the pond. The one who owned the pond had available to him the beavers that were there. For a hunter, there are two alternatives:

$\alpha_1 =$ hunt only those ponds marked as his;
$\alpha_2 =$ hunt any pond that he comes upon.

We can represent this problem, with a little imagination, in terms of a two-person noncooperative game. The game is reminiscent of the well-known prisoner's dilemma problem. The payoff matrix (see Figure 1) gives the number of beavers obtained per year by each hunter as beaver kill for I, beaver kill for II.

Figure 1

		II	
		α_1	α_2
I	α_1	(5,5)	(3,6)
	α_2	(6,3)	(4,4)

If both hunters follow α_1, no hutch is hunted twice and the equilibrium stock of beavers is sufficiently large to allow an annual kill of 10. The hunters would receive 5 each. If either person hunts both types of ponds, the equilibrium stock is reduced to a level where the annual kill is 9. The hunter following α_2 gets 6 and the other receives 3. With both following α_2 the stock is reduced (due to overkill) to a level which only yields a harvest of 4 each. Because of the prisoner's dilemma attribute of the game, the solution if both follow a dominance principle is (α_2, α_2).

General agreement arose, however, that if I violated the territory of II, then II had the right to exercise certain claims to property possessed by I—he could seek remedy. This change in the institutional structure surrounding the exchange relationship results in a different game. The new game is outlined below.

Let

α_1 = abstain from invading other's property; not,
β_1 = if invaded, exercise rights;
α_2 = invade;
β_2 = if invaded, abstain from the exercise of rights.

The strategies and payoffs are displayed in Figure 2.

The structural payoff of the game dictated by the *physical* technology is the same as before but the *institutional* technology has changed. The

Figure 2

		II			
		$\alpha_1\beta_1$	$\alpha_1\beta_2$	$\alpha_2\beta_1$	$\alpha_2\beta_2$
	$\alpha_1\beta_1$	(5,5)	(5,5)	(6,3)	(6,3)
	$\alpha_1\beta_2$	(5,5)	(5,5)	(3,6)	(3,6)
I	$\alpha_2\beta_1$	(3,6)	(6,3)	(4,4)	(6,2)
	$\alpha_2\beta_2$	(3,6)	(6,3)	(2,6)	(4,4)

game has been expanded according to the following rules governing property rights: the invaded has the right to take from the other individual an amount equal to his own "loss" (5 minus the harvest he actually obtained) plus a penalty cost of one beaver.

Notice the only Nash equilibrium of the game is for both players to use the strategy $(\alpha_1\beta_1)$. Thus, if we can depend upon the outcome being a Nash equilibrium, then by constructing penalties for rights violations—altering the rules of the game—we can make sure the efficient outcome is automatically achieved by the system. In other words, an appropriate specification of property rights has provided a resolution of the prisoner's dilemma.

The second example has features which are a little closer to those economists have dealt with under the heading of social costs and or externalities, even though some major features of the private goods case are present. We can, affectionately, call the games STEAL and PROPERTY. We first view a game the structure of which is outlined by the accompanying "tree" and normal form.

The first player is a seller, player I, who can make his goods available (act 0) or refrain from exposing them to the market (act N). The second player, II, is the customer. He has three alternatives in case the goods

Figure 3

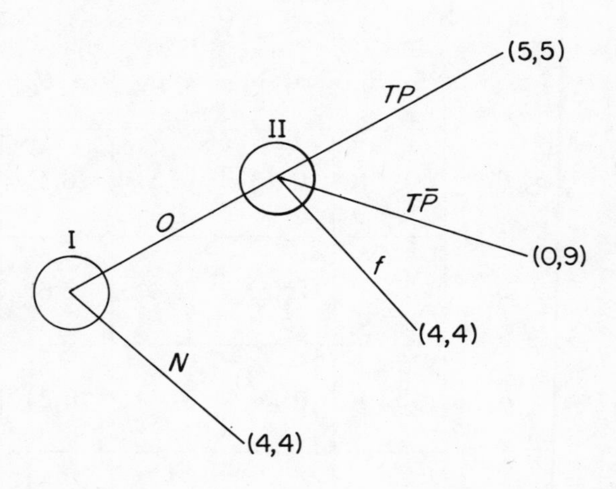

are available. He can appropriate them for himself and pay for them (act *TP,* which stands for take and pay), he can appropriate them for himself and *not* pay for them (act $T\bar{P}$), or he can fail to act (*f*). Now, if act *N* is chosen, or acts 0 and *f* are chosen, no "trade" takes place and player I keeps his goods (valued to him at 4) and player II keeps the money (valued to him at 4). If I plays 0 and II plays *TP,* I gets the money (valued to him at 5) and II gets the goods (valued to him at 5). Suppose, however, that I plays *0* and II plays $T\bar{P}$. Then I loses the goods (ending up with nothing, which he values at 0) and II gets goods and money (total

Figure 4

		II		
		TP	$T\bar{P}$	*f*
I	*O*	5,5	0,9	4,4
	N	4,4	4,4	4,4

value to him of 9) without worry, since, in this game, I cannot "seek remedy."

If we can be assured that individual II will always choose in accord with a dominance law and that I believes this also, then, from this design, we shall obtain a perfectly unambiguous outcome. Simply look at the game in normal form, as displayed. The dominant strategies are $(N, T\bar{P})$. Under this institutional arrangement the goods would never be offered. The structure of the institutions together with our behavioral law (the dominance principle) assures this. This system is "inefficient" [the Pareto optimals are (5, 5) and (0, 9)].

Notice we have a feature of particular interest. A social system organized along the lines modeled by the game STEAL will operate inefficiently—the outcomes are not Pareto optimal. Now, is there an "externality" involved? Is there a "divergence of social *benefits* from private *benefits*"? Yes. The *social benefit* of I's action is $1(5 - 4 = 1)$, while the *private benefit* is -4. In the traditional jargon, we would say that I would undertake a level of his activity below the "optimum amount" (he would offer nothing). Is there a divergence of *social* cost from *private* cost? The answer is again yes. The social cost of II's action is 4 (a benefit of -4 to I), while the private cost is 0. In the traditional jargon we would say that II undertakes too much of his activity (of taking without paying).

Now, let us alter the game of STEAL a little and turn it into a game of PROPERTY. Suppose we give I the *right* to seek remedy in the form of damages, should he want to, in cases where he has made his goods available to II, and II in turn has appropriated them without paying I. The game represented by the tree below is the same as STEAL with the exception of a new move which, depending upon previous moves, *may* become open to player I. This new move involves his right to seek (R), or not seek $(\bar{R})$, "remedy" for the previous actions of II. If II plays $T\bar{P}$ and I, in turn, seeks remedy, the "courts" will return the goods to I and fine II an amount equal to the money he has. The payoff is thus (4, 0). If $\bar{R}$ is chosen by I, then II keeps both the goods (valued to him at 5), and the money (valued to him at 4), while I gets nothing. The payoff then is (0, 9). See accompanying figures.

From the normal form we can see that OR is dominant for I. Player II, however, has no dominant strategy, so if the only admissible behavioral law is dominance, PROPERTY has no solution. It does have a solution if we can rely on II to follow a one-step conjectural variation. If he postulates that I will follow the dominant strategy, then II has a dominant strategy in this conjectured context. The solution is (OR, TP) with the Pareto optimal payoff of (5, 5). If we can rely on a simple behavioral law

Figure 5

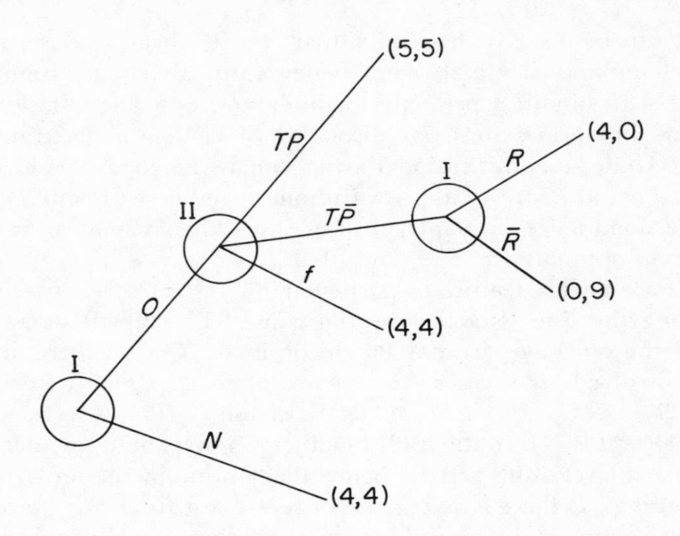

then we are able to use legal institutions to transform an inefficient mode of organization, STEAL, into an efficient mode of organization, PROPERTY.

There are several morals to the story. First, if there is a principle called the exclusion principle, it is likely to involve some interrelationship between behavioral laws and institutional structures. In the case displayed in Figures 5 and 6, the interrelation is a connection between budget constraints. Secondly, calling the problem one of a "lack of markets" due to a lack of sufficient number of participants, as Arrow [1] does, seems to be a gross oversimplification. In the case presented here, there was no apparent "small number" problem as a reader of Arrow might be led to believe. We simply needed to alter the institutional technology.

Finally, representing the problem as one of "transaction costs," as Demsetz tends to do, also seems to be glossing over some important features. There are several different ways by which one could introduce transactions costs into games such as PROPERTY so as to cause problems. For example, if the payments accruing to I as a result of remedy were reduced by a sufficiently large enforcement fee, the dominance principle could *not* be applied or used to guarantee an efficient outcome. Exactly how one goes about translating various altered organizational forms into something called transactions costs is not clear at all.

We do not intend to imply that there are no "costs" of reorganization.

Figure 6

		II	
	TP	$T\bar{P}$	*f*
OR	5,5	4,0	4,4
I $O\bar{R}$	5,5	0,9	4,4
N	4,4	4,4	4,4

There are likely to be costs which must be borne by someone and the covering of these costs does present a type of public goods problem. Both I and II benefit in the reorganization from STEAL to PROPERTY, but the pattern of resulting benefits is very sensitive to the organizational parameters. Consequently, it is not obvious that the altered form will be forthcoming. We are not addressing the problem of how the situation might thus evolve. We only note that lumping "organization" together with other things in a general category called "problems due to transactions costs" precludes a study of the independent features of each. Models that deal with the technology of externality relationships should be separated from models and theories that deal with the evolution of that technology.

Some Economic Quantities and Relationships

The examples above capture many aspects of our point of view. The idea is to take various aspects of "institutions," put them together in a manner which, cleverly taking advantage of a behavioral law, assures that the process outcome will lie in some class of outcomes which have been predesignated as acceptable. We turn, now, to a consideration of institutions. In particular we consider those applicable to "market"- or "economic"-related situations. We ignore situations which involve the problem of "publicness" of organization. That is, we do not address a theory of how organization evolves [14].

Now, as the reader probably suspected all along, we are not going to discuss real institutions. We shall study mathematical representations of institutions. Actually, we shall not even study the representation of any specific set of institutions at all; we shall study mathematical structures which (1) seem capable of capturing the "essence" of the major attributes of certain classes of institutions, (2) use forms typical of economic models, so that the concepts can be applied within standard models, and (3) appear to decompose the natural links among economic activities, allowing public goods to be viewed with respect to the link(s) where problems arise. It is at these links that we suspect the search for real "corrective measures" should concentrate.

Consumption and Appropriation

The links we seek first are those related to the consumer—the consumption and appropriation activities. It is convenient for us to depart from the standard model where consumption, ownership and use are not separated. In doing so, we identify a variable $c_i \geq 0$, $i = 1, 2, \ldots, n$, which we will call the *consumed quantity* commodity of i, and variables $x_i \geq 0$, $i = 1, 2, \ldots, n$, called the *appropriated quantity* of i. These variables are, at this point, scalars, and they should be thought of and labeled with respect to a particular individual.

Consumed quantities correspond to the idea of use or application while appropriated quantities correspond to the idea of ownership or claim. The distinction is important because many legal instruments are founded on such distinctions and because the relationship between the two concepts can serve as an interesting link in the externality chain.

In order to explore this distinction we postulate the existence of a consumption technology, $C(\mathbf{c}, \mathbf{x}) \leq \mathbf{0}$. The vectors $\mathbf{c}$, $\mathbf{x}$ are the consumed and appropriated amounts, respectively. The vector of consumed amounts affects utility, i.e., $U(\mathbf{c})$, and the consumer adjusts both $\mathbf{c}$ and $\mathbf{x}$ to maximize $U(\mathbf{c})$ subject to $C(\mathbf{c}, \mathbf{x}) \leq \mathbf{0}$ plus additional constraints (such as the budget constraint). The expression simply says that the limits of one's consumption activities are set by his activities of appropriation.

There are lots of different ways to think about the c's and the x's. A natural distinction might be a flow vs. stock distinction, where consumption is a flow.[3] A service vs. goods distinction could also be used. The idea of attributes as discussed in [11] being represented by the c's, while purchased commodities are represented by the x's is also appealing. Notice that for purposes of modeling one may or may not want to identify each of the c's with a particular x. For example, if c_1 is apple eating while x_1 is apple ownership, there seems to be some direct relationship between x_1 and c_1, perhaps independent from the other consumption activities and ownership activities. This closed system aspect is clearly not a necessary restriction. Suppose the consumer is interested in the beauty of his home as indexed by, say, the number of times per month that he receives compliments about it. Call this number c_1. The variables x_1, x_2, x_3 could reflect his contracts pertaining to the relative frequency of painting the house, washing the outside and supplementing any paint used with additives. The point of this example is that the seeming independence between the c and the x's is absent. A given level of c_1 may be attainable by several different combinations of the x's.

There are some potential advantages of dwelling upon the distinctions made here. First, the idea of ownership or appropriation can be connected with an *act* performed by someone. This connection will become important in the next section where the possibility of forced ownership is considered, and also later when we discuss the exclusion principle. Secondly, delineations found in legal institutions about the basis and extent of rights are hard to mirror within the standard public goods model unless one is willing to tolerate an endless proliferation of the dimensions of the commodity space. The distinction between consumption and appropriation might do a better job. Legal instruments sometimes base rights on theoretical distinctions about payment and possession as embodied in the concept of title. The act of acquiring ownership conveys certain rights to the owner. By virtue of being an owner, an individual has the right to take certain actions regardless of how others feel about it. Rights can also be founded on distinctions other than that of title. They can be based on the effects of actions. I may not have the right to take *any* action which causes you to experience certain types of discomforts regardless of your perhaps peculiar preferences and regardless of the pattern of ownership. The ownership of a gun does not carry with it the right to shoot people.

Rights which are defined in terms of the admissible actions afforded by ownership can be modeled as constraints on the consumption technology. These rights, when considered in conjunction with other institutions as well as the physical environment, would define, for every pattern of ownership (the amounts of the various goods appropriated), a consumption possibilities set. For example, if I own land I have the right to build any of the structures consistent with local building codes, regardless of how others feel about it. These form a consumption possibilities set. If I appropriate more land I expand alternative building possibilities accordingly. A change in building codes would be a change in rights and can be captured in the model by a change in the set of building possibilities associated with any given land ownership.

In reality it would seem that this class of rights is important since some attitudes are bound to be ignored in actual policy applications. Any small thing we do might offend someone. Dales [4] claims to have searched hard for a purely private good before settling on a real example —slippers. Even this homely example is contained in a paragraph where a paper titled "Costume as a Means of Social Aggression" is cited.

Rights which are conceptually intertwined with ideas of "effects," such as "damage" and or "influence" are related to the "taxes" and "bribes" proposed as solutions to the externality problem. Payment by person A should offset at the margin any undesirable influence exerted on person B who has the right to be free from such influences. A problem with this approach is that "utilities," which in some sense would provide a basis for all concepts of influence, are not observed. More seriously, they might not, and probably do not, in any meaningful sense, even exist. However, if we have $U(\mathbf{c})$ where $\mathbf{c}$ is observed—perhaps by convention where we, the court, or the law, can arbitrarily provide a definition—then, there exists something to work with. As we attempt to characterize the nature of problems and or model the effects of proposed remedies, the obvious magnitudes available for our use are the derivatives dc_i/dx_j and dx_i/dx_j. We can seek ways, perhaps by following the leadership of the courts, to place values on magnitudes like dc_i/dx_j. In the case of dx_i/dx_j we have the natural value of $P_i(dx_i/dx_j)$ when P_i exists, since it is a type of opportunity cost. Suppose I ruined some of your x_1, thereby causing you to adjust x_2; the damage, then, $P_i(dx_i/dx_j)$, could serve as a natural definition for the extent of my liability. Of course, if this number were used as the magnitude of liability in a decentralized pricing system, it could induce inefficiencies, but analysis of that problem introduces the different sphere of welfare economics in general.

We would like to pursue a general theory of consumption technology as influenced by legal parameters. The reason we do not do so is because, at this point, we have nothing profound to say on the subject. We do offer, however, one general characterization: the consumption technology is characterized by *total rejectability* in which case $\tilde{C}(\mathbf{c}, \mathbf{x}) \leq \mathbf{0}$ can be rewritten as $\tilde{C}(\mathbf{c}) - \mathbf{x} \leq \mathbf{0}$. The idea is similar to free disposability. If you are stuck with more of the product, you can get rid of it without cost. In cases of *total rejectability,* a person would never make an effort to own *less* of a variable (unless he could sell it at a positive price). He can simply proceed to "dump" the "trash" without effort or cost. An increase in ownership does not automatically affect the consumption pattern. There cannot be too much of anything. Consequently, total rejectability is, for purposes of finding externalities, not at all interesting since it is the lack of rejectability which can cause problems. However, for purposes of correcting externalities the idea is very important.

Now, clearly the concept of total rejectability is rather gross. It could easily be refined into concepts which allow for rejectability or lack of rejectability of certain select quantities. Perhaps it is possible to say something specific about institutions which foster nonrejectability or even develop a theory along these lines. We have not done so. One thing is clear, however. If all consumption technologies were characterized by perfect rejectability, there could never be a problem with external *dis*economies.

Before moving on, let us reflect on the concept of the appropriated quantity. This concept allows concepts pertaining to what one does at the consumption level to be separated from concepts and theories about how one appropriates or acquires ownership. Exactly what constitutes appropriation can be thought of as being determined by social conventions and or legal conventions. It sometimes involves overt actions. If I file a document proclaiming ownership of a property, then I have appropriated it. My signature on a contract may signify appropriation. If I attach a label, e.g., if I brand a cow, the appropriation is signified. On the other hand, the act of appropriation might be rather circumstantial. If I am in the theater during a show (whether or not I paid), it is assumed that I watched (appropriated). If I carry an item from a store (whether I paid or not), it is assumed that I appropriated it (claimed title)—even though I intended to bring it back.

Later, the concept of appropriation will play a fundamental role in our discussion of the exclusion principle. We will want to directly connect concepts of liability and or payment to the activity of appropriation. If I appropriate the pair of shoes or if I declare ownership over the pair of shoes, then I am automatically liable to the storekeeper for the amount posted on the price tag. Of course, in case I appropriated quietly, it might be difficult for the storekeeper to identify me as the one who is liable to him, but this is another story.

Appropriation and Availability

We now consider the activities of a producer or supplier. We will postulate two basic activities here. One is the act of *production* and to it is attached a produced quantity, y_k. The other is the activity of *distribution* or making available quantities, in terms of what is produced, to the other agents in the economy. The last concept, that of *available quantities,* $q_k = (q_k^1, q_k^2, \ldots, q_k^m)$ where $q_k^i \geq 0$, are interpreted with respect to, and are unique to, each of the consumers (m in number) or other economic agents. Notice now that we have distinguished between appropriated quantity, which, in principle, is controlled by the consumer in question, and the quantity available to that consumer, which, in principle, is controlled by the producer. We have also identified a variable termed the produced quantity. This latter variable will be examined in detail in the next section.

Let us pursue the relationship between appropriated quantities and available quantities. In general, the relationship $x_k^i \leq q_k^i$ must always hold. For example, if I am in a store, the amount of a commodity available to me is (presumably) the entire inventory. I can appropriate up to that amount, depending upon my own strategy in conducting my affairs. A park which is open 10 hours a day is available to me in that amount even though I might appropriate only one hour. The point is that appropriation and availability are, at points, independent concepts. The idea of access and opportunity, as opposed to ownership, can be found repeatedly in legal discussions. It is hoped that the distinction here will capture some of the complexities. We do not want to say that the available amount is necessarily what you pay for or is necessarily the amount you have appropriated. This would leave us no room within which we could characterize stealing, for example, or lack of rights enforcement or liability enforcement. Each of these involves the phenomenon of taking or appropriating without payment. Consequently, to the extent that these types of divergence between private and social costs are of interest, we need separate concepts.

Of particular interest is the case where a distinction between appropriation and availability cannot be made. The representation of this case is $x_k^i \equiv q_k^i$, and can be termed a situation of forced appropriation or lack of appropriation. This is the case where the concept of ownership begins to fade. If something is available to you, then it is yours. The example of smog seems relevant. We could also mention Mead's bees or Coase's rabbits. The point of both examples is that the courts decided that if the creatures are on your property then they are yours, even if you didn't want them. You acquired title through no act of your own. The ambiguity is underlined by the fact that we have not specified who controls x and who controls q, in the case of forced ownership. The agent who controls one must necessarily control both. We chose to discuss the issue on the presumption that the supplier, rather than the consumer, controls the variable. We could have just as easily called it forced availability, rather than forced appropriation, and postulated that the consumer controlled the variable; or we could always call it forced appropriation and label the agent who controls the variable the supplier.

As it turns out, the Mead and Coase examples are of interest because not only are they presented as though the relationship $\equiv$ holds between the appropriated and the available amounts, but also because nonrejectability is present. If, for example, the receptor could avoid the effects of smog by some (costless) means, some of the complexities of the problem might be avoided. The usual public goods cases can be viewed as involving both nonrejectability and forced appropriation.

A convenient feature of the distinction is that on the one hand it allows a characterization of the exclusion principle which will be based on the variable x_k^i, and on the other hand it makes possible certain types of jointness in supply which are based upon the $\boldsymbol{q}_k$. It is to these distinctions that we turn in the next section.

Availability and Production

We now consider the relationship between the vector of available amounts, $\boldsymbol{q}_k$, and the scalar amount produced, y_k. In general, we will represent the relationship by the vector valued function $\boldsymbol{q}_k = T(y_k, \boldsymbol{\alpha})$, where the vector $\boldsymbol{\alpha}$ is a representation of some arbitrary, nonspecified set of controls (real or abstract). The producer can control the amounts which become available to the various individuals by altering y_k and $\boldsymbol{\alpha}$.

Just a glance at the general form $T(.,.)$ indicates that a great number of forms of this mapping would make economic sense, and thus might be applicable for purposes of modeling. For example, one aspect of the usual private goods model where there are m individuals, can be captured by postulating controls of the form

$$\alpha_i \geq 0, \quad \sum_{i=1}^{m} \alpha_i = 1,$$

together with

$$T(y_k, \boldsymbol{\alpha}) \equiv (y_k)\boldsymbol{\alpha} \equiv (\alpha_1 y_k, \ldots, \alpha_m y_k) \equiv (q_k^1, \ldots, q_k^m).$$

This simply says that the supplier is physically able to make a given produced amount, y_k, available at the individual level in any pattern $(q_k^1, \ldots, q_k^m)$ such that

$$\sum_{i=1}^{m} q_k^i = y_k.$$

Actually, the private goods case above is a special case of what can be called *perfect selectivity of supply*. We say that $T(.,.)$ allows perfect selectivity of supply in case the range of the function (it maps on to) is the nonnegative orthant of the Euclidian m-space (a type of invertability assumption). That means that any point in the shaded area of the accompanying figure could be achieved by a proper adjustment of the controls. The producer is able to discriminate between receptors even though some discrimination may be costly. For example, the distribution (0, 5, 0, 1, 1) may take enormous amounts of y_k, while all other points are achievable via the private goods form of $T(.,.)$.

Figure 7

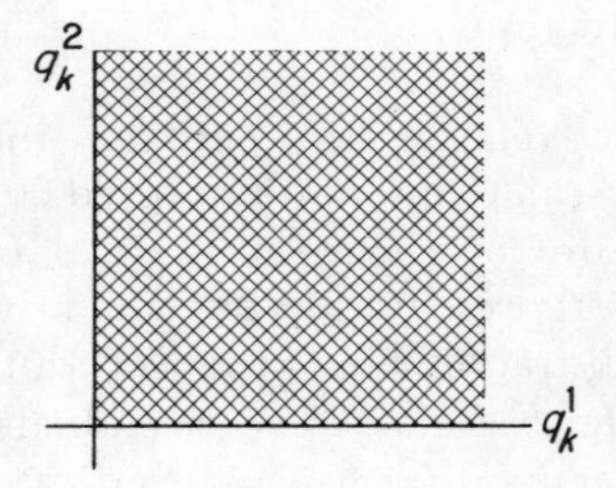

We underline the note that perfect selectivity of supply can operate independently of the other concepts. The case of a private goods world with free disposability can be viewed as a case of perfect rejectability, nonforced appropriation and perfect selectivity of supply. We can take the same example and do away with the free disposal. For example, consider something like Limburger cheese which must be stored at home. If you own it, you cannot simply ignore the effects of the ownership on your consumption activities—it simply stinks. Yet there is perfect selectivity of supply (the supplier is able to remove it as you approach the counter, even though it would be available to any other customer).

Another example might be interesting. Consider a big, fast fellow who supplies handbills on the street. He is fast, so has perfect selectivity in supply (he can reach anyone and or everyone). He is also big (with a slight whiff of ugliness), so there is forced appropriation. If he makes a handbill available to you, then you smile and take it. Now, if it is impossible to throw it away (without considerable effort), you also have nonrejectability. So, here is a seemingly private goods case which has many of the aspects of externalities. There are also many reasonable corrective measures for this problem which would not automatically evolve from our usual manner of modeling such processes: supply trash cans; make him stand in one spot; make him smile; give people tennis shoes so they can outrun him. This final suggestion, we are compelled to add, seems no less relevant to us than the several possible suggestions which involve the use of cost-benefit ratios.

Part of the existing models of public goods can be captured by another special form of $T(.\,,.)$. Suppose, where $\mathbf{1}$ is the vector of ones, we have

$$T(y_k, \boldsymbol{\alpha}) = (y_k)\mathbf{1} = (y_k, \ldots, y_k) = (q_k{}^1, \ldots, q_k{}^m).$$

Here we have something that looks like the traditional public goods case of $q_k{}^1 = q_k{}^2 = \ldots, = q_k{}^m = y_k$. The range of $T(.\,,.)$ as shown is simply the diagonal. We can call this a case of no selectivity of supply.

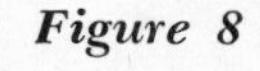

Figure 8

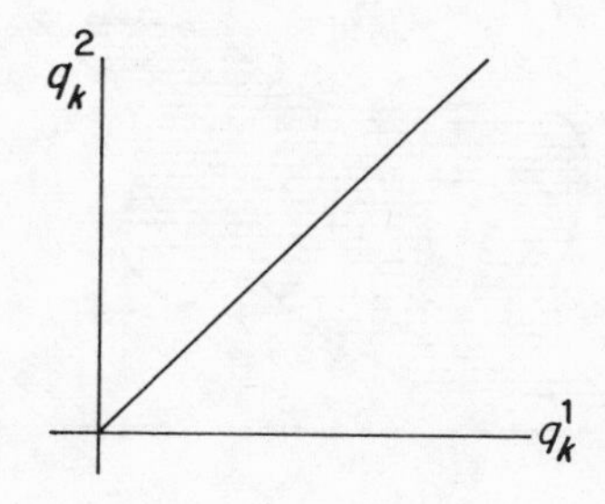

A potential example falling within this class is the case of the park. Park service is necessarily available to everyone in equal amount and that amount is the total supply. The park service can clearly be supplied free of forced availability, the case of $x_k{}^i \leq q_k{}^i$ (we do not consider crowding). We have then the idea that appropriation by one person does not reduce the amount available to others. Of course, this verbal description is frequently taken as a definition of "pure public goods." We can see from the accompanying structure, however, that there is no forced availability and no nonrejectability. In fact, in this case, it is possible for the exclusion principle to be operative as well. Even though the park has aspects of publicness, there is no necessary market failure.

The usual case of national defense should also be considered here. Presumably, there is no selectivity of supply but there is forced availability and (perhaps) nonrejectability. Although, it would seem to be easy enough (things being what they are) to expose oneself to any degree of risk necessary to perfectly offset safety supplied by defense. If this is true then nonrejectability would not apply.

Of course, the examples of $T(.,.)$ discussed so far do not exhaust the possibilities. Consider, for example, the range shown on Figure 9. The idea is that availability to some types of individuals cannot be separated. Consider an airline with two close routes (there is only one control parameter, α, that takes only 2 values) between two mountains. The people situated between the two routes are going to have noise available to them as long as there is any supply and regardless of what the company does. However, if the company chooses route 1, some people are far enough away to receive no noise at all. The same is true with route 2.

Figure 9

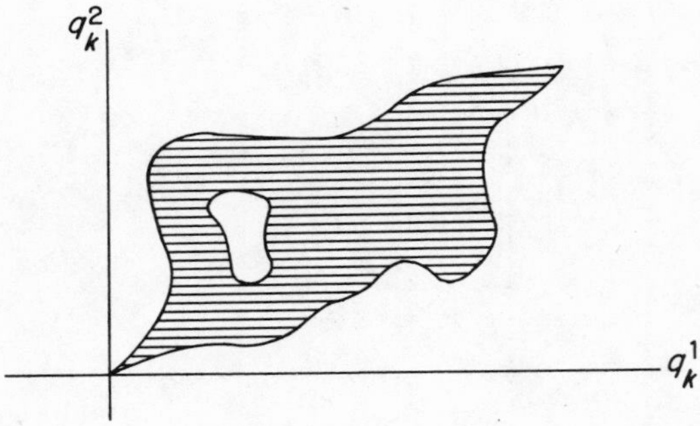

There appears to be no utility in pursuing these structures (which have been called cases of "parametric jointness" [16])[4] at this time. However, the alterations in the form of $T(.,.)$ could potentially provide vehicles for capturing the idea of transaction costs. There may be real costs (perhaps parameterized by α) in making a supply available to one person without making it available to others.

Before leaving the subject, we can offer two general concepts which can reduce speculation over admissible forms of $T(.,.)$. We say supply is *depletable* in case

$$T(y_k, \boldsymbol{\alpha}) \cdot \mathbf{1} \leq y_k \text{ for all } (y_k, \boldsymbol{\alpha})$$

and *nondepletable* in case for all y_k

$$T(y_k, \boldsymbol{\alpha}) \leq (y_k)\mathbf{1} \equiv (y_k, \ldots, y_k) \equiv (q_k{}^1, \ldots, q_k{}^m)$$

and for some $\boldsymbol{\alpha}^*$,

$$T(y_k, \boldsymbol{\alpha}^*) = (y_k)\mathbf{1}.$$

The idea of depletability is simply that the amount produced must be no less than the sum of the amounts available to individuals. The traditional private goods model serves as an example. Nondepletability means that the amount available to each individual is limited only by the total supply. As an example, consider the case of one evening of TV production and broadcast. Suppose the station controlled scramblers on each individual's set. In this case, there would be perfect selectivity of supply and nondepletability, since making the program available to an additional individual would not reduce the amount available to anyone else. The entire evening broadcast would be available to each individual.

The Exclusion Principle

The exclusion principle is probably the most elusive of all the ideas presented here. In fact, from time to time, we have wondered whether it is a principle at all, since we were unable to obtain a really satisfactory characterization. The game examples given in the introduction do give some hints when taken in conjunction with the concepts thus far elaborated.

Here, the dominance principle can be used to considerable advantage. The structure we will propose involves the postulation of a liability function which depends directly upon an individual's act of appropriation. If this function is properly constructed then we will be assured by virtue of the dominance principle that a given individual will act in the proper manner.

Before making the idea more precise, let us return to the game of PROPERTY. We have complicated it in order to make clear the analogy between it and our abstract concept. The seller, player I, will have the first move. He has the alternative to offer, 0, or not to offer, N, the goods for sale. If 0 is played at the first move, the second player has a move with two alternatives. He can appropriate (take) the goods, (T), and make a payment available to I, (A); or, he can take the goods, (T), and *not* make the payment available to person I, ($\bar{A}$). We ignore here the option of not appropriating any of the goods, so he has two alternatives: (TA) and ($T\bar{A}$). If he plays (TA), then player I has the option of collecting the payment (C) or not collecting ($\bar{C}$). The former induces a payoff of (5, 5), since I gets the money, valued to him at 5, and II gets the goods which are valued to him at 5. If $\bar{C}$ is chosen, then I loses the goods and II gets both the goods and the money.

If player II plays $T\bar{A}$ then I has the option of filing suit (R—for seek remedy) or not filing suit ($\bar{R}$). The former leads to a payoff of (5, 0)—player I receives damages of the foregone value of the sale (to him), and player II pays the damages and a fine. If player I does not file suit, then II gets everything.

Figure 10

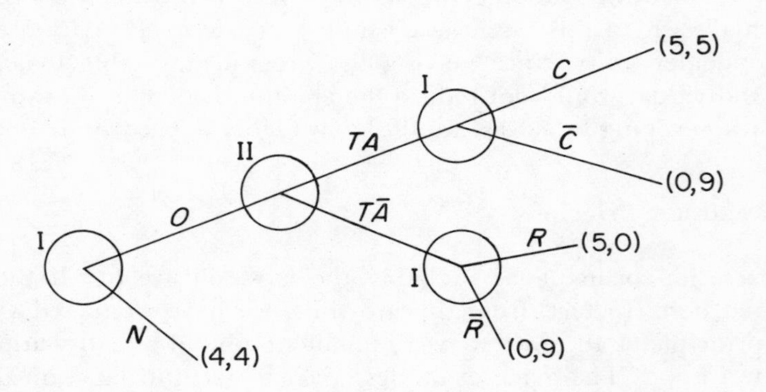

Player I has the dominant strategy of (OCR)—he offers the goods, collects when possible and files suit when collecting is not possible. Player II has no dominant strategy unless he conjectures that player I will play a dominant strategy. In this case, he always plays (TA). The outcome, of course, is Pareto optimal.

The key idea in Figures 10 and 11 is the interrelationship between the act of taking (appropriating) and the options that automatically become available to the other players as a result. In the traditional private goods model where P_x is the price of the good, the payment, $L(x) = P_x x$, is simply a linear function of the amount consumed (x). In our jargon, we would say, where x is the amount appropriated, that the amount $P_x x$ is money made available to the seller. That is, the act of appropriating some amount will automatically induce a liability amounting to $L(x) = P_x x$. Of course, if the seller "likes money" and also follows the dominance

Figure 11

		II	
		TA	$T\bar{A}$
I	OCR	5,5	5,0
	$OC\bar{R}$	5,5	0,9
	$O\bar{C}R$	0,9	5,0
	$O\bar{C}\bar{R}$	0,9	0,9
	N	4,4	4,4

principle, the individual can be assured that the seller will appropriate all of the money made available to him. If the individual can "steal" the commodity, we would say that $L(x) \equiv 0$. That is, the individual's act of appropriation was not accompanied by an induced liability. The seller, in this case, would have no alternatives available as a result of the first person's act of appropriation.

In general, and more abstractly, we suggest the use of a function $L(x_k^i)$ as a representation of the liability which results from appropriation. At a general level, $L(x_k^i)$ would simply identify some set of actions. At a more specific level, we suggest that $L(x_k^i)$ be interpreted in terms of a *reduction* in the amount of some given commodity (perhaps money) available to the appropriating party, individual i. Furthermore, for our purposes, we can simultaneously view the absolute amount of $L(x_k^i)$ as becoming available to the seller. We intend for this function, $L(x_k^i)$, to capture what we call the *circumstances of exclusion*.

The liability function is intended to describe, in a sense, the potential retaliations that one can experience as a result of his act of appropriation. It could, in reality, take any number of interpretations, even though we have suggested that it is a reduction in the amount of something available to him. It could, however, represent a potential increase in the amount of something which he is forced to appropriate and unable to reject: e.g., I am able to park in my neighbor's driveway (appropriate this parking space), but as a result I am forced to tolerate more of his children's climbing on my car. The example is reminiscent of reciprocal externalities.

We have no general theory about exclusion, nor general forms characterizing the circumstances of exclusion, but we can make several observations about particulars. First, the failure of exclusion might imply absolutely nothing about forced availability, nonrejectability, nondepletability, etc. The case of "theft" described above is an excellent example of a private good where exclusion is absent. It is also a case which, within the traditional model, could not be separated. We can also see that the property of nondepletability, i.e., that everyone can appropriate up to the produced quantity, need not be accompanied by the failure of exclusion. Here, we need only consider a park, picnic area, or beach which is privately owned. By suitable organizational arrangement, e.g., a big guard authorized to collect a fee if possible, and backed by laws governing trespassing, the owner is assured that anyone who appropriates some of the amount available to him will incur the proper liability. In other words, from those who enter, the owner is assured of a proper fee. We can worry about crowding, etc., but there appear to be ways of capturing that phenomenon without destroying the nondepletability attribute we have used in the characterization here.

There are, of course, problems with this formulation of exclusion in the cases of forced availability. In these cases, there is really no act with which liability can be associated. As was mentioned above, it is here that even the concept of appropriation and or ownership becomes unclear. However, there is no need to follow the usual path and extend complications that arise in this case to all of the other cases. Things are complicated enough as they are.

We can make the concept, $L(x)$, a little less abstract by discussing some features of processes which would be implied by its use. First, the supplier must be aware of $L(x)$. This implies that when $L(x)$ becomes available, it is accompanied by some type of signal. In the game theory jargon, the individual must be aware that it is his move. In some cases, the signal may occur almost naturally as part of the appropriation process. For example, large discount stores frequently place purchases in a bag, then staple the bag closed with the receipt of payment. Any customer who is observed carrying merchandise which is not so labeled serves as a signal to the supplier that certain options are available to him, e.g., collecting for the purchase. Coin-operated machines are excellent examples of cases where the interrelationships between appropriation and liability are so automatic as to be almost unnoticed. The coins in the machine at the end of the day, in a very trivial sense, signal their availability to the vendor. In some cases the knowledge about $L(x)$ may not be automatic. In fact, the knowledge may be acquired only after considerable search. For example, the actions available to you as a result of my appropriation of an idea on which you have a patent may never be known to you unless you "catch" me "using" it. In passing, we should observe that the knowledge of $L(x)$ itself is of primary importance, as opposed to the knowledge of who appropriated what and how much was appropriated. The knowledge of these other aspects can be viewed as either secondary or indirectly implied by the nature of $L(x)$.

The second feature of processes implied by the use of $L(x)$ is that $L(x)$ really represents some thing or some things which are actually available. After all, $L(x)$ does represent some alternative form of retaliation, so it cannot be vacuous or token. It does little good for someone to be liable to you if it is impossible for some reason or another for you to collect. A supplier who requires the deposit of a driver's license and/or credit card as part of a rental contract is attempting to assure himself that the agreed upon rate can actually be collected by him. In most cases discussed above, the thing represented by $L(x)$ was some kind of monetary payment. This is a very natural instrument to use, because it can be readily transferred into something the seller desires. Other things could be involved. For example $L(x)$ might involve the ability to initiate criminal proceedings. It could involve civil proceedings in pursuit of damages, injunctive relief or review of official decisions. It might be the case, as it is with most market models, that the terms of $L(x)$ are set by the supplier himself in the form of a price.

The point of the above discussions, for those interested in market corrections, is that slight perturbations in institutions can be the means by which the correction takes place. A small change in procedure, such as requiring a signature, might make the seller aware of his rights, and thus correct an externality: simply providing an unambiguous set of retaliatory measures might do the trick.

Concluding Remarks

What is a purely public good? In our jargon, it is characterized by nonrejectability, forced availability and no selectivity in supply. What is a purely private good? In our jargon, it is depletable, it may or may not involve nonrejectability, it may or may not have perfect selectivity of supply, and the exclusion principle applies. Do all permutations and combinations of our concept have a counterpart in reality? We have absolutely no idea.

There is no real divergence between what we have done and what has been developing in the literature. We have sought only to refine the concept of "cost internalization" and the elimination of "unnecessary" transactions costs. In doing so, we hope we have equipped the reader with a point of view which helps him recognize some modes of attack which may have otherwise gone unnoticed.

Perhaps we have been more preoccupied with noncooperative game formulations than some readers would like, but we have not found the characteristic function form of model useful when thinking about these problems, even when it is generalized as in [2] and [19]. Models using the characteristic function form of games collapse all variables, at the individual level, into one variable—the choice of coalition. The representation of institutions then must be introduced through the structure of the "possibilities set" or "payoff set" associated with a coalition and the closely associated description of an outcome of the game. Presumably, when one starts at the extensive form, such as we have done, the model can be collapsed to a characteristic function. But whether or not various standard institutions or technologies have a useful and straightforward natural representation as a characteristic function is simply not known. Some progress along these lines is evident. The interested reader should consult, for example, [15], [17], [21], [23] and, especially, [8], for examples where the technology is readily represented by a characteristic function.

Along with our tendency at the structural level, to avoid cooperative concepts, we have also avoided the use of solution concepts used typically in models of the cooperative form. This is somewhat implicit in what we have done, since the applicability of the concept we have used is limited to the operation of a dominance principle. It also placed limits on the type of formal apparatus we introduced. For example, we did not examine the type of institutions which would guarantee the existence of a core in any associated game model. We chose not to try to isolate institutions or technologies of this sort because we have serious reservations about whether or not the core can actually be depended upon as an operative behavioral proposition. We would aggressively pursue a theory of those institutions which would guarantee the existence of a core, if we were assured that existence of a core in the model was sufficient to guarantee that the outcomes would be restricted to it in reality.

We also diverge slightly from those, such as Drèze, Foley, Munch, Roberts, Samuelson, and others, who have been developing the structure of the Lindahl equilibria. We have made no attempt to systematically integrate our concepts with a set-theoretic general equilibrium model. Perhaps that can come later.

Footnotes

1. The research support for C. R. Plott provided by NSF grant No. GS-36214 is gratefully acknowledged. The word "technology" used in the title was adopted as a result of comments made by Robert Haveman.

2. The most recent direct attack at placing some structure on this problem of preference revelation can be found in Ledyard [12].

3. A model which capitalizes on distinctions along these lines can be found at [10]. They regard pollution as a stock which acts as a parameter on the flow of consumption—although they take "utility" directly to represent consumption rather than some intermediate variable like our variable, *c*.

4. This idea is also closely related to the type of jointness investigated by Smith [22].

References

1. Arrow, K. J. "The Organization of Economic Activity: Issues Pertinent to the Choice of Market *vs.* Nonmarket Allocation," *Public Expenditures and Policy Analysis,* edited by R. H. Haveman and J. Margolis. Chicago: Markham, 1972.
2. Aumann, Robert J. "A Survey of Cooperative Games Without Side Payments," *Essays in Mathematical Economics,* edited by M. Shubik. Princeton: Princeton University Press, 1967.
3. Buchanan, James M. *The Demand and Supply of Public Goods.* Chicago: Rand McNally & Co., 1968.
4. Dales, John H. "Rights and Economics," *Perspectives of Property,* edited by G. Wunderlich and W. L. Gibson, Jr. University Park: Institute for Research on Land and Water Resources, Pennsylvania State University Press, 1972.
5. Davis, O. A. and A. B. Whinston. "On the Distinction Between Public and Private Goods," *American Economic Review* (May 1967).
6. Demsetz, H. "The Exchange and Enforcement of Property Rights," *Journal of Law and Economics* (October 1964).
7. Drèze, J. H. and D. de la Vallée Poussin, "A Tâtonnement Process for Public Goods," *R. E. Studies* (April 1971).
8. Ellickson, Bryan. "A Generalization of the Pure Theory of Public Goods," *American Economic Review* (June 1973).
9. Foley, D. "Lindahl's Solution and the Core of an Economy with Public Goods," *Econometrica* 38 (January 1970), pp. 66–72.

10. Keeler, E., M. Spence, and R. Zeckhauser. "The Optimal Control of Pollution," *Journal of Economic Theory* (February 1972).
11. Lancaster, K. D. "A New Approach to Consumer Theory," *Journal of Political Economy* (April 1966).
12. Ledyard, J. O. "A Characterization of Organizations and Environments which Are Consistent with Preference Revelation," Center for Mathematical Studies in Economics and Management Science, Study Paper No. 5, Northwestern University, September 1972.
13. Meyer, R. A. "Private Cost of Using Public Goods," *Southern Economic Journal* (April 1971).
14. Krier, J. E. and W. D. Montgomery. "Resource Allocation, Information Costs and the Form of Government Intervention," Social Science Working Paper No. 11, California Institute of Technology, September 1972. (Forthcoming in *Natural Resources Journal*.)
15. Montgomery, W. D. "Markets in Licenses and Efficient Pollution Control Programs," *Journal of Economic Theory* (December 1972).
16. Plott, C. R. "Joint Production and the Multiproduct Firm," paper delivered at the 1968 meeting of the Midwest Economics Association, Chicago.
17. Reiter, S. and G. Sherman. "Allocating Indivisible Resources Affording External Economies or Diseconomies," *International Economic Review* (January 1962).
18. Roberts, D. J. "Existence of Lindahl Equilibrium with a Measure Space of Consumers," Department of Managerial Economics and Decision Sciences, Working Paper 99–72, Northwestern University, February 1972.
19. Rosenthal, R. "Cooperative Games in Effectiveness Form," *Journal of Economic Theory* (August 1972).
20. Samuelson, P. "Pure Theory of Public Expenditures and Taxation," *Public Economics,* edited by J. Margolis and H. Guitton. New York: St. Martin's Press, 1969.
21. Shapley, L. S. and M. Shubik. "On the Core of an Economic System with Externalities," *American Economic Review* 59 (1969), pp. 678–684.
22. Smith, V. L. "Dynamics of Waste Accumulation Disposal Versus Recycling," *The Quarterly Journal of Economics* (November 1972).
23. Wilson, R. B. "The Game-Theoretic Structure of Arrow's General Possibility Theorem," *Journal of Economic Theory* (August 1972).

COMMENT

Robert H. Haveman, University of Wisconsin

The paper by Plott and Meyer is designed to provide a new characterization of the standard concepts of private goods, public goods, and externalities. Although this characterization is based on the perspective of game theory and carries with it all of the complexities of that view, the paper also has the flavor of a layman's guide to externality control. As such it is either deceivingly simple or deceivingly complex.

The new characterization grows out of the perspective of game theory in which the behavioral law, represented by the dominance principle, prevails. Because such a framework and principle, together with a multi-person game structure, is theoretically equivalent to the standard economic model, Plott and Meyer offer no new theorems regarding market failures. In this sense, the paper is but old wine in new bottles.

Having said this, however, I confess that Plott and Meyer do seem to get limited mileage out of their analysis of these phenomena. The categories and concepts which they develop tend to focus attention on some fundamental attributes of the market failure phenomena which may be obscured by the standard neo-classical model. However, by dissecting the public goods-externality phenomena, their paper becomes a discussion of the technology of externalities and public goods. Their justification for this detailed look at mechanics is that good policy toward externalities requires a perception and evaluation of the full range of institutional or policy correctives for a market failure.

To convey the flavor of the Plott-Meyer approach, I will examine a few examples of standard phenomena which, when looked at through their glasses, somehow seem a bit richer than before. If this examination does some minor violence to their framework, I am sure they will forgive it.

First, the notion of *private goods*. To Plott and Meyer, a private good is one in which the following characteristics are present:

- *depletability*—the sum of the amounts of a good which are made available are no larger than the amounts produced. This is to be set against *nondepletability* in which the amount of a good available to every individual is limited only by total supply.
- *perfect selectivity*—the provider of a good can target the good (or bad) on whomever he desires.
- *no forced appropriation*—the individual does not have to take any good made available to him—or alternatively, if the good is a bad and he is forced to take it, he can dispose of it at no real cost.
- *a liability function exists*—appropriation of the good results in an immediate liability (it is this which substitutes for the exclusion principle in the Plott-Meyer framework).

At the other extreme, consider the notion of *public goods*. In the Plott-Meyer framework, such goods have the following characteristics, whose meaning can be inferred from the above definitions:

- *non-depletability*
- *no selectivity*
- *forced appropriations,* hence,
- *no liability function exists*
- *no rejectability*

Given these two polar cases, it is clear that the world of nonextreme externalities lies somewhere between them. Indeed, it appears that if any one of the characteristics of the private or public good cases are altered (or more than one appears in any combination or permutation), some phenomenon results which represents an externality-type market failure which may or may not have a counterpart in the real world. A couple of examples will illustrate this conclusion.

Consider first the phenomenon of *theft* as an external diseconomy. The producer-supplier in this case is the robber with the victim being the appropriator-consumer. In Plott and Meyer's jargon, there is *no liability function* present in this case or, in other words, exclusion is absent. Similarly, by the very nature of the case, there is no *rejectability*. Moreover, there is *perfect selectivity* by the supplier and *forced appropriation.* Finally, the good is *depletable.* Looking at the problem in this way, then, theft would be classified as a private good (or bad)—for example, there is selectivity—except that the liability function is absent. Correction of the market failure, then, requires an institutional change in which the opportunity for redress is imposed. Viewing the problem in this context, it is claimed, immediately focuses attention on the search for an institutional change to correct the absence of a liability function. It is the technology of this case which points directly to the corrective policy.

As a second example, consider the handbill distributor described by Plott and Meyer—a case similar to theft. Again there is *no liability function* even though there is *forced appropriation.* The pedestrian cannot make the distributor pay for the inconvenience of taking the handbill. Similarly, there is *perfect selectivity, no rejectability,* and *depletability.* In this case a number of policy remedies are available and the task is to choose the least costly of them (assuming equal effectiveness). A rule could be imposed denying the distributor the right to force appropriation. For example, he could be required to stand in one place. Or a rule could be imposed requiring him to incur a liability equal to the marginal inconvenience his action imposes. Or provision could be made for rejectability by placing a trash can just beyond him. All of these institutional changes would tend to turn an inefficient mode of social organization into an efficient one. In each case, the nature of the change falls directly out of the specific characteristics of the case, as distinguished by the Plott-Meyer framework.

Having so characterized the Plott-Mayer framework, the question of the benefits of this view of the world must be confronted. The first benefit claimed by Plott-Meyer has already been alluded to: By focussing on the detailed characteristics of the case of suspected market failure, a deeper understanding of the externality problem and a clearer comprehension of the full range of policy alternatives or institutional changes are obtained. This statement, it should be noted, is both a plea for a comprehensive search and evaluation of available policy options (*à la* Baumol-Oates) and a belief that their "nuts and bolts" framework will be helpful in undertaking that search and evaluation.

With respect to their plea, I have absolutely no quibble. To insist on a full understanding of the institutions surrounding a market failure problem prior to taking corrective action is unexceptional. Few economists would advocate anything but the need for a comprehensive search for policy options and a careful examination of the benefits and costs of each. Such a procedure would seem to follow directly from standard welfare economic analysis and would be advocated by most careful public administrators. Plott and Meyer's framework would seem to add little to this proposition.

With respect to Plott and Meyer's *belief* regarding the efficacy of their framework, I do have a quibble. Surely, their perspective is helpful in illuminating the case of small numbers externality problems. Through their framework, insights are conveyed—for example, in the theft and handbill examples discussed above—which a more gross statement regarding externalities would not be likely to reveal.

However, I am not at all convinced that the Plott-Meyer framework provides illumination in the case of large numbers externality problems, such as those which dominate discussions regarding environmental pollution. Nor am I convinced that an economist or a policy maker confronted with a task of the framing, say, federal water pollution control policy would find the insights from their framework more helpful than those stemming from the standard welfare economics framework. Clearly the range of possible policies and policy mixes is very large—charges, subsidies, regulations, prohibitions, assignment of property rights, and combinations of any of these. Presumably, careful analysis within the standard welfare economic framework can point to the correct policy mix. The Plott-Meyer concepts of forced appropriation, rejectability, selectivity and so on add little insight if any to the design of optimal externality policy where large numbers interactions exist. It is revealing that, in their paper, Plott and Meyer deal with no concrete environmental problem nor illustrate any large number externality problems for which their framework would be either a complement or a substitute for the standard analysis.

A second potential benefit of the Plott-Meyer framework is as a contribution to the formal externality literature and to the modeling of environmental problems. If I am correct regarding the lack of relevance of their framework to most large number externalities problems, it is not likely that it will contribute substantially to the success of modeling efforts. As a contribution to the formal analysis of externalities, their framework is enlightening. Their critique of Arrow's conclusion that the problem is a lack of markets due to too small a number of participants follows directly from their game theory framework. Similarly, their point regarding Demsetz' focus on the role of transaction costs as the prime cause of market failure is attributable to their framework and is helpful.

Finally, their framework does seem to illuminate some of the more subtle characteristics of externality problems which are not typically noted. In the terms of their framework, external diseconomies are a problem only if nonrejectability is present. Similarly, rejectability must be thought of as a continuous variable so that the closer the world is to the pole of nonrejectability, the more serious the problem. Can we then think of technological change to expand the possibilities for rejectability or to reduce the cost of rejectability? In much the same way, viewing the liability function as a continuous variable ranging from zero liability to full liability is helpful in understanding the meaning of the exclusion principle and in contemplating possibilities for "privatizing" goods.

In conclusion, then, I found the Plott-Meyer paper to be a mixed bag. The framework which it sets out does convey some insights regarding the process of externality generation. It is a frustrating paper, however, in several regards. The tie to the standard externalities literature is never made. For example, how do the notions of consumption and production externalities fit into the Plott-Meyer framework? Is their framework helpful in distinguishing real from merely pecuniary externalities? Moreover, it is not clear that the framework can really enlighten some real world, real life externality cases—perhaps even in the environmental area. The authors have yet to demonstrate the effective relevance of the framework. As a final point, if various constellations of Plott-Meyer characteristics yield a large, yet finite, number of externality cases between pure public and pure private goods, it should be possible to map corrective institutional changes onto the set of cases. While this might be a worthwhile effort, it has not been undertaken.

The Instruments for Environmental Policy

Wallace E. Oates, Princeton University and
William J. Baumol, New York University
and Princeton University

In their part of the continuing dialogue on environmental policy, economists have quite naturally stressed the role of policy tools operating through the pricing system. The case for heavy reliance on effluent charges to internalize the social costs of individual decisions is, at least in principle, a very compelling one. However, a cursory survey of potential policy instruments reveals the existence of a wide spectrum of methods for environmental control ranging from outright prohibition of polluting activities to milder forms of moral suasion involving voluntary compliance.

In spite of the economist's predilection for a central role for direct price incentives, we suspect that even he recognizes that a comprehensive and effective (and even the "optimal") environmental policy probably involves a mix of policy tools with the use of something more than only effluent fees. The purpose of this paper is a preliminary exploration of the potential and limitations of the various policy tools available for environmental protection; our concern here is what we can say in a systematic way about the particular circumstances under which one type of policy is more appropriate than another and how various policy tools can interact effectively. We stress the word preliminary, because this paper is, in effect, an interim report on a study of environmental policy.

In the first section, we enumerate and classify the available policy instruments. In the following three sections, we present a simple concep-

NOTE: We are grateful to the National Science Foundation whose support has greatly facilitated our work on environmental policy.

tual framework for the analysis of environmental policies and a discussion of what *in principle* would appear to be the appropriate roles for the various policy tools. We turn in the fifth section to an empirical examination of the effectiveness of the different environmental policies. Our work here is in its early stages; we have at this point some admittedly fragmentary and piecemeal evidence on the efficacy of available policy instruments. In some cases, we have had to rely upon evidence that is indirect, occasionally derived from experiences other than environmental programs, to obtain some insight into the likely effectiveness of a particular policy tool.

Policy Tools for Environmental Protection

Before examining the various active policy options available for the control of environmental quality, we want to acknowledge the case for a policy of no public intervention: we could rely wholly on the market mechanism as an instrument for the regulation of externalities, unimpeded by public programs designed to protect the environment. In fact, as Ronald Coase has shown in his classic article, it is actually possible, under certain conditions, to achieve an efficient pattern of resource use through private negotiation that internalizes all social costs or benefits. This can, at least in principle, result from the incentive for parties suffering damage from the activities of others to make payments to induce a reduction in these activities.

The difficulties besetting the Coase solution are well known, particularly the free rider problem and the role of transaction costs. The main point we wish to make here is that the Coase argument is plausible only for the small group case, for only here is the number of participants sufficiently small for each to recognize the importance of his own role in the bargaining process.[1] Note, moreover, that this requires small numbers on *both sides* of the transaction; even if the polluter is a single decision-maker, a Coase solution is unlikely if the damaged parties constitute a large, diverse group for whom organization and bargaining is costly. A quick survey of our major environmental problems—air pollution in metropolitan areas, the emissions of many industries and municipalities into our waterways—indicates that these typically involve large numbers.

1. Even in the small group case, the use of certain bargaining strategies or institutional impediments to side payments may prevent efficient outcomes.

This would suggest that the Coase solution is of limited relevance to the major issues of environmental policy.[2]

Turning to the remaining policy alternatives, we present in the following list a classification of policy tools that is admittedly somewhat arbitrary. We will examine four classes of policy instruments. The first category includes measures that base themselves on economic incentives, either in the form of taxation of environmentally destructive activities or, alternatively, of subsidization of desired actions. Under the second heading, we group programs of direct controls consisting of quotas or limitations on polluting activities, of outright prohibition, and of technical specifications (e.g., required installation of waste treatment devices). Third, we consider social pressure with no legal enforcement powers so that compliance on the part of individual decision makers remains voluntary. Finally, the fourth set of programs consists of an actual transfer of certain activities from the private to the public sector.

Tools for environmental policy:

1. Price Incentives[3]
    a) Taxes
    b) Subsidies
2. Direct Controls
    a) Rationing
    b) Prohibition
    c) Technical Specifications
3. Moral Suasion: Voluntary Compliance
4. Public Production

We stress at the outset that, while the list seems simple enough, it does conceal the vast number of ways in which these policy tools may be employed. Taxes, for example, may vary with time and/or place, may apply to particular inputs, or, alternatively, outputs or byproducts of productive activities, and so forth. Similarly, direct controls on polluting activi-

2. In certain instances, no intervention may, of course, be optimal for totally different reasons: not because the market will resolve the externalities itself, but because in that particular case the damage happens to be small while the social cost of regulation is large. Here we fail to intervene not because the disease will cure itself, but because the cure is worse.

3. The auctioning of pollution rights could be added here. However, considering the major environmental problems before us, the practicality of this proposal seems to us rather limited.

ties can take an enormous variety of forms, involving the courts or special regulatory agencies, permitting and sometimes encouraging citizen lawsuits, and so forth. This list is neither exhaustive nor composed of mutually exclusive policy measures. Programs of taxes and regulations, for example, can be combined to control waste emissions; we will, in fact, consider such policy mixes shortly.

Forms of Environmental Damage

In this section, we consider, in general terms, the various forms that insults to the environment may take. More specifically, we are interested in different types of environmental damage functions. As we will argue later, the damage function that characterizes a particular type of polluting activity may be of central importance in determining the policy instrument appropriate for its control.

The first distinction is between the situation in which the current level of environmental quality is a function of the *current* level of the polluting activity and the case where it depends on the history of *past* levels of the activity. The state of purity of the air over a metropolitan area, for example, depends largely on the quantities of pollutants currently being emitted into the atmosphere. This we will call a *flow* damage function.

Alternatively, past levels of activity may build up a stock of pollutant. Therefore, the extent of environmental damage depends on the history of the activity. This we call a *stock* damage function. Such damage functions are typically associated with nondegradable pollutants, such as mercury and DDT. The pollutant accumulates over time and thus constitutes an ever increasing environmental threat. The stock and flow damage functions are pure, polar cases. In reality there is a spectrum of damage functions in which historic levels of polluting activity assume varying degrees of importance in determining the present level of environmental quality.[4] However, the distinction is a useful one for certain policy purposes.

Of equal importance is the particular form of the damage function. Economists are familiar with cost functions which exhibit monotonically increasing marginal costs; a familiar example in the literature is the case

4. For an interesting theoretical study using a more general damage function which incorporates both stock and flow elements, see C. G. Plourde.

of crowding on highways. Once costs of congestion set in, the time loss to road users resulting from the presence of an additional vehicle rises rapidly with the number of vehicles. Many environmental phenomena, however, appear to involve more complex damage functions; some exhibit important discontinuities or threshold effects. When, for example, waste loads in a river become sufficiently heavy, the "oxygen sag" may become so pronounced that the assimilative capacity of the stream is exceeded. The dissolved-oxygen content may in such cases fall to zero, giving rise to anaerobic conditions. In such cases, the cost of exceeding the threshold level of the activity may be exceedingly high. There may, moreover, exist a series of thresholds so that the damage function can be exceedingly complex. In addition, the precise form of the damage function itself may be problematic, thus injecting an important element of uncertainty into the situation.

The uncertainty element in the damage function is not a haphazard affair, but arises out of the very nature of the relationship. It is essential to recognize that damage functions are multivariate relationships, functions of a vector of variables many of them entirely outside the control of the policy maker. The effects of a given injection of pollutants into the air depend on atmospheric conditions. The damage caused by a waste emission into a stream is determined largely by the level of the water flow: it may be relatively harmless when poured into a stream that is near its crest, but very dangerous when put into the same stream when depleted by drought. Externalities in urban affairs will be more or less serious depending on the state of racial tension, the level of narcotics use, and a variety of other crucial influences.

Expressed somewhat more formally, the function describing the determination of environmental quality at time s, q_s, may be written

$$q_s = f(m_s, E_s), \tag{1}$$

where m_s is the level of waste emissions and E_s is a vector whose components are environmental conditions, such as the direction and velocity of the wind, the quantity of rainfall, and so forth. The important thing about E_s is that it includes variables over which we have little, if any, control. The exogenous variables describing the vector, E_s, are themselves likely to be random variables, or at least subject to influences which can best be treated as random.

The environmental damage function may be defined as

$$z_s = g(q_s) = h(m_s, E_s). \tag{2}$$

While q_s indicates the state of environmental quality (e.g., the sulfur dioxide content of the atmosphere or the dissolved oxygen level of a waterway), z_s denotes the social cost associated with the value of q_s. For example, higher levels of sulfur dioxide in the air people breath appear to induce a higher incidence of respiratory illnesses and mortality (see Lave and Seskin); the costs associated with these repercussions are represented by z_s.[5]

The introduction of uncontrolled determinants of environmental quality and the associated uncertainty creates some difficult policy problems. For example, environmental control policy may have a combination of several objectives such as (a) the achievement *on average* of a level of environmental quality, q_s, such that the cost of environmental insults is acceptable; and (b) prevention of the attainment of some threshold level of q_s at which there is discontinuity in the damage function, thus causing social costs to soar to unacceptably high levels.

If the values for the components of E_s were known precisely for all future periods, we could set values of m_s for each period s so as to achieve these objectives, and we would look for the least cost methods of holding emissions to these specified levels. Unfortunately, we frequently do not know the values of E_s in advance. Normally however, we can make some predictions about them. In fact, we almost have a kind of probability distribution for variables such as weather conditions. Often the dispersion of the distribution becomes much smaller as the pertinent point in time approaches (e.g., we have a better idea about tomorrow's weather than next week's weather).

Even so, the policy maker cannot control most of the variables in the vector, E, and even his ability to foresee their values remains highly limited. The science of meteorology has not yet reached a stage at which forecasts can be made with a high degree of certainty. Meteorologists are unable to determine the timing of next year's or even next month's atmospheric inversions or rainfall patterns so that plans for the intermittent crises that are likely to result may be made in advance. This phenomenon can be extremely important in the selection of policy tools. It may be that, because of limited attention to this issue in the economics

5. More realistically, we can regard q_s and m_s as vectors whose components represent, respectively, various measures of environmental quality and levels of discharges of different types of wastes. This, however, seems to add little to the analysis. Note that z_s is a scalar, not a vector, for it represents the social cost, measured in terms of a numeraire, of the level of environmental deterioration (q_s) generated jointly by m_s and E_s.

literature, we have tended to overlook the merits of policy instruments usually favored outside the profession.

Matching Policy Tools with Environmental Conditions

Before proceeding to a more detailed empirical analysis of policy tools, we want to consider under what circumstances one policy tool is likely to be more appropriate than another. As a frame of reference, let us assume a set of standards or targets for environmental quality with an eye toward devising an effective environmental policy to realize these standards.[6]

In the case of stock damage functions with costs directly related to the accumulated quantity of the pollutant, a positive level of the polluting activity implies that the *level* of environmental damage will increase continually over time. The stock of pollutants will increase over time with the flow of emissions from one period to the next. Environmental quality will thus continue to deteriorate. Any damage thresholds may eventually be exceeded, and clearly the target level of environmental quality will not be achieved. In these cases there would appear to be a strong case for outright prohibition of polluting activities, for simply reducing the level of the activity will serve only to slow the cumulative process of environmental deterioration.[7] Outright prohibition would, therefore, seem to be an appropriate policy measure where damage functions are of the stock form. The recent ban on the use of DDT in the United States is a case in point.

Where, in contrast, environmental quality depends primarily on the current level of polluting activities, prohibition may be excessively costly. Achievement of the target level of environmental quality requires adjustment of the current levels of activities to those consistent with the target.

6. We could specify alternative types of objective functions. For example, we could assume standard utility and cost functions and, following the usual maximization procedures, derive our first order optimality conditions requiring that environmental quality be improved (or polluting activities curtailed) to the point where benefits and costs are equal at the margin. The major problem here is the difficulty of measuring benefits and costs. On this issue, see, for example, Baumol and Oates. Most of the discussion in the present paper applies, incidentally, to both of these approaches to environmental policy.

7. It might be desirable to curtail the flow of emissions gradually over time if the costs of rapid adjustment are high. This raises the interesting problem of the optimal path of reduction in the rate of flow, a problem which we note but which goes beyond the scope of this paper.

These required levels, in many cases at least, can be expected to be nonzero. A variety of the policy instruments included in our earlier list may then be appropriate to influence levels of polluting activities.

What *in principle* can we say about the relative effectiveness of these policy instruments? The efficiency-enhancing properties of taxes (effluent charges) are widely recognized and need little discussion here.[8] In terms of our objective, the realization of a set of specified standards of environmental quality, we have shown elsewhere (Baumol and Oates) that, assuming cost-minimizing (not necessarily profit-maximizing) behavior by producers, effluent charges are the least-cost method of attaining the target: the proper effluent fee will generate, through private decisions, the set of activity levels which imposes the lowest costs on society. Any other set of quotas determined by regulatory authorities and consistent with the specified environmental standards will thus involve a higher opportunity cost.

This would appear to establish a presumption at the conceptual level in favor of price incentives over regulatory rationing, and to make a system of fees an ideal standard with which others should be compared and judged as more or less imperfect substitutes. However, the proof of the superiority of the tax instrument involves a number of simplifying assumptions (and typically utilizes a static analytic model); there are several other critical considerations without which it is impossible to understand fully the inclination toward other policy instruments on the part of many noneconomists who are demonstrably well informed and well intentioned.

Once we enumerate these elements, their relevance is obvious. We will show that on economic grounds they may often call for measures other than the tax instruments that receive primary attention in the economic literature. This list includes the following.[9]

8. See, for example, Kneese and Bower, and Upton.

9. We might consider adding to this list the "political acceptability" of the program. This is not without an important economic dimension. Suppose we are given two programs *A* and *B* the first of which is shown capable of yielding an allocation of resources slightly better than that which would be produced by the latter. However, suppose that *B* can be "sold" to a legislature with little expenditure of time and effort, while the enactment of *A*, if it can be secured at all, would require a highly costly and time consuming campaign. In such a case, *purely economic considerations* may favor the advocacy of *B* in preference to *A*, if we are willing to take the predisposition of the legislature as a datum in exactly the same way we take the production function for a particular product as given for the problem of determination of outputs.

1. Administrative and enforcement costs (playing a role analogous to transactions costs elsewhere in theoretical analysis).

2. Exclusion or scale problems, which may make it difficult for the private sector to provide activities appropriate for the protection of the environment. (If one wishes, this can be classified as a special case of the problem of high administrative costs, the costs of collecting payment for an environmental service or of assembling the large quantity of capital needed to supply it efficiently.)

3. Time costs. Here we include not only the interval necessary to design a program and put it into effect, but also the period of adjustment of activities to the program.

4. Problems of uncertainty.

Let us now explore how these considerations, in the context of the objective of allocative efficiency, influence the choice among the basic types of policies listed in "Tools for Environmental Policy."

Pollution taxes

Beginning once more with the tax measures we see that, in addition to their desirable allocative properties, effluent charges possess a further major attraction: their enforcement mechanism is relatively automatic. Unlike direct controls, they do not suffer from the uncertainties of detection, of the decision to prosecute, or of the outcome of the judicial hearing including the possibility of penalties that are ludicrously lenient. Like death, taxes have indeed proved reasonably certain. Few are the cases of tax authorities who neglect to send the taxpayer his bill, and that is the essence of the enforcement mechanism implicit in the tax measures. They require no crusading district attorney or regulatory agency for their effectiveness.

However, once we leave this point, we are left with considerations in terms of which tax measures generally score rather poorly. We will defer the issue of time costs to a later point where its role will be more clear. It is true that *enforcement* costs are likely to be relatively low, although like any other taxes we can be confident that they will provide work for a host of tax attorneys employed to seek out possible loopholes. Perhaps more important in many cases are high monitoring or metering costs. One of the major reasons additional local telephone calls are supplied at zero charge to subscribers in small communities is the high cost of devices that record such calls, and the same is apparently true of communities in which water usage is not metered universally. This is particularly to the point when we recognize that allocative efficiency requires tax charges to vary by season of the year, time of day, or with unpredictable changes in environmental conditions (e.g., the charge on smoke emissions should presumably rise sharply during an atmospheric "inversion" that produces a serious deterioration in air quality). Moreover, in many cases there is no one simple variable whose magnitude should be monitored. Waste emissions into waterways should ideally be taxed according to their BOD level, their content of a variety of nondegradable pollutants, their temperature, and perhaps their sheer volume. Obviously, the greater the number of these critical attributes, the more costly will be the monitoring program required by an effective tax policy. This, of course, increases the complexity of other types of regulatory programs as well.[10]

A special problem may arise from the structure of the polluting industry. Under pure competition, fees will, in principle, work ideally; in addition, it is easy to show that they tend to retain their least-cost properties in any industry in which firms minimize cost per unit of output. However, under oligopoly or monopoly, management's interests may conflict with such a goal, and taxes on polluting activities may fail to do their job with full effectiveness. If an industry routinely shifts virtually all of the cost of such fees without attempting to reduce waste emissions in order to lower its tax payments, much of the intended effect of the tax program will be lost.

From all this we do not conclude that economists have been ill-advised in their support of tax measures. On the contrary, we continue to believe strongly that in many applications they will in the long run prove to be the most effective instrument at the disposal of society. However, it is clear that certain environmental and industrial characteristics can impair

10. The technology of monitoring industrial waste emissions appears still to be in its infancy; metering devices which provide reliable measures of the composition and quantities of effluents at modest cost are (to our knowledge) not yet available. Environmental officials in New Jersey, for example, rely heavily on periodic samples of emissions which they subject to laboratory tests, which involve costly procedures. However, there is a considerable research effort underway to design effective and inexpensive metering mechanisms. This may well reduce substantially the administrative costs of programs whose effectiveness depends on measurement of individual waste discharges. In this connection, William Vickrey has stressed, in conversation with us, the dependence of the cost of metering on the degree of accuracy we demand of it. In many cases, high standards of accuracy may not be defensible. As Vickrey points out, a ten-hour inspection of an automobile will undoubtedly provide a more reliable and complete description of its exhaust characteristics than a half-hour test, but it is surely plausible that the former exceeds the standard of "optimal imperfection" in information gathering!

their effectiveness. This, as we will suggest shortly, may point to the desirability of a mixed policy of fees and controls.

Subsidies

An obvious alternative to taxes is the use of subsidies to induce reductions in the levels of these activities; what can be accomplished with the stick should also be possible with the carrot. Kneese and Bower, for example, have argued that "Strictly from the point of view of resource allocation, it would make no difference whether an effluent charge was levied on the discharger, or a payment was made to him for not discharging wastes" (p. 57). However, in addition to some extremely important differences at the operational level between taxes and subsidies, Bramhall and Mills have pointed out a fundamental asymmetry between the effects of fees and payments. While it is true that the price of engaging in a polluting activity can be made the same with the use of either a tax or subsidy, the latter involves a payment to the firm while taxes impose a cost on the firm. As a result, the firm's profit levels under the two programs differ by a constant. We have shown formally that, in long-run competitive equilibrium, subsidies (relative to fees) will result in a larger number of firms, a larger output for the industry, and a lower price for the commodity whose production generates pollution. Moreover, it is plausible the net effect will be an *increase* in total industry emissions over what they would be in absence of *any* intervention. Subsidies tend to induce excessive output. Thus, at least at a formal level, taxes are to be preferred.[11]

Direct controls

Direct controls often seem to score poorly on most of our criteria, in spite of their appeal to a curiously heterogeneous group composed largely of activists, lawyers, and businessmen. They are usually costly to administer,

11. Subsidies may be desirable if there is reason to suspect that direct controls constitute the only alternative that is feasible politically. Two reasons for this are obvious to the economist: a) direct controls are likely to allocate pollution quotas among polluters in an arbitrary manner while taxes *or* subsidies will do this in a manner that works automatically in the direction of cost minimization; b) a direct control that prohibits a polluter from, say, emitting more than x tons of sulfur dioxide per year, under threat of punishment, offers that polluter absolutely no incentive to reduce his emissions one iota below x even though the private cost of that reduction to him is negligible compared to its social benefits. Thus, subsidies may sometimes be preferable to direct controls even though both of them produce misallocations.

because they involve all the heavy costs of enforcement without avoiding entirely the costs of monitoring in whose complete absence violations simply cannot be detected. We have already noted their tendency to produce a misallocation of resources. Moreover, experience suggests that their enforcement is often apt to be erratic and unreliable, for it depends largely on the vigor and vigilance of the responsible public agency, the severity of the courts, and the unpredictable course of the public's concern with environmental issues.

Yet direct controls do possess one major attraction: *if enforcement is effective,* they can induce, with little uncertainty, the prescribed alterations in polluting activities. We cannot expect controls to achieve environmental objectives at the least cost, but they may be able to *guarantee* substantial reductions in damages to the environment, a consideration that may be of particular importance where threats to environmental quality are grave and time is short. This points up two limitations of effluent charges: first, the response of polluters to a given level of fees is hard to predict accurately, and second, the period of adjustment to new levels of activities may be uncertain. If sufficient time is available to adjust fees until the desired response is obtained, the case for effluent charges becomes a very compelling one. However, environmental conditions may under certain situations alter so swiftly that fees simply may not be able to produce the necessary changes in behavior quickly (or predictably) enough. Where, for example, the air over a metropolitan area becomes highly contaminated because of extremely unfavorable weather conditions, direct controls (perhaps involving the prohibition of incineration or limiting the use of motor vehicles) may be necessary to avoid a real catastrophe.

There *may* be a further role for direct controls in industries dominated by a few large firms whose market power enables them to pass forward taxes on polluting activities without much incentive to undertake major adjustments in production techniques to reduce environmental damage.[12] This is frankly a difficult case to evaluate. Perhaps the best example is the ongoing attempt to impose technical standards for exhaust discharges on new automobiles. Because of the highly concentrated character of the auto industry, it is not clear that taxes on motor vehicles (perhaps graduated according to the level of exhaust emissions) would have much effect

12. Of course, it is normally desirable that some portion of the tax be passed forward in the form of price increases, as a means to discourage demand for the polluting output. The issue is that an oligopoly whose objectives are complex may not always minimize the costs of producing its vector of outputs.

on automobile design or usage.[13] A more promising approach may consist of legislated emission standards that will compel alterations in the design of engines so as to reduce the pollution content of vehicle discharges. However, the use of standards also involves difficult problems: witness the protracted "bargaining" between auto-industry representatives and federal legislators over the level of the standards and the timing of their implementation. Moreover, there is always the danger of adopting standards approaching complete "purity" that impose enormous costs; the reduction of polluting activities typically involves marginal costs that increase rapidly as the required reductions in waste discharges approach 100 per cent. The setting of emission standards without adequate regard for the costs involved may produce some highly inefficient results.

Hybrid programs

Even those policy makers who have come to recognize the merits of a system of charges as an effective instrument of control seem normally unwilling to rely exclusively on this measure. Rather they typically prefer a mixed system of the sort in which each polluter is assigned quotas or ceilings which his emissions are in any event never to be permitted to exceed. Taxes are then to be used to induce polluters to do better than these minimum standards and to do so in a relatively efficient manner.

While this may at first appear to be a strange mongrel, some of the preceding discussion suggests that, under certain circumstances, such a mix of policies may have real merit. If taxes are sufficiently high to cut emissions well below the quota levels, the efficiency properties of the tax measure will be preserved. Moreover, it retains the advantage of the pure fiscal method in forcing recognition of the very rapidly rising cost of further purification as the level of environmental damage is reduced toward zero. It is all too easy to set quotas at irresponsibly demanding levels, paying no attention to the heavy costs they impose. But it is hard not to take notice when tax rates must be raised astronomically to achieve still further improvements in environmental quality.

On the other hand, the quota portion of the program can make two important contributions, safety and increased speed of adjustment and implementation. Suppose, for example, there is a threshold in the damage

13. As Roger Noll points out, the case for effluent fees is the weakest "when regulators must deal with firms with considerable market power, and, at the other extreme, individuals with very little freedom of choice arising either from a lack of economic power, lack of knowledge, or lack of viable technical options" (pp. 34–5).

function so that a form of environmental abuse imposes a serious threat, but only beyond some point that is fairly well known. In this case, a hybrid policy can make considerable sense, since the quotas it utilizes can be employed to make reasonably certain that damages never get beyond the danger point. Taxes can be unreliable for this purpose, since, as noted earlier, the tax elasticities of pollution output are generally not well known and these fees may not induce changes in activity levels with sufficient rapidity. Thus, reliance on tax incentives alone may impose unacceptable risks, which can be prevented by a set of direct controls that set ceilings on levels of polluting activities.

Controls can, moreover, introduce additional flexibility into an environmental program. In terms of our illustrative case, urban air pollution, we noted that authorities may be able to invoke temporary prohibition, or at least limitations, on polluting activities when environmental deterioration suddenly reaches extremely serious levels.

Hybrid programs of taxes and controls thus represent a very attractive policy package. The tax component of the program functions to maintain the desired levels of environmental quality under "normal" conditions at a relatively low cost and also avoids the imposition of uneconomically demanding controls. The controls constitute standby measures to deal with adverse environmental conditions that arise infrequently, but suddenly, and which would result in serious environmental damage with normal levels of waste emissions.[14] Such a mixed program should not involve notably higher administrative costs than a pure tax policy, since much of the monitoring structure used for the latter should also be available for enforcement of the controls. In sum, where threshold problems constitute a serious environmental threat and where levels of polluting activities may require substantial alteration on short notice, which is not a rare set of circumstances, a hybrid program using both fees and controls may be preferable to a pure tax-subsidy program.

Moral suasion: voluntary compliance

We come next to the cases in which it seems appropriate to rely on appeals to conscience and voluntary compliance. As economists, we tend to be somewhat skeptical about the efficacy of long-run programs which

14. In this volume, Lave and Seskin report evidence that the mortality danger of air pollution crises may have been exaggerated. Nevertheless, it remains true that, during periods of stagnant air, the social cost of a given emission level will be high, because a great proportion of the polluting element remains over the city for a protracted period.

require costly acts of individuals but offer no compensation aside from a sense of satisfaction or the avoidance of a guilty conscience. In fact, the appeal to conscience can often be a dangerous snare. It can serve to lure public support from programs with real potential for the effective protection of the environment. Later, we will provide some evidence that suggests this to be a real possibility.

There is nevertheless an important role for voluntary programs. In particular, in an unanticipated emergency there simply may be no other recourse: the *time cost* of most other instruments of control may be too high to permit their utilization under such circumstances. A sudden and dangerous deterioration of air quality allows no time for the imposition of a tax or for the drawing up and adoption of other types of regulatory legislation. There may be no time for emergency controls, particularly if they have not previously been instituted in standby form, but there can be an immediate appeal to the general public to avoid the use of automobiles and incinerators until the emergency is passed. Moreover, as we shall indicate in a later section, there is evidence to indicate that the public is likely to respond quickly and effectively to such an appeal. Perhaps social pressures and a sense of urgency lie behind the efficacy of moral suasion in such cases.[15]

Casual observation suggests that the sense of high moral purpose is likely to slip away rather rapidly and thus implies little potential for long-term programs that rest on no firmer base than the public conscience. However, that is no reason to reject this instrument where it can prove effective, particularly since no effective alternative may be available. We suspect that we have not yet experienced the last of the unforeseen emergencies and, in extremis, time cost is likely to swamp all other costs in the choice of policy instruments.

Public provision of environmental services

The direct public "production" of environmental quality may be justified in two types of situations. The first is the case where the current

15. There is another precondition for the efficacy of moral suasion, even in an emergency. We can usually expect a few individuals not to respond to a public appeal. Thus, voluntarism cannot be relied upon in a case where universal cooperation is essential, as during a wartime blackout where a single unshielded light can endanger everyone. However, in most environmental emergencies as long as a substantial proportion of the persons in question are willing to comply with a request for cooperation, a voluntary program is likely to be effective. For example if, during a crisis of atmospheric quality, an appeal to the public may lead to a temporary reduction in automotive traffic of some 70 or 80 per cent, that may well be sufficient to achieve the desired result.

quality of the environment is deemed unsatisfactory (i.e., falls below the specified standard) as a result of "natural" causes and where this cannot be corrected through market processes because the particular environmental service is a public good. It is hard to find a perfect illustration, but natural disasters such as periodic droughts or flooding come close. Here the problem is not one of restricting polluting activities on the part of the individual; it is one of providing facilities such as dams and reservoirs to prevent these catastrophies. The private sector of the economy may handle such situations adequately if the commodity needed to avert the disaster is not a public good—that is, if exclusion is possible (or, more accurately, not too costly) and consumption is rival. However, where exclusion is difficult and/or consumption is joint, as in the case of protection from flood damage, the public sector may have to take direct responsibility for the provision of the good.

The second type of situation in which direct public participation *may* be appropriate is that involving large economies of scale and outlays. An example may be the case of a large waste treatment facility used by a multitude of individual decision makers. The reduced cost of treatment of effluents made possible by a jointly used plan may not be realized if left to the private sector.

This example, incidentally, suggests a further type of environmental service that the public sector must provide, namely the planning and direction of systems for the control of environmental quality. The need of reaeration devices, for instance, depends upon water flows (influenced by reservoir facilities), the levels of waste emissions (determined in part by current fees or regulations), and so forth. The point is that the control of water quality in a river basin or atmospheric conditions in an air shed requires systematic planning to integrate effectively the use of quality-control techniques. Kneese and Bower stress the need for river basin authorities to plan and coordinate a program of water-quality management. Urban areas require similar types of authorities to develop integrated air quality programs. Thus, public agencies must not only directly provide certain physical facilities, but must also exercise the management function of coordinating the variety of activities and control techniques that serve jointly to determine environmental quality. Such agencies need not be federal, but must be sufficiently large so that their jurisdiction includes those activities that influence environmental conditions in a given area. This implies jurisdictions sufficiently large to encompass systems of waterways and areas whose atmospheric conditions are dependent on the same activities.

Optimal Mixed Programs: A Simple Model

The logic of the argument in the preceding section for the use of hybrid programs in the presence of random exogenous influences can be more clearly outlined with the aid of a simple illustrative model. Such a model can indicate not only the potential desirability of such a hybrid as against a tax measure or a program using direct controls alone, it can also illustrate conceptually how one might go about selecting the optimal mix of policy instruments.

A relationship apparently used frequently in the engineering literature to describe the time path of environmental quality is (in a much simplified form)[16]

$$q_s = k_s q_{(s-1)} + m_s, \tag{3}$$

where:

q_s is a measure of environmental quality during period s,
k_s is a random exogenous variable (call it "average wind velocity") during time s, and
m_s is the aggregate level of waste emissions in period s.

In the presence of a tax program, the level of waste discharges will presumably be determined in part by the tax. Let us define

m_{is} = waste emissions of firm i in period s,
$c_i(m_{is}, \ldots)$ = the total cost function of firm i, and
t = tax per unit of waste emission.

Then, if the firm minimizes its costs, we will presumably have in equilibrium

$$\frac{\partial c_i}{\partial m_{is}} = -t. \tag{4}$$

That is, the firm will adjust waste discharges to the point where at the

16. Other forms of this relationship are obviously possible. For example, k_s and $q_{(s-1)}$ may be additive rather than multiplicative. The facts will presumably vary from case to case, but within wide limits the choice of functional form does not affect the substance of our discussion.

margin the cost increase resulting from a unit reduction of emissions (e.g., the marginal cost of recycling) is equal to the unit emission charge. Using the cost function for the firm and its cost-minimizing emission condition, (4), we can derive a relationship expressing the level of waste discharges of the ith firm as a function of the unit emission tax:

$$m_{is} = h_i(t_s). \tag{5}$$

Aggregating over all i firms, we get an aggregate waste-emission function

$$m_s = h(t_s) = \sum h_i(t_s). \tag{6}$$

From equation (6), we can thus determine the total level of waste discharges into the environment in period s associated with each value of t, the effluent fee.

Next, suppose we know the probability distribution of k_s, our random and exogenous environmental variable ("average wind velocity") in equation (3). For some known value of environmental quality in period $(s-1)$, we can then determine the distribution function of environmental quality in time s associated with each value of the emission tax, t. Figure 1 depicts some probability distributions corresponding to different tax rates.

Figure 1

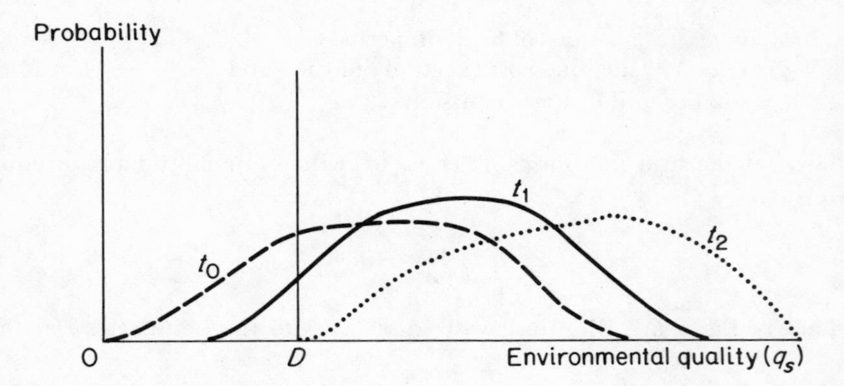

We see that a reduction in the emission tax from t_1 to t_0 shifts the distribution leftward. Once a lower tax rate is instituted, higher levels of waste emissions become profitable, thereby increasing the likelihood of a period of relatively low environmental quality.

Assume, moreover, that the environmental authority cannot readily change t in response to current environmental conditions so that t is essentially fixed for the period under analysis.[17] Let there also be some accepted "danger standard" (i.e., a minimum acceptable level of environmental quality). We designate this danger standard as D in Figure 1 and assume that the environmental authority is committed to maintaining the level of environmental quality above D at *all* points in time.

How can the authority achieve this objective at the least cost to society? One method of guaranteeing that q_s will never fall below D is to set the tax rate so high that waste emissions can never, regardless of exogenous environmental influences, reach a value sufficiently high to induce environmental quality to deteriorate to a level less than D.[18] In terms of Figure 1, this would require an emissions tax of t_2, which shifts the environmental probability distribution rightward until its horizontal intercept coincides with D. However, as we suggested earlier, this method of achieving the objective may be an excessively costly one, because it is likely to require unnecessarily expensive reductions in waste discharges during "normal" periods when the environment is capable of absorbing these emissions without serious difficulty. It may be less costly to set a lower emission tax (less than t_2 in Figure 1) and to supplement this with periodic introductions of controls to achieve additional reductions in waste discharges during times of adverse environmental conditions (periods of "stagnant air").

In Figure 2, we illustrate an approach to the determination of the optimal mix of emission taxes and direct controls. Let the curve TT' measure the *total* net social cost associated with each value of t. There are two components of this social cost. The first is the added costs of production that higher taxes impose by inducing methods of production consistent with reduced levels of waste emissions. This cost naturally tends to rise with tax rates and the associated lower levels of waste discharges. However, we must subtract from this "production" cost a negative cost (or social gain) which indicates the social benefits from a higher level of environmental quality. Over some range of values for t (up to t_0 in Figure 2), we might expect the sum of these costs to be negative, that is, the social benefits from improved environmental quality may well exceed the in-

17. Alternatively, we can assume that the response of waste emissions to changes in t is not sufficiently rapid for the tax adjustments in period s to influence significantly waste discharges during that period.

18. It may, of course, be impossible to achieve such a guarantee with any finite tax rate, no matter how high.

Figure 2

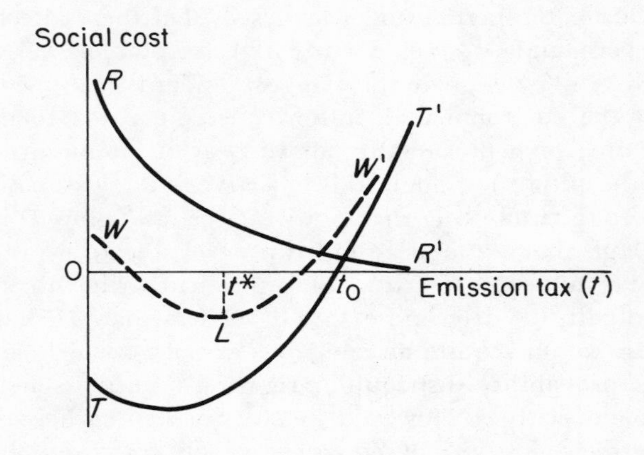

creased costs of production. However, as tax rates rise and waste discharges decline, the *marginal* net social cost will typically rise. The marginal production cost of reductions in waste emissions (equated in value to t) will obviously increase, while we might expect diminishing social gain from positive increments in environmental quality.[19] The TT' curve will, therefore, typically begin to rise at some point and, for values of t in excess of t_0 in Figure 2, the net social cost of the tax program becomes positive.

We recall that the environmental authority is committed to the maintenance of a level of environmental quality no lower than the danger point, D. We will thus assume that, whatever the level of the emission tax, environmental officials will introduce direct controls whenever necessary to maintain q above D. One relationship is immediately clear: the higher the emission tax, the less frequently will environmental quality threaten to fall below D and hence the less often (and less "intensely") will the use of direct controls be required. Controls, like taxes, impose increased costs of production by forcing reductions in waste emissions. Therefore, the more frequent and extensive the use of direct controls,

19. We have drawn TT' with a "smooth" shape (a continuous first derivative), but there could easily be flat portions of TT' corresponding to ranges of values of t over which the level of waste emissions remains unchanged. Note, however, that even in this instance TT' would still exhibit the general shape depicted in Figure 2 and, most important, would still possess a well defined minimum for some value (or continuous range of values) of t.

the greater the increment in production costs they will generate. We depict this relationship in Figure 2 by the curve RR', which indicates that the higher the tax rate the less the reliance and, hence, the lower the costs associated with the periodic use of direct controls to maintain q above D.[20]

When we sum TT' and RR' vertically, we obtain the net social cost (WW') associated with each level of the emission tax (t) *supplemented* by a program of direct controls which prevents environmental quality from ever falling below the danger point (D). In Figure 2, we see that the lowest point (L) on the WW' curve corresponds to the cost-minimizing or optimal tax rate (t^*) and determines residually the optimal use of direct controls.[21]

We stress that the treatment in this section is purely illustrative. It indicates an approach to the determination of the optimal mix of emission taxes and direct controls. A rigorous solution to this problem requires an explicit recognition of the stochastic element in the curves in Figure 2. The social costs generated by a given tax program depend in part on the values taken by our random exogenous environmental variable ("wind velocity"), so that the curves in this diagram must be regarded in some sense as "averages." More formally, the solution involves the minimization of a stochastic social cost function subject to the constraint that $q \geq D$. Elsewhere we will show how this can be formulated as a nonlinear programming problem, whose solution yields the optimal mix of effluent taxes and direct controls.

Environmental Policy Tools in Practice

In this section, we want to present some preliminary evidence on the effectiveness of the various tools of environmental policy. Since evidence in the form of systematic, quantifiable results is scarce, we have had to resort in some instances to case studies suggesting only in qualitative

20. Unlike the tax-cost function (TT'), the social cost of direct controls does not include a *variable* component related to the benefits from varying levels of environmental quality. Direct controls in this model are used solely to maintain q above D. We can treat the social benefits derived from the guarantee that environmental quality never falls below D as a constant (independent of the level of t), and we can, if we wish, add this constant to RR' (or to TT' for that matter). The essential point is that we can expect RR' to be a function that decreases monotonically in relation to t.

21. Note that the curve WW' may possess a number of local minima. It need not increase monotonically to the right of L.

terms the nature of the response to the programs. Many of the findings, however, do seem roughly consistent with the preceding discussion.

Price incentives

While economic theory suggests an important role for price incentives, particularly effluent fees, for environmental control, we really have limited experience with their use. The opposition to proposals for effluent charges has been strong, in some measure, we suspect, because people realize they will be effective and wish to avoid the inevitable costs of environmental protection.[22] Nevertheless, there has been some use of charges, and what evidence is available suggests that effluent fees have in fact been quite successful in reducing polluting activities.

The most striking and important case appears to be the control of water quality in West Germany's Ruhr Valley. The site of one of the world's greatest concentrations of heavy industry, the rivers of the Ruhr Valley could easily have become among the most polluted rivers in Europe. However, since the organization of the first *Genossenschaft* (river authority) in 1904 along the Emscher River, the Germans have been successfully treating wastes in cooperatives financed by effluent charges on their members. There are presently eight *Genossenschaften*. Together they form a closed water-control system which has maintained a remarkably high quality of water. In all but one of the rivers in the system, the waters are suitable for fishlife and swimming. Together, the eight cooperatives collect approximately $60 million a year, mainly from effluent charges levied on their nearly 500 public and private members. The level of charges is based largely on a set of standards for maintaining water quality, although the formulas themselves are rather complicated. As Kneese and Bower point out, the fee formulas do not correspond perfectly to the economist's version of effluent fees ("they violate the principle of marginal cost pricing," p. 251).[23] Nevertheless, the charges, in conjunction with an integrated system of planning and design for the entire river basin, "is a pioneering achievement of the highest order" (Kneese and Bower, p. 253).

There has been a scattered use of effluent fees for environmental protection in North America, and these, to our knowledge, exclusively for

22. For an excellent survey and evaluation of the most frequent arguments directed against programs of effluent fees, see Freeman and Haveman.

23. For a more detailed discussion of the Ruhr experience, see Kneese and Bower, Chapter 12.

the control of water quality. However, this evidence does again point to the effectiveness of fees in curtailing waste emissions. Kneese and Bower cite three instances in which the levying of local sewer charges induced striking reductions in waste discharges.[24] C. E. Fisher reports similar responses to a local sewerage tax in Cincinnati, Ohio. Fees were established in 1953 with the proviso that a rebate would be given to anyone who met a specified set of standards by a certain date. Subsequently, some 23 major companies invested $5 million in pollution control in less than two years to meet these standards.

There also exist three more systematic studies of industrial responsiveness to sewerage fees. Löf and Kneese have estimated the cost function for a hypothetical, but typical, sugar beet processing plant in which cost is treated as a function of BOD removal from waste water. Their results suggest, assuming the firm stays in business, that a very modest effluent charge would induce the elimination of roughly 70 per cent of the BOD contained in the waste water of their typical plant. Likewise, a recent regression study by D. E. Ethridge of poultry processing plants in different cities imposing sewerage fees indicates substantial price responsiveness on the part of these firms. In a total of 27 observations from five plants, Ethridge found that "The surcharge on BOD does significantly affect the total pounds of BOD treated by the city; the elasticity of pounds of BOD discharged per 1,000 birds with respect to the surcharge on BOD is estimated to be −0.5 at the mean surcharge" (p. 352).

The most ambitious and comprehensive study of the effects of municipal surcharges on industrial wastes in U.S. cities is the work of Ralph Elliott and James Seagraves. Elliott and Seagraves have collected time-series data on surcharges, waste emissions, and industrial water usage for 34 U.S. cities. They have put these data to a variety of tests and their findings indicate that industrial BOD emissions and water consumption do indeed appear to respond negatively to the level of surcharges on emissions. In one of their tests, for example, they have pooled their cross-section and time-series observations and, using ordinary least squares, obtained the following estimated equations:

$$T = 13.1 - \underset{(8.5)}{14.6S} - \underset{(79.3)}{120.0G} + \underset{(22.6)}{36.2P} \tag{7}$$

$$R^2 = .17 \qquad N = 190,$$

24. These involved sewerage fees in Otsego, Michigan, in Springfield, Missouri, and in Winnipeg, Canada. See Kneese and Bower, pp. 168–70.

$$W = 2.2 - \underset{(2.9)}{5.2S} - \underset{(24.7)}{36.8N} + \underset{(7.2)}{8.6P} + \underset{(26.0)}{75.1F} \qquad (8)$$

$$R^2 = .32 \qquad N = 179,$$

where;

T = pounds of BOD per $1,000 of value added in manufacturing;
S = surcharge per pound of BOD in 1970 dollars;
G = price of water (per 1,000 gallons) in 1970 dollars;
P = the real wage rate (per hour) in 1970 dollars;
N = net cost of additional water (per 1,000 gallons) in 1970 dollars;
F = proportion of value added in manufacturing in the city contributed by food and kindred products.

The coefficients on the surcharge variable (S) possess the expected negative sign and are statistically significant using a one-tail test at a .05 level of confidence. Using typical values for the variables, the authors estimate the elasticity of industrial BOD emissions with respect to the level of the surcharge to be -0.8, and the surcharge elasticity of water consumption at -0.6.[25] We are thus beginning to accumulate some evidence indicating that effluent fees can in fact be quite effective in reducing levels of industrial waste discharges into waterways.

In contrast, our experience with charges on waste emissions into the atmosphere is virtually nil. However, there is one recent and impressive study by James Griffin of the potential welfare gains from the use of emission fees to curtail discharges of sulfur dioxide into the air. Using engineering cost data, Griffin has assembled a detailed econometric model of the electric utility industry.[26] The model allows for desulfurization of fuel and coal, substitution among fuels, substitution between fuel and capital (using more capital allows more energy to be derived from a unit of fuel), and for the substitution away from "electricity-intensive" products by consumers and industry. Griffin then ran a series of nine alternative simulations involving differing effluent fees and other assumptions

25. The explanatory power (R^2) of the Elliott-Seagraves' equations is not extremely high. Among other things, this reflects the difficulties of accounting for varying industrial composition among cities and for intercity differences in the fraction of waste emissions that enter the municipal treatment system. Ethridge's equations, which use observations on only a single industry (poultry-processing), have much higher R^2 (of about .5).

26. In 1970 "power plants contributed 54% of the nation's sulfur dioxide emissions" (p. 2).

based on the estimates provided by the Environmental Protection Agency of the social damage generated by emissions of sulfur dioxide. In all the simulations, substantial net welfare gains appeared. The results were somewhat sensitive to assumptions concerning the availability and cost of fuel gas desulfurization processes about which there is some uncertainty. However, with such techniques available at plausible costs, Griffin's average annual welfare gains ranged from $6.5 to $7.7 billion, and these estimates do not allow for possible shifts to nuclear power sources.

The evidence thus does suggest that effluent fees can be an effective tool in reducing levels of waste emissions. This, of course, is hardly surprising. We expect firms and individuals to adjust their patterns of activity in response to changes in relative costs. It has often been observed that in less developed countries, where wages are relatively low, more labor intensive techniques of production are typically adopted than in higher wage countries. Moreover, in a regression study of the capital labor ratios across the states in the U.S. for 16 different manufacturing industries, Matityahu Marcus found that factor proportions did indeed vary systematically in the expected direction with the relative price of capital in terms of labor. There does seem to be sufficient substitutability in relevant production and consumption activities for modest effluent charges to induce pronounced reductions in waste emissions.[27]

What would be even more interesting is some measure of the relative costs of other control techniques (for example, the imposition of uniform percentage reductions in the waste discharges of all polluters). Evidence on this is scarce. However, one such study has been made, a study of the costs of achieving specified levels of dissolved oxygen in the Delaware River Estuary.[28] A programming model was constructed using oxygen balance equations for 30 interconnected segments of the estuary. The next step was to specify five sets of objectives and then to compare the costs of achieving each of these objectives under alternative control policies. Although effluent charges were not included specifically as a policy alternative in the original study, Edwin Johnson headed a subsequent study using the same model and data. This made possible the comparison of four alternative programs for reaching specified levels of dissolved oxygen in the estuary. The results for two D.O. objectives are presented

27. For a useful summary of estimates of price elasticities for polluting activities, see the paper by Robert Kohn.

28. Federal Water Pollution Control Administration, *Delaware Estuary Comprehensive Study: Preliminary Report and Findings* (1966); a useful summary of this study is available in Kneese, Rolfe, and Harned, Appendix C.

in Table 1, where LC is the least-cost programming solution, UT is a program of uniform treatment requiring an equal percentage reduction in discharges by all polluters, SECH is a program consisting of a single effluent charge per unit of waste emission for all dischargers, and ZECH is a zoned charge in which the effluent fee is varied in different areas along the estuary. As indicated by Table 1, the substantial cost savings of a program of effluent fees relative to that of uniform treatment is quite striking. Moreover, it should be noted that the least-cost programming solution involves a great deal more in the way of technical information and detailed controls than do the programs of fees. The reduced costs from the use of fees instead of quotas thus appear to be potentially quite sizable.

TABLE 1
Cost of Treatment Under Alternative Programs

D.O. Objective (ppm)	*Program*			
	LC	*UT*	*SECH*	*ZECH*
	(million dollars per year)			
2	1.6	5.0	2.4	2.4
3–4	7.0	20.0	12.0	8.6

Source: Kneese, Rolfe, and Harned, p. 272.

As we mentioned in the preceding section, effluent fees are, in theory, a more efficient device for achieving standards of environmental quality than subsidies. Fees appear, moreover, to possess a number of practical advantages as well. The design of an effective and equitable system of subsidies is itself a difficult problem. If a polluter is to be paid for reducing his waste emissions, it then becomes in his interest to establish a high level of waste discharges initially; those who pollute little receive the smallest payments.

In practice, subsidies have been used far more extensively in the United States than fees. The federal government has relied heavily on a program of subsidization of the construction of municipal waste treatment plants and on tax credits to business for the installation of pollution control equipment. The serious deficiencies in the first program are now a matter of record in the 1969 Report of the General Accounting Office. The failure to curtail industrial pollution; the subsidization of plant construction but not operating expenses (resulting in many instances of incredibly ineffective use of the facilities); and the inappropriate location of many plants have resulted in the continued deterioration of many major U.S. waterways despite an expenditure of over $5 billion.[29]

Although we have been unable to find any direct evidence on the tax credit program, there is a simple reason to expect it to have little effect. As Kneese and Bower (pp. 175–78) point out, a firm is unlikely to purchase costly pollution control equipment which adds nothing to its revenues; the absorption of k per cent (where $k < 100$) of the cost by the government cannot turn its acquisition into a profitable undertaking.

Thus both theory and experience point to the superiority of effluent charges over subsidies as a policy tool for environmental protection. Finally, we might also mention that, from the standpoint of the public budget, fees provide a source of revenues, which might be used for public investments for environmental improvements, while subsidies require the expenditure of public funds.

Direct controls

As James Krier points out, "Far and away the most popular response by American governments to problems of pollution—and indeed, to *all* environmental problems—has been regulation . . ." (p. 300). Three general types of regulatory policies for environmental control: quotas, prohibition, and the requirement of specified technical standards are stated in the list of tools for environmental control. However, this classification does not indicate the vast number of ways in which these direct controls may be implemented. The directive for polluters to cease certain activities or to install certain types of treatment equipment may come from an empowered regulatory authority, may result from a court order, or might be forced by the citizenry itself through a referendum. Even this is an oversimplification. There are, for example, several methods by which action through the Courts may be initiated (see Krier). Our category of "direct controls" thus encompasses an extremely broad range of policy options. It is beyond the scope of this paper to examine in detail, for instance, the potential of various forms of litigation for effective environmental policy. We shall rather examine somewhat more generally the success or failure of each of these approaches with particular attention to the circumstances which appear to bear on their effectiveness.

The record of regulatory policies in environmental control is not very impressive. This stems at least as much from administrative deficiencies

29. For further documentation of the ineffectiveness and abuses under this subsidy program, see Marx, and Zwick, and Benstock.

in the application of regulatory provisions as in the establishment of the provisions themselves. A successful regulatory policy generally requires at least three components.

(1) A set of rules that, if practiced, will provide the desired outcome. In this case, satisfactory levels of environmental quality achieved at something reasonably close to the least cost.

(2) An enforcement agency with sufficient resources to monitor behavior.

(3) Sufficient power (the ability to impose penalties) to compel adherence to the regulations.

The design of an efficient set of rules is itself an extremely difficult problem. As mentioned earlier, effluent charges have important efficiency enhancing properties. Moreover, the specification of an efficient set of regulatory provisions will generally require at least as much, and frequently more, technical information than the determination of schedules of fees.[30] In addition, experience suggests that substantial transaction costs in terms of resources devoted to bargaining (as noted earlier in the case of the continuing controversy over auto emission standards) may be involved in the rule selection process.

Even an effective set of regulations can only achieve its objective if it is observed. Unfortunately, the history of environmental regulation in the United States is not encouraging on this count. Regulatory agencies have frequently been understaffed and unable, or unwilling, to enforce antipollution provisions. An interesting historical example is the River and Harbors Act of 1899 which prohibits the discharge of dangerous substances into navigable waterways without a permit from the Army Corps of Engineers. As of 1970, only a handful of the more than 40,000 known dischargers had valid permits. Moreover, the newspapers abound with accounts of huge plants which have paid trivial sums (sometimes a few hundred dollars) for serious violations of pollution regulations. Many of the provisions simply have not given the agencies the power they require for enforcement.

Action through the courts has also not proved very effective. Environmental lawsuits, where a plaintiff can be found, have often stretched over years or even decades without resolution. However, even if judicial proceedings were prompt, it is difficult to envision how suits by individual plaintiffs for damages could lead to an efficient environmental policy.

30. In an interesting paper, Karl Göran-Mäler has shown recently that the determination of an efficient set of effluent standards (or quotas) among activities requires at least as much information as that necessary to solve for an optimal set of effluent charges.

Kneese and Bower, while acknowledging the potential of some support from the judicial process, conclude simply that ". . . efficient water quality management cannot be achieved through the courts" (p. 88).

Nevertheless, where enforcment is effective, and it surely has been in a significant number of cases, direct controls can lead to substantial reductions in polluting activities. A variety of regulations in various metropolitan areas have generated large reductions in waste discharges into the atmosphere. The banning of backyard incineration and of the use of sulfur bearing fuels over several months of the year led to significant reductions during the 1950's in smoke, dust, and sulfur oxide discharges into the air shed over the Los Angeles basin. Likewise, tough new regulations in Pittsburgh during the 1940's, requiring the switch from coal fuels to natural gas for heating purposes, resulted in notable improvements in air quality. Strong regulations combined with aggressive enforcement *can* clearly raise the level of environmental quality.[31] The difficulties, of course, are that the improvements may come at an unnecessarily high cost, or, alternatively, may come not at all, if the regulations are themselves inadequate or are ineffectively enforced.

Moral suasion and voluntary compliance

We suggested earlier that, while moral suasion is likely to be an ineffective policy tool over longer periods of time, it may prove quite useful in times of emergency. An interesting illustration of this pattern of response involves voluntary blood donations. In September of 1970, New York City hospitals were facing a blood crisis in which reserves of blood had fallen to a level insufficient for a single day of operation. The response to a citywide plea for donations was described as "fantastic" (*New York Post,* September 4, 1970, p. 3); donors stood in line up to 90 minutes to give blood. The statements by some of the donors were themselves interesting:

"I've never given blood before, but they need it now. That's good enough reason for me."

"I was paying a sort of personal guilt complex."

"It's the least I could do for the city."

31. Direct controls in the form of "technical specifications" for polluting activities may be the only feasible policy instrument, where the monitoring of waste emissions is impractical (or, more accurately, "excessively costly"). For example, if difficulties in metering sulfur dioxide emissions into the atmosphere were to preclude a program of effluent fees (or quotas, for that matter), it might well make sense to place requirements on the quality of fuel used, on the technical characteristics of fuel burners, etc.

And yet within a few months (*New York Times,* January 4, 1971, p. 61), the metropolitan area's blood stocks were again down to less than one day's supply. It was also noted that many donors who promised to give blood had not fulfilled their pledges.[32]

A somewhat similar fate seems to have characterized voluntary recycling programs. Individuals and firms greeted these proposals with substantial enthusiasm and massive public relations efforts. Many manufacturers agreed to recycle waste containers collected and delivered by nonprofit volunteer groups. While the initial response was an energetic one, it seems to have tailed off significantly. "Many (of the groups) disbanded because of a lack of markets or waning volunteer interest" (*New York Times,* May 7, 1972, p. 1 and p. 57). The Glass Manufacturers Institute announced that used bottles and jars returned by the public were being recycled at a rate of 912 million a year, but this represents only 2.6 per cent of the 36 billion glass containers produced each year. Similar reports from the Aluminum Association and the American Iron and Steel Institute indicated recycling rates of 3.7 per cent and 2.7 per cent respectively for metallic containers. The reason for the failure of these programs to achieve greater success is, according to several reports, "that recycling so far is not paying its own way" (*New York Times,* May 7, 1972, p. 1). Experience with recycling programs also points to a danger we mentioned earlier: that these types of programs will be instituted instead of programs with direct individual incentives for compliance. There are a wealth of examples of businesses providing active support for voluntary recycling as parts of campaigns *against* fees or regulations on containers. The *New York Times* (May 7, 1972, p. 57), for instance, cites a recent case in Minneapolis in which the Theodore Hamm Brewing Company and Coca-Cola Midwest, Inc. announced that they would sponsor "the most comprehensive, full-time recycling center in the country." This pledge, however, was directed against a proposed ordinance to prohibit local usage of cans for soft drinks and beer.

A final example of some interest involves a recent attempt by General Motors to market relatively inexpensive auto-emission control kits in Phoenix, Arizona. The GM emission control device could be used on most 1955 to 1967 model cars and could reduce emissions of hydrocarbons, carbon monoxide, and nitrogen oxides by roughly 30 to 50 per cent. The cost of the kits, including installation fees, was about $15 to $20.

32. Other cases we are currently investigating are the formation of car pools both in emergency and "normal" periods to cut down on auto emissions, and the extent of voluntary reductions in usage of electricity during periods of power crises.

Despite an aggressive marketing campaign, only 528 kits were sold. From this experience, GM has concluded that only a mandatory retrofit program for pre-1968 cars, based upon appropriate state or local regulation, can assure the wide participation of car owners that would be necessary to achieve a significant effect on the atmosphere. The Chrysler Corporation has had a similar experience. In 1970 Chrysler built 22,000 used car emission control kits. More than half remain in its current inventory. In fact after 1970 Chrysler had experienced "negative" sales. About 900 more kits were returned than shipped.

The role of moral suasion and voluntary compliance thus appears to promise little as a regular instrument of environmental policy. Its place (in which it may often be quite effective) is in times of crisis where immediate response is essential.

Concluding Remarks

Our intent in this paper has been a preliminary exploration of the potential of available tools for environmental policy. There is, as we have indicated, a wide variety of options at the policy level with differing instruments being appropriate depending upon the characteristics of the particular polluting activity and the associated environmental circumstances. The "optimal" policy package would no doubt include a combination of many approaches including the prohibition of certain activities, technical specifications for others, the imposition of fees, etc. We hope that the analysis has provided some insight into the types of situations in which certain policy instruments promise to be more effective than others.

Our own feeling, like that of most economists, is that environmental policy in the United States has failed to make sufficient use of the pricing system. Policies relying excessively on direct controls have not proved very effective in reversing processes of environmental deterioration and, where they have, we would guess the objective has often been achieved at unnecessarily high cost. Moreover, to the extent that environmental authorities have used price incentives, they have typically adopted subsidies rather than fees. These subsidy programs have often been ill-designed, providing incentives only for the use of certain inputs in waste treatment activities and by absorbing only part of the cost so that investments in pollution reducing equipment continue to be unprofitable. We still have much to learn at the policy level about the proper use of price incentives in environmental policy.

What emerges from all this is the conclusion that there is considerable validity to the standard economic analysis of environmental policy. There is good reason for the economist to continue to emphasize the virtues of automatic fiscal measures whose relative ease of enforcement, efficiency enhancing properties, and other special qualities are too often unrecognized by those who design and administer policy.

On the other hand, we economists have often failed to recognize the legitimate role of direct controls and moral suasion, each of which may have an important part to play in an effective environmental program. These policy tools may have substantial claims in terms of their efficiency, particularly under circumstances in which the course of events is heavily influenced by variables whose values are highly unpredictable and outside the policy-maker's control. In environmental economics we can be quite certain that the unexpected will occur with some frequency. Where the time costs of delay are very high and the dangers of inaction are great, the policy-maker's kit of tools must include some instruments that are very flexible and which can elicit a rapid response. A tightening of emission quotas or an appeal to conscience can produce, and has produced, its effects in periods far more brief than those needed to modify tax rules, and before any such change can lead to noteworthy consequences. Where intermediate targets, such as emission levels, may have to be changed frequently and at unforeseen times, fiscal instruments may often be relatively inefficient and ineffective.

In sum, as in most areas of policy design, there is much to be said for the use of a variety of policy instruments, each with its appropriate function. Obviously this does not mean that just any hybrid policy will do, or that direct controls are always desirable. Indeed, there are many examples in which their use has provided models of mismanagement and inefficiency. Rather, it implies that we must seek to define particular mixes of policy that promise to achieve our environmental objectives at a relatively low cost.

References

1. Baumol, W. and W. Oates. "The Use of Standards and Prices for Protection of the Environment," *The Swedish Journal of Economics* 73 (March 1971), pp. 42–54.
2. Bramhall, D. and E. Mills. "A Note on the Asymmetry Between Fees and Payments," *Water Resources Research* 2 (Third Quarter 1966), pp. 615–616.

3. Coase, R. "The Problem of Social Cost," *Journal of Law and Economics* 3 (October 1960), pp. 1–44.
4. Comptroller General of the United States. *Examination into the Effectiveness of the Construction Grant Program for Abating, Controlling, and Preventing Water Pollution.* Washington, D.C.: Government Printing Office, 1969.
5. Elliott, R. and J. Seagraves. "The Effects of Sewer Surcharges on the Level of Industrial Wastes and the Use of Water by Industry," *Water Resources Research Institute* Report No. 70 (August 1972).
6. Ethridge, D. "User Charges as a Means for Pollution Control: The Case of Sewer Surcharges," *The Bell Journal of Economics and Management Science* 3 (Spring 1972), pp. 346–354.
7. Fisher, C. "Cincinnati Industry Reduces Sewer-Surcharges," *Sewage and Industrial Wastes* 28 (September 1956), pp. 1186–1187.
8. Freeman, A. M. and R. Haveman. "Residual Charges for Pollution Control; A Policy Evaluation," *Science* 177 (July 28, 1972), pp. 322–329.
9. Griffin, J. "An Econometric Evaluation of Sulphur Taxes," *Journal of Political Economy* (forthcoming).
10. Johnson, E. "A Study in the Economics of Water Quality Management," *Water Resources Research* 3, No. 2 (1967).
11. Kneese, A. and B. Bower. *Managing Water Quality: Economics, Technology, Institutions.* Baltimore: Johns Hopkins, 1968.
12. Kneese, A., S. Rolfe, and J. Harned, eds. *Managing the Environment: International Economic Cooperation for Pollution Control.* New York: Praeger, 1971.
13. Kohn, R. "Price Elasticities of Demand and Air Pollution Control," *Review of Economics and Statistics* 54 (November 1972), pp. 392–400.
14. Krier, J. *Environmental Law and Policy.* New York: Bobbs-Merrill, 1971.
15. Lave, L. and E. Seskin. "Acute Relationships Among Daily Mortality, Air Pollution, and Climate," *Economic Analysis of Environmental Problems,* edited by Edwin S. Mills. New York: National Bureau of Economic Research, 1974.
16. Löf, G. and A. Kneese. *The Economics of Water Utilization in the Beet Sugar Industry.* Baltimore: Johns Hopkins, 1968.
17. Mäler, K. G. "Effluent Charges versus Effluent Standards," Working Paper (May 20, 1972).
18. Marcus, M. "Capital-Labor Substitution Among States: Some Empirical Evidence," *Review of Economics and Statistics* 46 (November 1964), pp. 434–438.
19. Marx, W. *Man and His Environment: Waste.* New York: Harper and Row, 1971.
20. Noll, R. "Institutions and Techniques for Managing Environmental Quality," 1970. Mimeographed.
21. Plourde, C. "A Model of Waste Accumulation and Disposal," *Canadian Journal of Economics* 5 (February 1972), pp. 119–125.

22. Upton, C. "Optimal Taxing of Water Pollution," *Water Resources Research* 4 (October 1968), pp. 865–875.
23. Zwick, D. and M. Benstock. *Water Wasteland.* New York: Grossman, 1971.

COMMENT

Charles Upton, The University of Chicago

The authors consider a variety of policy instruments for regulating environmental quality. For a variety of reasons which they—and I—find compelling, they reject proposals such as subsidies and moral suasion and suggest instead a mixture of pollution taxes and direct controls. They conclude that although "environmental policy in the United States has failed . . . to make sufficient use of the pricing system . . . economists have too often failed to recognize the legitimate role of direct controls . . . which may have an important part to play in an effective environmental program."

According to Oates and Baumol, an important reason for using direct controls is the stochastic nature of environmental quality. Since a fixed tax will result in periods of low environmental quality, direct controls should—again, according to the authors—be employed on those occasions. Yet this argument is an invalid comparison between controls which can be varied and a tax structure which cannot. Their argument essentially rests on the quite strong assumption that pollution taxes cannot be changed to deal with "emergencies," but the level of direct controls can be changed.

But one can change the level of taxes. Indeed, one should. For example, air quality in urban areas is usually lower in winter than in summer, suggesting the use of a two-part emissions tariff, and not a uniform emissions tax throughout the year supplemented by direct controls during the winter and summer months.

The notion of a differential tax can be further extended to other cases. For example, air quality drops during "thermal inversions" and so presumably does the optimal level of emissions. Oates and Baumol call for direct controls under such circumstances. But a temporary rise in the emission tax—of sufficient magnitude—could achieve the same reduction in emissions as direct controls. To be sure, one could not impose a "thermal inversion surcharge" until it could be determined that an inversion had occurred. Hypothetically if this took one day, taxes would not be useful on the first day. But then one could not impose emergency controls until it was determined that an inversion had occurred.[1]

In sum, it is important to distinguish between cases in which one knows that a parametric shift in the environmental quality has or will occur and those which are unpredictable. In the first case, either taxes or controls can be used; the well-known efficiency properties of taxes to which the authors allude suggest that taxes are appropriate. In the second case, there is no emergency policy which will be able to affect emissions.

As the authors admit, their argument rests on the assumption that the taxes cannot be changed as rapidly as direct controls can be imposed. However, they do not address the question of how one could institute controls and have them effective if it is impossible to use taxes.[2]

Integral to Oates and Baumol's discussion of the uncertainty issue is their analysis of the environmental authority's objective, which they take to be meeting some prescribed standard of environmental quality. The true objective of the authority is to maximize the value of environmental quality net of emission treatment costs. Since this rule may prove difficult to implement, it may sometimes be useful to adopt as a proxy an objective of meeting a prescribed environmental quality standard.[3] However, if the shifts in the parameters such as wind conditions that affect environmental quality are truly stochastic and emission levels cannot be changed in response, it may be impossible to meet any standard with certainty.

Even if it is possible to change emissions to meet a given environmental quality standard, optimal social policy may be to accept variations

1. One could however announce a tax schedule which would be applied whenever inversions occur, even though there might be a delay in determining that an inversion had occurred. But unless one assumes that firms have superior ability to recognize the start of an inversion, this plan will not make the tax any more effective since firms will respond only when they believe an inversion has begun.

2. Two caveats on this point. First, firms will set the short run marginal cost of reducing pollution equal to the tax. So if only a short-run reduction in emissions is desired, one may require a tax higher than the one which would be required if a permanent reduction in emissions was desired (assuming, of course, that the short-run marginal cost of reducing emissions is higher than the long-run marginal cost).

Second, there is another problem if the regulatory authority is unsure of the effects any given tax or control schemes will have on emissions. This is a difficult problem which has no simple solution. However, this is not the problem taken up in the paper.

3. To be more precise, it is sometimes useful to analyze pollution control as the dual tasks of meeting a quality standard at minimum cost and determining the optimal standard.

Figure 1

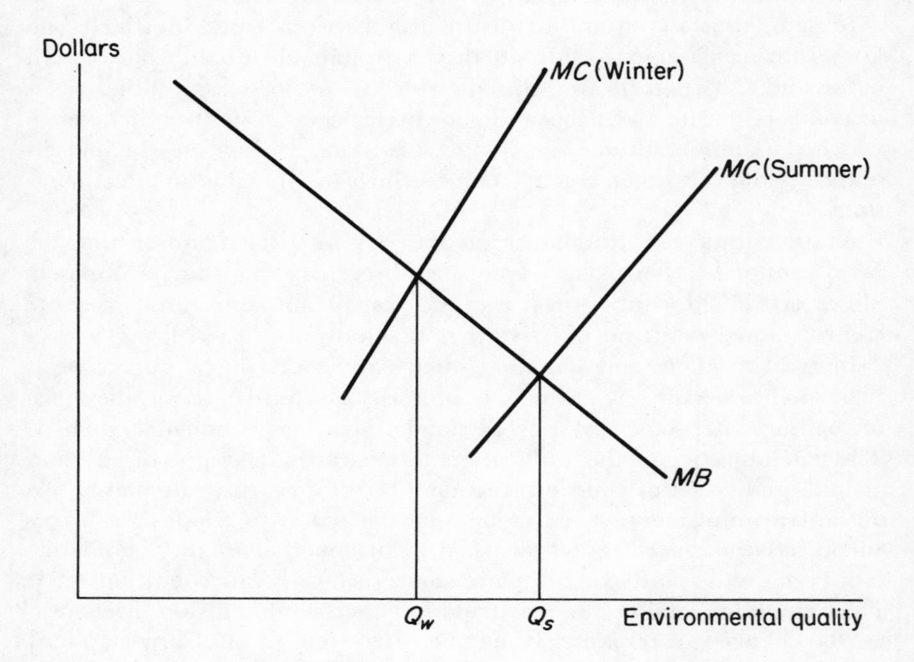

in environmental quality. For a simple example, consider again the case of winter and summer months. One factor behind the difference is the wintertime demand for heating. If we interpret this as meaning that the marginal cost curve for environmental quality shifts to the left in winter months and if, for simplicity, we assume that the marginal benefit curve is the same for both seasons, it is optimal to have seasonal changes in environmental quality standards. As Figure 1 illustrates, the optimal level of environmental quality is Q_w in the winter and Q_s in the summer.

Another difficulty with emission taxes raised by Oates and Baumol lies in their application to oligopolies. Since the authors attach only minor importance to this issue and since an oligopoly is an ill-defined concept, my comments will be brief. First, note that in the simple case of a profit-maximizing monopolist, emission taxes are more efficient than direct controls. An emission tax will induce any profit-maximizing firm to reduce emissions and substitute hitherto more costly factors of production, thus minimizing the total social cost of producing output. Direct controls probably will not do that. To be sure, a monopoly will not necessarily pass along the full cost of emission control to the consumer (as would a competitive firm), but this will be true whether the costs arise from direct controls or control via emission taxes. So, on balance, the differences lie in favor of emission taxes.

Two additional cases are those of the regulated industry and an oligopoly which has objectives other than maximizing profits. However, there is a question of their relevance: to what extent does regulation matter and do oligopolies exist? But these are old issues and there seems little point in repeating the arguments here.

Another regulatory device which Oates and Baumol do not consider is nonintervention. Indeed they begin their paper by specifically ruling out this possibility, claiming that the transactions costs involved in the provision of environmental quality by private action make such a solution impossible. The transactions costs involved in private action do not constitute an absolute barrier to the provision of public goods by private action; they mean public goods might thereby be undersupplied. Although little is known about the economics of political processes, it is possible that political control of environmental quality could mean an oversupply of environmental quality. If so, a policy of nonintervention which results in an undersupply of environmental quality may well be preferable to a policy of government intervention which provides an oversupply. The expected cost of an undersupply must be weighted against the costs of a possible oversupply and inefficient production of environmental quality possible with a nonmarket solution.[4]

It is even more difficult to reject *a priori* a policy of nonintervention by the federal government when one considers the possibility of local control. Regional differences in factor endowments suggest that there should be regional differences in the provision of environmental quality. Indeed, even were there no differences in factor endowments, differences in individual tastes would argue for cities providing different levels of environmental quality.

Surely, almost all of the externalities from, for example, Pittsburgh's air pollution are internalized within Pennsylvania, and there would seem little necessity for federal intervention to set air quality standards. To be sure, there are some cases like the Chicago SMSA where problems cross state lines. However, the number of negotiators required to inter-

4. Or to put it another way: most economists would agree *a priori* that there is some inefficiency in a water pollution control act which called for zero effluents, and it is conceivable that the social welfare would be lower than it would be under a policy which permitted unlimited discharges of pollutants.

nalize interstate externalities is sufficiently small (the governors of Illinois and Indiana) that the Coase solution seems appropriate.

Thus, it is difficult to rule out a policy of nonintervention at the federal level. Although a case can be made for economies of scale implicit in federal control,[5] these gains must be weighted against the welfare loss from the provision of uniform levels of environmental quality (which seems implicit in federal control).

5. A common example of these economies is the possible cost to the automobile industry of dealing with fifty state automobile emission standards.

The Resource Allocation Effects of Environmental Policies

G. S. Tolley, The University of Chicago

Once, if you asked an economist what to do about externalities, the answer was sure to be: tax them. A number of questions have been raised about the traditional tax approach, and nontax approaches have continued to find more favor in actual policy. These developments help explain why interests of economists have widened to direct limitations on outputs and inputs, zoning, salable rights, legal recourse and a variety of other formal and informal arrangements (see Bohm, Buchanan, Ciriacy-Wantrup, Clarke, Dales, Kamien, Kneese, Mishan, Tideman, Tolley, Turvey, Upton, Wolozin, Wright, Zerbe).

The traditional economics literature on taxes and most of the recent literature on nontax policies have been qualitative. How to measure the benefits and costs of the policies has been neglected. The measurement task is often taken to be the obvious gathering of facts, not recognizing deficiencies in concepts needed for their collection and interpretation. Previous literature has tended to deal with one policy at a time. Different forms of control on polluting firms, procrusteanism of imposing uniform requirements, and spatial arrangements have been particularly neglected. A framework is needed for systematically comparing policies and indicating how effects depend on underlying demand and production conditions.

With these concerns in mind, the first section of this paper considers benefits from reducing a single negative externality. Results are obtained on how to use information on physical effects of the externality, on de-

NOTE: Helpful comments were made by Gardner Brown, Charles Upton, Richard Zerbe and University of Chicago urban economics workshop participants.

fensive acts of those harmed, and on factor rewards. Several needs for modifying benefit estimation practices emerge.

The second section considers the costs of reducing an externality through (a) emission regulation, (b) requirement of emission control equipment, (c) restrictions on inputs and (d) restriction on output. General cost expressions are developed, the policies are compared using algebraic forms, and applications of current interest are discussed.

After a third section on how to bring together benefits and costs with identical factors, the fourth section considers losses from identical requirements where there are uncertain multiple externalities with nonuniform factors. This section gives most attention to nonuniformity within a shed where physical effects are interrelated. Quantitative restrictions, taxes, salable rights and zoning—all of which are the same for a single externality under certainty—are compared. The final major section deals with location of activity between sheds giving attention to land bids needed for optimum location incentives.

Damages and Defensive Acts

Firms

If reducing effluent will lower production costs of downstream firms, one part of benefits is the lowering in costs of producing the prevailing level of output downstream. Since the change in the total cost of producing the output is the sum of changes in marginal costs, this part of the benefit is equal to the sum of changes in the marginal cost of production from zero output up to the prevailing output downstream. If the demand curve facing the downstream firms is not completely inelastic, the lowering of marginal cost curves will increase the output at which marginal cost equals price. On each increment of increased output, there is a net gain equal to the difference between the demand value of the increment and marginal cost of production. The part of the benefit resulting from increased output of the downstream firms is the sum of the differences on each increment between the old and new output.

In Figure 1, H is output of the downstream firms prevailing before the reduction in the externality. The part of the benefits which is the change in cost of producing the prevailing output is the difference in marginal costs from zero up to H, or the area $ABCE$. As a result of reducing the externality, output expands to where the new marginal cost curve intersects the demand curve at output I. The part of the benefit due to additional output is the sum from H to I of incremental differ-

Figure 1

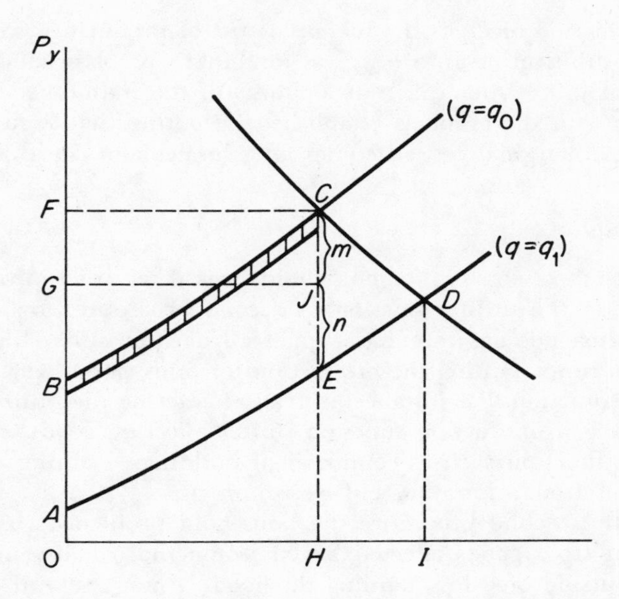

ences between demand value and marginal cost, or the area *CDE*. The total benefit from reducing the externality is the total of the areas *ABCE* and *CDE*, or *ABCD*. This is the standard result that benefit is equal to the change in producer plus consumer surplus [10].

Let the demand curve facing producers of a commodity y which is adversely affected by pollution be

$$p_y = F(y). \tag{1}$$

The production function is

$$y = y(z, q), \tag{2}$$

where z refers to inputs controlled by the producer. The variable q is a public good such as quality of water or air and is not controllable by the producer of y. The system is completed familiarly by equating marginal value of y to input price times inputs required to produce an extra unit of y:

$$p_y = p_z/y_z(z, q) \tag{3}$$

where $y_z(z, \hat{q})$ is the partial of (2) with respect to z and is the marginal product of z.

The right side of (3) is the marginal cost of producing y. Solving (2) for z and substituting into (3) gives marginal cost of producing y as a function of y itself, for different amounts of the public good q. Correspondence with the graph is established by noting that Figure 1 shows two of these marginal cost schedules and the demand curve (1).

Households

A first procedure possible for households would be to let environmental quality enter the utility function. A second procedure, to be followed here, is to exclude environmental quality from the utility function and let it be an input shifting the production for other goods which do enter the utility function. For instance, instead of entering the utility function, air quality is a production function shifter affecting goods which enter the utility function such as condition of buildings, clothing cleanliness, and freedom from respiratory and eye symptoms.

Under this second procedure the household problem is to maximize satisfaction from goods affected by environmental quality. Air quality affects inputs devoted to obtaining the goods. Among several advantages of this procedure, the analysis of benefits from improving environmental quality for the household becomes identical to that just given for the firm, permitting an institutionally neutral approach not arbitrarily affected by whether activity, such as laundering, takes place in the firm or household.

As applied to the household, Figure 1 shows how a lowering of environmental quality raises the marginal cost curve for attaining the goods on the x-axis which are affected by environmental quality. Defensive measures and other time and money responses to pollution are cost outlays devoted to the goods. The total cost outlay is the sum of the marginal costs up to the output achieved, or $OADI$ at the higher level of environmental quality and $OBCH$ at the lower level. The costs include housewife time in the case of cleanliness, and they include medical bills and time lost from work in the case of health. Even under adverse environmental conditions, medical bills and time lost from work are subject to choice since options would be not to have medical treatment and not to stay away from work, in which case health would be reduced below the low best level OH achievable with the reduced environmental quality.

To derive equations (1), (2) and (3) for the household, note that in contrast to the firm problem max $p_y y - p_z z + \mu_y[y - y(z, q)]$, the problem for the household is max $U(y, z') + \mu_y [y - y(z, q)] + \mu_z(Z - p_z z - p_z' z')$ where Z is total wealth and z' is all other goods. For the firm, given the demand curve (1) and the production function (2), the derivation of (3) by Lagrangian maximization is straightforward. For the household, the production function (2) is given, but no price of y or demand curve are given. Letting p_y be the internal demand price or amount of money the household is willing to give up to get an extra unit of y, this amount of money must be such that the utility from spending an extra dollar on y and z' are the same, $U_y/p_y = U_{z'}/p_{z'}$ (see Becker), or rearranging $p_y = (U_y/U_{z'})p_{z'}$. Using the budget constraint to substitute out z' and taking $p_{z'}$ as given, the foregoing price condition gives the internal demand curve (p_y as a function of y) which is equation (1) for the household. Using the price condition together with the Lagrangian solution to the household maximization problem gives $p_y = p_z/y_z$ which states that marginal valuation is equated to marginal cost of producing y and is equation (3), completing the demonstration that the same system is obtained for the firm and household.[1]

The expenditure approach

The effects of air quality on household expenditures are often estimated to gain an idea of the benefits of air pollution reduction (see Ridker).

1. The Lagrangian solution to the household maximization problem in the text can be written $U_z = U_y y_z$. Indicating the variables appearing in each function and using the production function $y = y\ (z, q)$ to eliminate y, the equilibrium condition for the text formulation—where environmental quality does not enter the utility function—is $U_{z'}(z', z, q) = U_y(z', z, q) y_z(z, q)$.

If environmental quality does enter the utility function, expenditures on things z affected by environmental quality are considered to be expenditures on goods with utility, instead of being expenditures on inputs. The formulation of the household maximization problem becomes max $U(z', z, q) + \mu_z(Z - p_z z - p_z z')$ for which the equilibrium condition is $U_{z'}(z', z, q) = U_z(z', z, q)$.

Compare the right sides of the equilibrium conditions under the two different formulations. If environmental quality enters the utility function as in the formulation just given in this footnote, the marginal utility of things affected by environmental utility is seen, in terms of the text formulation, to be a product whose unobserved components are the marginal utility of the output affected by environmental quality times the marginal productivity of inputs in producing the output.

If environmental quality enters the utility function, activities which are responses to pollution must enter as related goods. They have to be analyzed in terms of "substitutability" with environmental quality, which seems arbitrary and prevents consideration of the more ultimate household satisfactions y. Because of the suppression of ultimate satisfactions, information on health and other physical measures of well-being cannot be used in benefit estimation using the formulation, given in this footnote, where environmental quality enters the utility function.

The change in expenditures is the difference between $OADI$ and $OBCH$. Since the two costs have $OAEH$ in common, the change in expenditure is $ABCE$ minus $HEDI$. $ABCE$ is the change in costs necessary to maintain the level of y at H, and is equal to $ky\Delta C$ where ΔC is the vertical shift in marginal cost at H and k is the ratio of the average of vertical shifts at all the previous values of y relative to the shift at H. Expressed as a percentage of the value of $p_y y$, $ABCE/p_y y$ is $k\Delta C/C$ since at the margin price p_y equals marginal cost C. The area $HEDI$ is $p_y \Delta y$ minus the area EJD, which in turn is $n\Delta y/2$. Making use of the fact that the elasticity δ of the marginal cost curve is $(\Delta y/n)(C/y)$ and again expressing results as a per cent of $p_y y$, $HEDI/p_y y$ equals $(\Delta y/y)[1 - (\Delta y/y)/2\delta]$. To find $\Delta y/y$, making use of the fact that the elasticity of demand β is $(-\Delta y/m)$ (p_y/y) and of the expression for δ, obtain $m + n = \Delta C$ as a function of Δy. Solving for Δy and dividing by y gives $(\Delta y/y) = (\Delta C/C)$ $\{1/[(1/\delta) - (1/\beta)]\}$. These results may be combined to obtain changes in expenditures as a percentage of value

$$\Delta E/p_y y = (\Delta C/C)\{k + \beta[1 + \beta(\Delta C/C)/2(\delta - \beta)]/[1 - (\beta/\delta)]\}. \quad (4)$$

The special case of a horizontal marginal cost curve is

$$\Delta E/p_y y = (\Delta C/C)(1 + \beta) \quad \text{if} \quad \delta = \infty \quad \text{and} \quad k = 1. \quad (4\overline{C})$$

Extra effort over a wide range should continue to yield substantial effects on physical characteristics defining cleanliness, thus suggesting that marginal cost is fairly constant for attaining these attributes and thereby proving ($4\overline{C}$) a good approximation for cleanliness. A commonly reported finding is that higher pollution does not lead housewives to devote more effort to cleaning. Contrary to the inference one might be tempted to draw that there are no cleanliness benefits, a possibility is that the elasticity of demand for cleanliness is unity ($\beta = -1$) since this is the only condition making the right side of ($4\overline{C}$) zero. Even with error in answers, the lack of perceptible expenditure response under extreme pollution conditions suggests a downward response of cleanliness demanded to a rise in its cost ($\beta < 0$).

The area $ABCE$ is $-k\Delta C/C$ as already noted. The additional benefit area CDE is $(\Delta y)(\Delta p)/2 - (\Delta y)(\Delta C - \Delta p)/2$ or $-(\Delta y)(\Delta C/2)$. Adding the two areas, making use of the solution for $\Delta y/y$ and dividing by $p_y y$ gives benefits as a fraction of product value:

$$\Delta B(y)/p_y y = -(\Delta C/C)\{k + (\Delta C/C)/2[(1/\delta) - (1/\beta)]\}, \quad (5)$$

which reduces to

$$\Delta B(y)/p_y y = -(\Delta C/C)[1 - \beta(\Delta C/C)/2] \quad \text{if} \quad \delta = \infty \quad \text{and} \quad k = 1. \qquad (5\overline{C})$$

Cleanliness

Comparing ($4\overline{C}$) and ($5\overline{C}$) makes clear that zero change in expenditure ($\beta = -1$) does not indicate that benefits are zero. Under conditions that seem typically satisfied of rises in marginal costs less than one hundred per cent and absolute value of elasticity of demand of one or less, benefits in ($5\overline{C}$) are the same order of magnitude as the use in marginal cost.

Suppose marginal cost of maintaining household cleanliness is raised twenty-five per cent due to heavy pollution in a neighborhood. Assuming $\beta = -1$, $\delta = \infty$, and $k = 1$, ($4\overline{C}$) indicates change in expenditures is zero while ($5\overline{C}$) indicates costs (negative benefits) are 28.1 per cent of the total expenditures for cleanliness. If the yearly value of materials and time expended on cleanliness is $1,000 per household, the pollution costs are $281 per household of $2.81 million per year for a neighborhood of 10,000 households showing that pollution costs may be substantial even in the absence of an observed expenditure response.

Medical services

Instead of being horizontal, marginal cost curves may be upward sloping and may be shifted nonuniformly. For a disease, the abscissa is an index of freedom from the disease symptoms. In the absence of pollution, rising marginal costs might be encountered only at a health level far to the right. With air pollution, the marginal cost curve would be shifted up and could become more steeply sloped at a lower level of health. For a disease with high treatment costs or debilitating effects, the relative rise in marginal cost $\Delta C/C$ may be high at H, and change in marginal cost at H may be greater than average change in marginal cost on the units of x to the left of H. At the lower level of cost, which Figure 1 indicates to be the relevant cost curve for the calculation, the supply curve might be highly elastic. The fact that expenditures are observed to increase is suggestive that the demand elasticity is less than one. If $\Delta C/C = .10$, $k = 2$, $\delta = 7.5$ and $\beta = -.5$, (4) and (5) give $\Delta E/p_y y = .153$ and $-\Delta B/p_y y = .205$.

At the extreme, if no defensive expenditures are possible, the marginal cost curves become vertical lines. With no observed changes in expenditures, the benefits are determined entirely by the slope of the demand curve ignored in the expenditure approach.

Mortality

The model of (1) – (3) can guide studies of physical effects of pollution. The benefit of a one-unit change in environmental quality is the sum of the effects on marginal costs of all units of x up to the observed level, illustrated as the sum of the small quadrangles in Figure 1. In view of (3), the sum is $\int_0^y [d(p_z/y_z)/dq]dY$. Carrying out the differentiation under the integral sign, substituting in $p_y = p_z/y_z$ and making use of $dY = y_z dZ$ to change the variable of integration gives as the sum of quadrangles $\int_0^z p_y y_{zq} dZ$ which equals $p_y y_q$ and says that the benefit from a one-unit change in environmental quality is the value of a unit of y times the effect of the environmental change on y. Another way of representing the benefit area $ABCE$ plus CDE thus is

$$B(y) = \int_{q_0}^{q_1} p_y y_q \, dQ, \tag{6}$$

which suggests how measures of pollution effects y_q on physical attributes should enter benefit estimation. Note that y_q is a marginal productivity concept holding all other inputs z constant.

Suppose the only health effects of air pollution are small effects on probability of survival, which probability is the good measured as the abscissa. Suppose the change in probability is so small that the marginal value of survival is not affected (demand curve flat over the range being considered) and there are no defensive measures (marginal cost curves perfectly vertical). Then equation (6) indicates the appropriate measure of benefits is the observed change in survival expectancy times p_y, a measure of the value of life.

Morbidity

If the demand curve is not flat or if defensive expenditures are undertaken, as is the rule for morbidity, in applying (6) one must first allow for changes in marginal value p_y along the demand curve. Econometric studies are conceivable estimating sacrifices people are willing to make to avoid physical effects as a way of facing this valuation problem. Second, the effect of the expenditures on physical attributes needs to be subtracted out to obtain the sole effect y_q of pollution or physical attributes. The observed association between morbidity and pollution understates the benefits from pollution reduction since morbidity is reduced by defensive expenditures. Clinical data might throw light on effects of defensive measures and might also be used to directly estimate y_q if situations can be found of the same defensive measure under different pollution levels. For damages to materials, as opposed to human beings, controlled observations are promising.

Land and labor returns

The problems of goods definition encountered in analyses of expenditures and physical effects do not arise in the factor rewards approach. Since any environmental effect which is less than nationwide can be escaped by moving, given consumer knowledge the shaded benefit area *ABCE* plus *CDE* can be expected to show up as a factor reward difference, estimable without the conceptual problems surrounding Figure 1. The idea that air quality differences within a city are reflected in land values, provides a rationale for benefit estimates based on econometric studies of pollution effects on residence values (see, for example, Crocker and Anderson).

Environmental effects pervading an entire city are not mutable by a residence change within the city. However, because they are mutable by moving between cities, they can be expected to show up in differences in wages between cities. In contrast to work on land values, there has been little estimation of environmental effects on wages. To indicate possibilities, a preliminary result by Oded Izraeli is a regression of deflated wages of laborers in SMSAs on human capital, public expenditure and environmental variables. The R^2 is .81. Regarding air pollution, the elasticity of wages with respect to sulfates is .09 and with respect to particulates is .01. Both signs are as expected, and the coefficient of sulfates is significant above the 5 per cent level.

Productivity of Pollution

Turning from benefits to costs, the costs of pollution reduction consist of losses in satisfaction from commodities whose production causes pollution. In the absence of incentives to control pollution, pollution can be ignored as a consideration in production of these commodities. The traditional theory of production without controls suffices. If pollution is reduced from the point of no control, losses may be incurred because less of the product is produced and it is produced in a higher cost way. While the existence of pollution control costs has been recognized, the reasons for losses have not often been considered explicitly. At most, even in theory, a cost schedule for reducing emissions is usually assumed as a starting point without being derived.

To find out why and by how much the costs of different methods of control differ, in addition to needing to know about product demand and the traditional production function for product output, knowledge is needed about an additional production function indicating how pollutant emissions depend on producer decisions. Specifically, emissions depend on waste producing inputs and pollution control inputs. In this section, it will be shown that the production function for emissions is a key determinant of differences in policy costs. Under an emission regulation policy, producers can choose between adjusting waste producing inputs and pollution control inputs. Because they can choose, this policy is least costly. Under requirement of pollution control devices, producers have incentives to reduce emissions using the devices but not to adjust waste producing inputs; whereas under regulation of waste producing inputs, these incentives are reversed. The relative costs of the latter two policies depend on the marginal effects on emissions of pollution control devices and waste producing inputs. The most costly policy of all is restriction of product output, under which the only reason for emission reduction is a fall in output, with no action being taken to reduce emissions caused by any given output.

Policy effects can be analyzed as responses to incremental exogenous changes. The marginal emission benefit is achieved by allowing emissions to increase one unit through incremental changes in a policy, as for example, the benefit from relaxing restriction on waste producing inputs just sufficiently to allow emissions to increase by one unit. The cost of a policy (measured as benefit foregone) is the sum of marginal benefits from allowing emissions to increase from their level under the policy up to the uncontrolled level. Since uncontrolled emissions are pushed to the point where they have no further value, marginal benefit is zero from allowing emissions to increase at the no control equilibrium under any policy. The magnitude of total benefits foregone depends on how rapidly marginal benefits decline in approaching the no control equilibrium. Thus comparing policies requires comparing *change in marginal benefits* as emissions are allowed to increase. After presenting the no control model, a model of producer decision will be set up for each policy, from which will be derived marginal benefits, change in marginal benefits and the resulting policy costs.

No control

Let the demand curve for a commodity whose production causes pollution be

$$p_x = D_x(x) \tag{7}$$

where p_x is price or value of an extra unit of x. For a producer having no effect on price, p_x is given implying the slope D_{xx} is zero. If output affects price, as for a local utility, it will be assumed that regulation enforces marginal cost pricing, leaving for future analysis other pricing policies. If the polluting entity is a household, the price is marginal valuation within the household. The incentive is then to maximize the area under the curve less foregone expenditures in producing the commodity.

In the absence of expenditures to reduce emissions, the only physical relation of concern to the producer is the traditional production function explaining product output:

$$x = x(u, f), \tag{8}$$

where f consists of inputs such as coal or gasoline which are polluting and u consists of all other inputs that increase the production of x. The assumed demand conditions imply familiar incentives to make output price times marginal physical product equal to input price. The problem is max $\int_0^x D(X)dX - p_f f - p_u u + \lambda_x[x - x(u, f)]$, whose solution by Lagrangian maximization gives:

$$p_x x_u = p_u, \tag{9}$$

$$p_x x_f = p_f, \tag{10}$$

where x_u and x_f are the partials of (8) and p_u and p_f are the input prices.

Equations (7)–(10) determine commodity price, output, and the two inputs in the absence of efforts to control pollution. They describe market behavior toward pollution assuming there are free rider and other impediments to private negotiations. To consider how changes will affect this system, a generalized displacement can be represented by taking the differential of each equation. The resulting coefficients of differential changes are:

$$\begin{matrix} dx & dp & du & df \\ \end{matrix} \qquad dE$$

$$\begin{bmatrix} D_{xx} & -1 & 0 & 0 \\ -1 & 0 & x_u & x_f \\ 0 & x_u & px_{uu} & px_{uf} \\ 0 & x_f & px_{fu} & px_{ff} \end{bmatrix} = \begin{bmatrix} e_x \\ e_p \\ e_u \\ e_f \end{bmatrix}. \qquad \begin{matrix} (7') \\ (8') \\ (9') \\ (10') \end{matrix}$$

The determinant on the left hand side will be denoted M. On the right hand side dE refers to any exogenous change. The coefficients e_x, e_p, e_u and e_f indicate the effect, if any, of the change in each equation. With no controls, exogenous changes refer to shifts in demand function, production function or factor prices. With controls, the exogenous changes can also refer to incremental changes in a policy control.

Emission regulation

The production function specifying emissions is:

$$s = s(f, c) \tag{11}$$

where c refers to inputs devoted to controlling emissions. The polluting inputs f increase emissions, and control inputs c decrease them.

One set of policies of interest theoretically and practically operates on emissions s; e.g., s is limited to some maximum amount. This type of policy induces producers to use less polluting inputs and to incur emissions control expenditures. With a given limit of allowable emissions, the marginal cost to the producer of adding a unit of polluting input is the input price *plus* the cost of controlling the emissions from the extra input. The extra emissions are given by the partial of the emission relation (5) with respect to f, or s_f. For example s_f is pounds of smoke resulting from an extra ton of coal. Since adding one unit of precipitator inputs will reduce pounds of smoke emitted by $-s_c$, precipitator inputs required per pound of smoke reduction are $-1/s_c$. Multiplying the precipitator inputs required per pound of smoke by the extra pounds of smoke gives $-s_f/s_c$, the control inputs required to keep emissions from increasing. The magnitude $-s_f/s_c$ is the marginal rate of substitution between control inputs and polluting inputs and will be denoted σ. The cost of controlling emissions from an extra unit of polluting inputs is this amount times the price of control inputs or $p_c\sigma$. In contrast to (10), the conditions governing use of polluting inputs becomes

$$p_x x_f = p_f + p_c\sigma. \tag{10s}$$

Figure 2

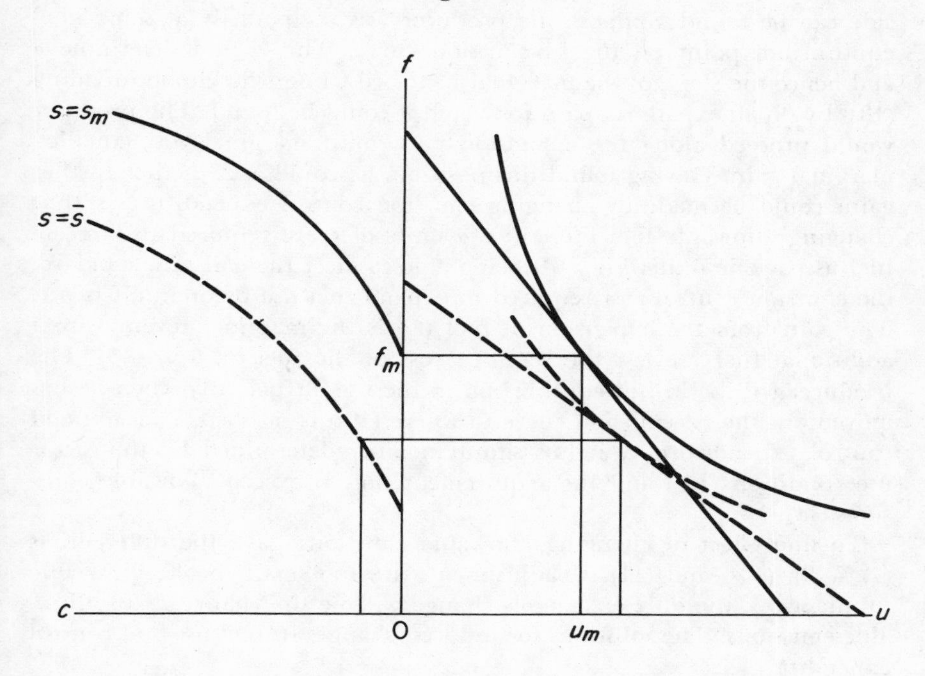

The control costs $p_c\sigma$ add to the marginal cost of using polluting inputs, giving incentives to use less of them.

Equations (7)–(9), (10s) and (11) describe the system under the regulation controlling s. As compared with the free market system, there is an additional endogenous variable c. In the free market system, there are no incentives to use control inputs ($c = 0$). If the regulation is effective, c will take on a positive value.

Free market and the control situation are compared in Figure 2. The right side contains iso-product curves for x. The free market inputs f_m and u_m are determined in the usual manner by tangency between an iso-product curve and factor cost line having slope $-p_u/p_f$. With an emission standard, the slope of the factor cost line is the dashed line $-p_u/p_f + \sigma p_c$. The left side of Figure 2 contains iso-emission curves. Taking the differential of (11) holding s constant and solving for df/dc gives slope of iso-emission curve $-s_c/s_f$, the reciprocal of σ. If allowed emissions are lowered from the free market level s_m to $\bar{s}$, the producer contemplates positions along the new iso-emission curve, each position implying a different slope of marginal factor cost line on the right side. For any choice of c on the left side, an optimum production decision for x on the right side can be found. Suppose the producer was temporarily at some non-equilibrium point on the iso-emission curve. This would determine σ and hence the slope of the marginal factor cost line whereupon, dividing (10s) by (9) an expansion path for x and u could be found. The producer would proceed along the expansion path until marginal cost equalled marginal gain. Having found this position, he could ask whether further gains could be made by changing emission control expenditures c thus changing allowable fuel use. Since σ units of c are required to increase fuel use by one unit while still being able to meet the emission standard, the emission control cost required to expand fuel use by one unit is $p_c\sigma$. The gain from the expansion of fuel use is the marginal revenue from additional fuel use less the resource cost of the fuel or $p_x x_f - p_f$. The producer will be in full equilibrium in the use of fuel only when he has moved out the iso-emission curve to where (10s) is satisfied. Fuel use and control expenditures are thus simultaneously determined by the factor use condition (10s) and the requirement not to exceed allowable emissions (11).

To find effect of changing allowable emissions, take the differentials (7) – (9), (10s) and (11). If adjustments are too small to affect variable input prices, the only exogenous change will be the change $d\bar{s}$ in allowable emissions. The solutions for induced changes in fuel use and control expenditures are

$$df/d\bar{s} = p_c\sigma_c M_{ff}/M_s, \tag{12s}$$

$$dc/d\bar{s} = (M - p_c\sigma_f M_{ff})/M_s, \tag{13s}$$

where $M_s = s_c M + (s_f\sigma_c - s_c\sigma_f)p_c M_{ff}$. The first subscript of a double subscript for M indicates the deletion of a row, and the second indicates deletion of a column.

The benefits from producing x are $b(x) = \int_0^x D(X)dX - p_u u - p_f f - p_c c$, that is, the consumption benefits less the input costs. The change in benefits from imposing an incremental adjustment in s is obtained by differentiating benefits with respect to s to obtain $b(x)_s^{\bar{s}} = p_x(dx/ds) - p_u(du/ds) - p_f(df/ds) - p_c(dc/ds)$. This expression can be simplified by inserting the derivative of the production function for commodity output (8) with respect to s, $(dx/ds) = x_u(du/ds) + x_f(df/ds)$, into the change in benefits to eliminate (dx/ds), giving $(p_x x_u - p_u)(du/ds) + (p_x x_f - px_f - p_f)(df/ds) - p_c(dc/ds)$. Substituting in the marginal productivity conditions (9) and (10s) further simplifies the change in benefits to $b(x)_s^{\bar{s}} = p_c\sigma(df/d\bar{s}) - p_c(dc/d\bar{s})$. The first term on the right side is the excess $p_c\sigma$ of the marginal benefits from fuel use over the marginal resource cost of fuel, times the change in fuel resulting from a one-unit change in allowed emissions $\bar{s}$. The second term is the resource cost of emission controls c resulting from a unit change in s. The simplifications leading to $b(x)_s^{\bar{s}}$ make use of the idea that in a no control equilibrium the total benefits in the production of x are maximized implying marginal benefits are zero; i.e., extra resources devoted to x are just worth the benefits obtained. A change in benefits when s is changed occurs only if the marginal conditions are not fulfilled. The change in benefits is the difference between the marginal resource costs incurred for those inputs not being used so as to maximize benefits in the production of x.

The change in benefits can be simplified further because the two terms in the centered expression for $b(x)_s^{\bar{s}}$ just given are control cost effects. The term $-p_c(dc/d\bar{s})$ is the direct change in control costs as a result of a change in allowable emissions and would be the entire change in benefits if there were no induced change in f. On the other hand, if there were no change in control costs and the entire adjustment was to change f, adjustments in control costs would be avoided. A reduction in fuel use of one unit reduces emissions by s_f making it possible to avoid reducing control inputs by s_f/s_c. Since $\sigma = -s_f/s_c$ the saving on control costs is $p_c\sigma$. Differentiation of the emission relation (11) with respect to $\bar{s}$ gives as a necessary condition between fuel and control input changes $1 = s_c(dc/d\bar{s}) + s_f(df/d\bar{s})$ indicating that the sum of the emissions changes due to control input and fuel adjustment must equal the total emissions change. Rearranging, the change in fuel is $df/d\bar{s} = [1 - s_c(dc/d\bar{s})]/s_f$ or the part of the emission change not met through control costs divided by change in emissions per unit of fuel change. Substituting this change in fuel into the expression for $b(x)_s^{\bar{s}}$ gives

$$b(x)_s^{\bar{s}} = -p_c/s_c. \tag{14s}$$

The benefit resulting from a change in allowed emissions reduces the control cost saving that would be made possible by allowing a one-unit emissions change, holding fuel constant. Comparing with the previous expression for $b(x)_s^{\bar{s}}$ the benefit is not the actual control cost change but rather is what the control cost change would be if the entire adjustment in emissions were achieved via a change in control inputs.

The slope $b(x)_{ss}^{\bar{s}}$ of the marginal benefit schedule, needed to evaluate the cost of an emission regulation, can be found by differentiating (14s) with respect to s to obtain

$$b(x)_{ss}^{\bar{s}} = (p_c/s_c^2)[s_{cf}(df/d\bar{s}) + s_{cc}(dc/d\bar{s})] \tag{15s}$$

where $df/d\bar{s}$ and $dc/d\bar{s}$ are given by (12s) and (13s).

Pollution control devices

A second type of policy would not control emissions directly but would require producers to undertake emission control expenditures, making $\bar{c}$ exogenous. There are then no incentives to hold down fuel use. The producer model consists of (7) – (10) plus the condition that c is exogenous, which is the same as the free market model except that c is nonzero. The cost of this policy is simply the emission control expenditure. The effect on benefits (negative of costs) of a one-unit change in emissions achieved through altering control inputs is input price p_c times the $1/s_c$ emission control inputs required to reduce emissions by one unit.

$$b(x)_s^{\bar{c}} = -p_c/s_c. \tag{14c}$$

The right hand sides of (14s) and (14c) are identical because benefit change (14s) under the $\bar{s}$ policy can be expressed as a hypothetical control cost expenditure that would be necessary. In (14c) the change in expenditure is actual.

The *change in marginal benefits* with respect to *emission control inputs* is the derivative of (14c) with respect to c, or $-p_c s_{cc}/s_c^2$. The slope being sought is the change in marginal benefits with respect to *emissions* and is this derivative divided by the associated change in emissions $ds/d\bar{c}$. Since there are no incentives to change f, $ds/d\bar{c}$ is obtained by differentiating (11) with respect to c holding f constant or s_c. Thus the slope of the marginal benefit schedule under the policy of controlling c is

$$b(x)_{ss}^{\bar{c}} = p_c s_{cc}/s_c^3. \tag{15c}$$

Restricting waste producing inputs

The simplest example of a policy operating through waste producing inputs is a direct control on an amount of a fuel. Instead of choosing fuel according to (4) or (4s), fuel f becomes exogenous. The producer model then is (7) – (9) determining price of output p_x, output x and nonfuel inputs u. Since no incentive is given to make emission control expenditures, $c = 0$.

Differentiate benefits $\int_0^x D(X)dX - p_u u - p_f f - p_c c$ with respect to f, substitute in the derivative of the production function with respect to f, and make use of (9) and the condition that $c = 0$ to ascertain that the change in benefits with respect to f is $p_x s_f - p_f$, or the difference between marginal revenue and marginal cost from the extra unit of f, which is reasonable since the other inputs are either in equilibrium or are zero. Since the amount of fuel needed to reduce emissions by one unit is $1/s_f$, the effect on benefits of a unit change in emissions achieved through reducing fuel inputs is

$$b(x)_s^{\bar{f}} = (p_x x_f - p_f)/s_f. \tag{14f}$$

It was possible to express benefit change under general emission control (14s) in terms of control costs because the difference between marginal revenue and marginal cost of fuel $p_x x_f - p_f$ was equal to addition to control costs required due to adding fuel, i.e., from (10s) $p_x s_f - p_f = p_c \sigma$. The latter equality does not hold under the fuel restriction policy. As f is reduced, the divergence between marginal revenue and marginal cost will grow. The value of $\sigma = -[s_f(f, 0)]/[s_c(f, 0)]$, or the control inputs that would be required to keep emissions from increasing when f is changed, might be altered little if at all. Thus (14f) must remain as stated with no conversion to equivalent control cost.

To obtain the *change in marginal benefits* with respect to *fuel*, differentiate (14f) with respect to f to obtain $[p_x x_{fu}(du/d\bar{f}) + p_x x_{ff} + x_f(dp_x/d\bar{f})]/s_f$. This approximation holds as long as the second term in the differentiation is zero $\{-[(p_x x_f - p_f)/s_f^2]s_{ff} = 0\}$, which is necessarily so at the free market equilibrium where $p_x x_f - p_f = 0$. The approximation remains good as long as the fuel restriction is not so severe as to raise $p_x x_f - p_f$ to a significantly large value. Another defense of the approximation is the likelihood that s_{ff} will be small. The reasonable assumption that, with zero emission controls, emission will tend to be proportional to fuel input, implies s_{ff} is zero, making the term in question drop out. This assumption is used in the functional form examples later.

Take the differentials of (7) – (10) letting f change exogenously, and solve the linear system to obtain $du/d\bar{f} = -M_{fu}/M_{ff}$ and $dp_x/d\bar{f} = M_{fp}/M_{ff}$. Substitute these results into the change in marginal benefits resulting from a change in fuel given at the beginning of the previous paragraph, factor out $1/M_{ff}$ from the bracket, and note that the bracket then equals M. The change in marginal benefits from a unit change in *emissions,* achieved via an input policy such as fuel restriction, is obtained by dividing $ds/d\bar{f}$ $(= s_f)$:

$$b(x)_{ss}^{\bar{f}} = (1/s_f^2)(M/M_{ff}). \qquad (15f)$$

A policy giving the producer equivalent incentives to adjust the amount of fuel would be a tax on fuel equal to $p_x x_f - p_f$, i.e., a tax making an equivalent divergence between marginal revenue and marginal resource cost of coal. From (14f) it is seen that the marginal benefits are proportional to the amount of this tax. The *change* in marginal benefits is then proportional to the change that would occur in such a tax. M/M_{ff} on the right side of $b(x)_{ss}^{\bar{f}}$ is the reciprocal of the response of fuel use to a change in fuel price and is thus, in fact, equal to the change in tax that would be necessary to bring about a unit change in fuel use, which is then converted to an emissions basis by the $(1/s_f^2)$ term.

Restricting output

A fourth type of policy seeks to control emissions even more indirectly, through affecting the producer's decision as to amount of x produced. The simplest example is a direct restriction making x exogenous. In the model of producer decision, the demand relation (7) is dropped since the regulation of x prevents the producer from adjusting output to demand. The model then consists of the production function (8) and the factor demand relations (9) and (10) in which price of output p is replaced by the marginal cost of output λ. The producer adjusts factors to minimize the cost of a given output but is unable to carry output to where $p = \lambda$. In the other models, where x is not controlled, marginal cost equals price making it unnecessary to distinguish between p and λ.

Since the derivative of benefits with respect to x is $p_x - p_u(du/dx) - p_f(df/dx)$, since (3) and (4) permit the substitutions $p_u = \lambda x_u$ and $p_f = \lambda x_f$, and since the derivative of the production function (2) with respect to x gives the substitution $1 = x_u(du/ds) + x_f(df/ds)$, the marginal benefit from a change in x reduces to $p_x - \lambda$ which, reasonably, is the value of an extra unit of x minus the cost of producing it. The *marginal benefit* with respect to *emissions,* achieved through the exogenous changes in x, is obtained as in the other cases by dividing by the change in emissions resulting from the change in x:

$$b(x)_s^{\bar{x}} = (p_x - \lambda)/s_f(df/dx). \qquad (14x)$$

Using the same logic as for the fuel restriction policy, the *change in marginal benefits* with respect to *emissions* when x is changed, $b(x)_{ss}^{\bar{x}}$, is $(1/s_f^2)[(dp_x/dx) - (d\lambda/dx)]$ divided by $(M_{xf}/M_{xx})^2$ which is the square of the fuel change resulting from a change in x obtained from solving (8′) – (10′) with x exogenous. Also from (8′) – (10′), $d\lambda/dx = -M_{xp}/M_{xx}$. From the demand relation (1), $dp_x/dx = D_{xx}$. Making these substitutions in the expression for $b(x)_{ss}^{\bar{x}}$, factoring out $1/M_{xx}$ from the bracket and noting that the bracket then equals M, gives as the slope of the marginal benefit schedule for the case where output is controlled

$$b(x)_{ss}^{\bar{x}} = (1/s_f^2)(M_{xx}/M_{xf})(M/M_{xf}), \tag{15x}$$

which can be interpreted as the change in tax on output required to change emissions by a unit $M/M_{xf}s_f$, divided by the change in emissions per unit change in output s_fM_{xf}/M_{xx}.

Comparison of the four policies

The curved lines in figure 3 are total benefits from x and are a maximum $b(x)^m$ at the free market level of emissions s_m. Benefits from x are reduced as one moves to lower emissions. The dark curved line shows the benefits from x under one particular policy. The dark straight line is the marginal benefit curve from x for this policy. The light lines in Figure 3 pertain to an alternative policy. The cost of a policy is the difference between free market benefits and benefits under the policy, or $b(x)^m - b(x)^\pi$ where π denotes the policy. This cost is the shaded area under the marginal benefit schedule in figure 3.

The cost is the sum of marginal benefits in going from free market emissions s_m to emissions s_π under the policy, or $[b(x)^m - b(x)^\pi] = -[\int_{s_\pi}^{s_m} b(x)_s^\pi ds]$ $b(x)_s^\pi ds$. Marginal benefit at the free market solution is marginal benefit at any lower level of emissions s plus the sum of changes in marginal benefits going from the lower level up to the free market level, or solving for the marginal benefits at the lower level $b(x)_s^\pi = b(x)_s^m - \int_s^{s_m} b(x)_{ss}^\pi\, dS$. Substituting this result into the expression for cost and assuming free market marginal benefits from pollution are zero, the cost of any policy π is:

$$b(x)^m - b(x)^\pi = \int_{s_\pi}^{s_m} \left[\int_s^{s_m} b(x)_{ss}^\pi\, dS\right] ds. \tag{16}$$

Figure 3

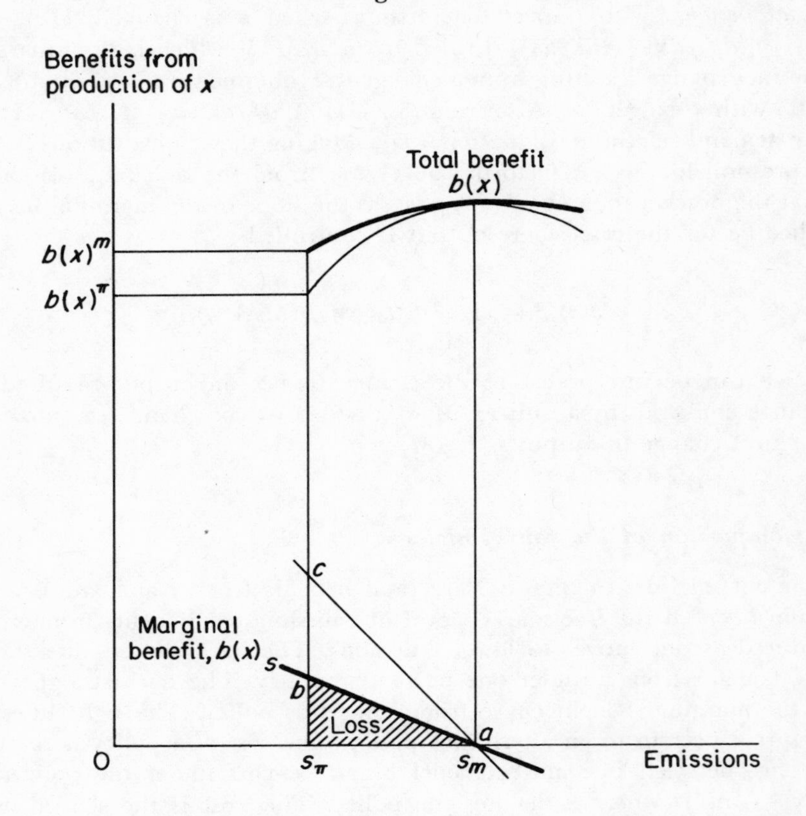

If the marginal benefit schedule can be approximated as linear, $b(x)_{ss}{}^{\pi}$ is constant giving as the cost of any policy π:

$$b(x)^m - b(x)^\pi = b(x)_{ss}{}^{\pi}(s_m - s_\pi)^2/2, \qquad (16L)$$

verifiable as the shaded area by inspection. For a given emission reduction, the costs of the policies are thus proportional to the slopes $b(x)_{ss}{}^{\pi}$ of the marginal benefits schedules (15s), (15c), (15f) and (15x).

A possible functional form for the emission relation (11) is $s = mfe^{-kp_cc/p_ff}$, where m is emissions per pound of fuel if there are no control inputs and k is the percentage reduction in emissions per pound of fuel resulting from an extra dollar of expenditures on control inputs relative to fuel. Evaluating (15s) with this functional form, inserting the result into (16L) and dividing by $p_f f$ reveals that the estimated cost, relative to fuel expenditures, of a policy of regulating emissions is

$$[1/(-k + \eta_f)](a^2/2), \qquad (16s)$$

where a is the reduction in emissions as a per cent of total emissions and η_f is the own price elasticity of demand for fuel. Similarly, the cost relative to fuel expenditures of a policy of requiring emission control inputs is

$$(1/-k)(a^2/2). \qquad (16c)$$

The cost relative to fuel expenditures of restricting fuel inputs is

$$(1/\eta_f)(a^2/2). \qquad (16f)$$

Finally, the cost relative to fuel expenditures of restricting the firm's output is

$$(1/-v\eta_x)(a^2/2), \qquad (16x)$$

where v is the change in value of fuel inputs per unit of output accompanying a change in output and η_x is the elasticity of demand for fuel with respect to the price of output.

With regard to the last policy, the elasticity η_x in (16x) is a firm scale effect. The only reason that fuel use is affected by the price of x is that there is a product output response which changes all inputs. On the other hand, the elasticity η_f in (16f) contains both a scale effect and a substitution effect. In addition to giving incentive to change the scale of output, a change in the price of fuel gives incentives to substitute between fuel and other inputs, indicating that the cost of a fuel restriction policy relative to fuel expenditures is less than that of restricting the firm's output.

Comparing (16f) and (16c) indicates that whether a fuel restriction policy is cheaper than requiring emission control inputs depends on whether η_f is less than k. Since the formulas express costs as a per cent of fuel expenditures, the cost comparison also depends on the absolute level of fuel expenditures. The least costly of the four policies relative to fuel expenditures is emission regulation, which (16s) reveals to be a combination of the fuel restriction and control input policies. If the latter two policies happen to be equally costly, the emissions regulation will be half the cost of either of them.

As a further application, if the firm faces a perfectly elastic product demand and has a CES production function for output, evaluation of M and its cofactors gives $\eta_f = [1 - \epsilon + (\epsilon - \gamma)/(1 + p_f f/p_u u)]/(1 - \epsilon)(\gamma - 1)$ and $\eta_x = 1/(1 - \gamma)$ where the elasticity of substitution is $1/(1 - \epsilon)$ and the scale parameter γ is the percentage change in output that would result from a simultaneous 1 per cent increase in inputs u and f. For a short run situation, suppose the elasticity of substitution is zero ($\epsilon = \infty$). Suppose that expenditures on fuel and other variable inputs are each a third of the value of output, the total of the shares being substantially less than one due to short run fixity of many inputs. Assuming the shares add to the elasticity of output with respect to the inputs implying $\gamma = 2/3$, the costs of a fuel restriction and an output restriction policy are identical because of the zero elasticity of substitution assumption and are $a^2/3$ of fuel expenditures. For a long run situation, suppose that the elasticity of substitution is one ($\epsilon = 0$), fuel is a third the value of output, and other inputs are one-half the value of output with $\gamma = 5/6$. As a per cent of fuel expenditures, the costs of a fuel restriction policy are then $a^2/6$ and the costs of output restriction policy are $5a^2/24$. If k is 5, costs relative to fuel expenditures of a policy of requiring emission control inputs are $a^2/10$. The costs of emission regulation relative to fuel costs are $a^2/13$ in the short run and 1/8 in the long run. These examples illustrating how factor substitution and scale effects determine policy costs are consistent with the idea that costs rise with increasing rapidity as emission reduction approaches 100 per cent, in view of the a^2 term.

Relevance

This section has dealt with production theory for a firm under restrictions, in contrast to previous studies in which information about specific control devices and fuels has been used to estimate dollar costs at a point assuming no substitutions, for example the two studies done by the U.S. Environmental Protection Agency in 1963 and 1970. In future work it would be most useful to draw on details in engineering and physical science studies to estimate emission and output production functions, thus obtaining refined measures of substitution and scale effects.

Each policy type has many examples, all in need of the analysis contained in this section. The proposed tax on sulfur dioxide emissions is an example of the least costly of the four policy types. The major approach to air and water pollution followed in practice is of the same general type in that it deals with emissions. In the emission relation (11), there is a positive relation between polluting inputs and emissions. The common practice is to vary emission standards in line with this relation, allowing larger plants to pollute more than small ones so as to constrain every plant to emissions proportionally below the uncontrolled level. The standards are designed to force firms to adopt the more expensive control devices presently manufactured for each scale of output. This practice could be optimal economically in the special case where plants regardless of size are forced to equally high marginal costs of control and where damages from an extra pound of smoke are the same regardless of plant size. While the special case is probably never encountered exactly, the common practice is almost surely more optimal than setting standards invariant to scale.

For automobiles, the tax on leaded gasoline, requirement of catalytic mufflers, and emission standards or tests for cars, are examples of fuel input, control devices and emission policies. Limiting auto use in central business districts is an output policy, specifically restricting auto travel output.

The static case considered in this section provides a starting point for dynamic extensions in which policy costs are conceived as a present value. In view of incentives to move toward new factor combinations, enhanced inducements to discovery will lead to research and development responses to policies neglected in previous estimates of costs. Since substitutions depend on the wearing out of equipment, the optimal timing of pollution reduction depends on capital replacement decisions. Capital replacement analysis is needed for variances granted in judicial and administrative proceedings that allow delay in meeting standards. London laws banning coal in space heating provide examples of input policies with dynamic dimensions. Coal has been declining as a household fuel due to relative cost changes. Replacement of existing furnaces determines the timing. Part of the costs of reducing coal air pollutants is the present value of switching out of coal sooner rather than later.

Net Benefits

Best

The production functions considered so far have included equation (2) $y = y(z, q)$ explaining the output of commodities affected by pollution, equation (8) $x = x(u, f)$ explaining output of commodities whose production causes pollution and equation (11) $s = s(f, c)$ explaining pollution emissions. The system is completed by another production function

$$q_i = q_i(s_1, \ldots, s_n), \tag{17}$$

showing how pollution emissions are transformed into changes in the public good causing pollution damages. This function differs for air, water and solid wastes, and for pollutants within any one of the waste forms. It can indicate how treatment facilities affect q and how emissions s at one time affect the public good at later times. For the air pollution examples in this paper, (17) is an air dispersion model. As will be shown later, the effect of an emission s_j on air quality q_i within the same shed varies depending on the location of j and i.

If there is only one x and y producer each or if producers are identical with no locational differentiation within the shed, (17) reduces to $q = q(s)$. This equation can be used in expressions for y benefits such as (5) and (6) to replace q with s. One obtains benefits from y as a function of s which can be compared with the benefits from x as a function of s derived in the preceding section. In Figure 3, the y benefits as a function of s reach a maximum to the left of the maximum for the x good. Proceeding from the y maximum, the marginal y benefits are negative as s increases. The marginal y benefits with changed sign are marginal costs which may be plotted in the same quadrant as the marginal benefits from x. The marginal cost schedule is upward sloping and crosses the marginal benefit schedule for x still to the left of s_m. If the two marginal schedules are linear, the gain from moving from the free market situation s_m to the maximum net benefit point, say s^*, where the two marginal benefit schedules cross, is the area between the schedules or $-[b(y)_s{}^{s_m}]^2/2[b(y)_{ss}{}^s - b(x)_{ss}{}^s]$, where $b(y)_s{}^{sm}$ is the y benefit from a unit increase in emissions at the free market level. The foregoing gain is the maximum potential gain from an environmental policy. Solving for the level of emissions where marginal benefits equal marginal costs and subtracting from free market emissions gives reduction in emissions necessary to achieve the maximum gain.

Absolutely inferior

If emissions are reduced beyond what is necessary to achieve the maximum gain, a level $\bar{s}$ will eventually be reached below which net benefits are less than at the free market level. With linearity, the critical emission reduction is twice the reduction necessary for maximum potential gain, or

$$s_m - \bar{s} = 2b(y)_s{}^{s_m}/[b(y)_{ss}{}^s - b(x)_{ss}{}^s], \tag{18}$$

that is, twice the marginal effect of emissions on y benefits at the free market levél, divided by the sum of the slopes of the marginal schedules. Nonlinearities might increase negatively on the slope of the marginal benefits schedule for y at higher emissions and increasing marginal costs of control for x at lower emissions. Since the nonlinearities have opposite effects on s^* and $\bar{s}$, they could conceivably be offsetting. These considerations indicate information needed to eliminate policies, that may be put forth in the course of policy deliberations, which are worse than no policy.

Waste interfaces

An illustration brings out the economics of the much discussed possibility that reduction of one externality may increase another. Suppose a choice is being made whether to get rid of garbage by incinerator or landfill. Direct costs per ton of garbage are $10 for incineration and $6 for landfill. To estimate external costs, suppose the volume to be handled of 500 tons per day, if incinerated, would result in 10 tons per day of particulate emissions. If damages from the particulates increase by $10,000 for each increase of one ton in daily particulate emissions, the external cost of the incinerator is $100,000 per year. Assuming a total of 125,000 tons are handled during a year, the external cost for incineration is $.80 per ton of garbage. A landfill will impose external costs on surrounding residences due to unsightliness, smell and noise. For this volume of waste, assume a landfill would impose an average property volume loss of $2,000 on 1,000 residences or $2 million capital loss which implies a yearly loss of perhaps $200,000. The external costs for landfill are then $1.60 per ton. Direct plus external costs per ton are $10.80 for incineration and $7.60 for landfill. The external costs do not reverse the ranking based on direct costs in this example, but, of course, they ought in other cases.

The landfill with external costs concentrated on a few residences in a vocal outlying community, might generate greater public opposition than the incinerator causing the landfill to be rejected in spite of its cost advantage. Requiring one payment for all compensation could make incentives of damagers and damagees coincide with incentives to maximize net benefits. If compensation of $2 million to surrounding property owners were required on opening the landfill (and likewise compensation to those damaged by air pollution were required if the incinerator were built) the external costs would be borne by those disposing of wastes. The damagéd parties being fully compensated would lose their economic incentives for opposing the facilities.

Uniformity Losses

Within a metropolitan area, identical polluters cause different damages. Figure 4 shows SO_2 changes within the Chicago area that would result from a new power plant. The damages from the plant depend on densities of activity along each isopleth. Locating the plant differently would

Figure 4

Increase in SO_2 Levels from a New Generating Plant ($\mu g/M^3$)

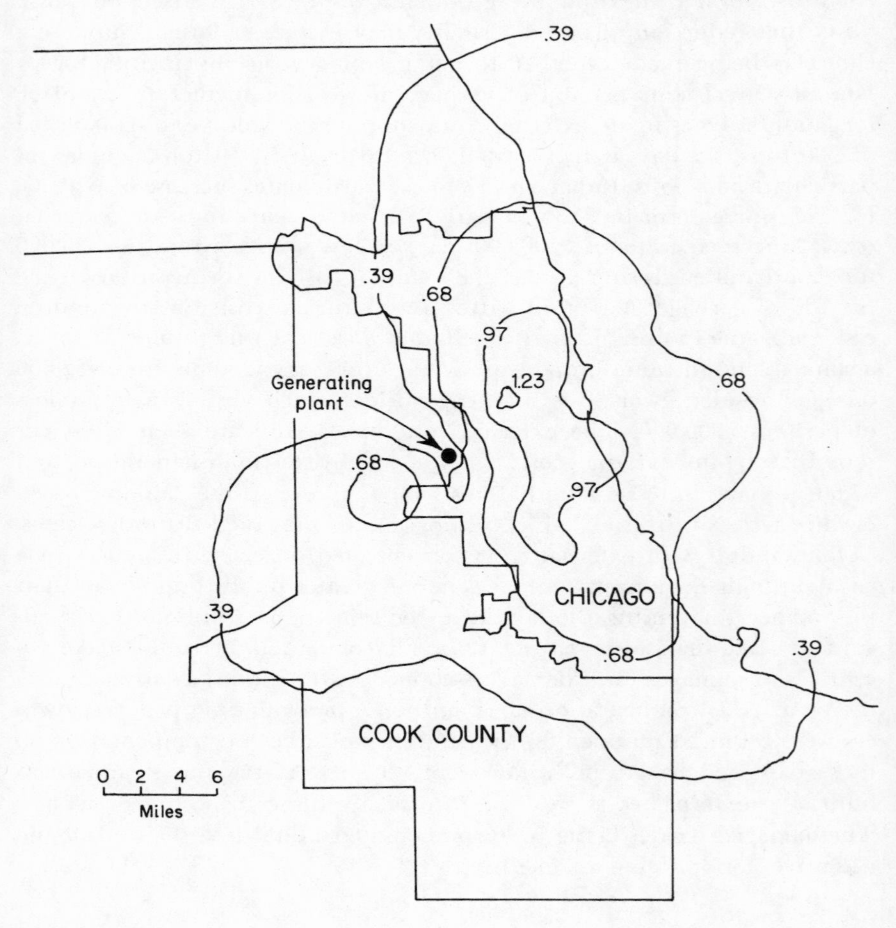

change the isopleths leading to different damages. In outlying low density areas, the damages would be smaller.

Imposing uniform restrictions on emissions regardless of location imposes identical control on different externalities. All the current policies mentioned in the section on pollution control costs have this defect. Control may result in positive net benefits for some sources, while for other sources emissions are restricted below $\bar{s}$ with control costing more than worth. If quantitative restrictions are to be used, a case can be made for administrative or court regulation set case by case according to damages, rather than having a uniform emission standard. A uniform tax on emissions also imposes identical incentives on different externalities. To avoid the uniformity losses, the tax rate on emissions should vary at different points in the city. Salable emission rights also encounter the problem. If sales of rights to emit were allowed unrestricted within the metropolitan area, their trading value would tend to be the same as the tax necessary to attain the same level of emissions. With salable emission rights, *transfer fees* or *rebates* might be desirable equal to differences in damages caused by a change in the location of emissions.

Uniformity losses could be restricted by establishing zones within cities, within which uniform incentives would be imposed. Zoning could avoid a failure, possible under the foregoing policies, to find gains from eliminating pollution from some areas altogether. Joe Reid has emphasized the implications of nonconvexities, indicating a formal case for zoning to exclude some types of pollution altogether if the marginal damages are decreasing rather than increasing. The stability conditions are then violated, and the market—or policies trying to correct the market marginally—may find a local polluting optimum where net benefits are less than with no pollution. The normal concave situation of increasing marginal damages probably applies to materials damages and health. The convexity argument applies to esthetic and recreational uses of land, whereby a little bit of blight is sufficient to reduce drastically the benefits and with further blight not having much effect after the initial impact.

None of the policies mentioned so far avoids the free rider problems of getting people to bid either for reduced pollution or for the generation of knowledge about health and other adverse effects of pollution. Schemes have been proposed, so far at an abstract level, for including people to reveal their preferences for a public good by making the supply curve facing them be the actual supply curve less other people's bids (Clarke, Tideman). These schemes face the same problems connected with nonuniformities as do other policies.

The issue of losses from uniformity extends beyond spatial variation within one metropolitan area. When the wind is blowing and the weather is mild, reducing demand for fuels for heating, the marginal damage caused by a given level of emissions may be small, in contrast to great damages in still, cold weather, and even greater damages during an inversion. If a single emission standard must be observed at all times, as most plans require, different external costs are being treated alike over time as well as over space. Emergency measures required during episodes are a step in the direction of recognizing uniformity losses.

Administrative costs and regulatory behavior partly determine how much uniformity should be imposed, but the benefits and costs being considered in this paper also influence the choice. Because there are costs of varying the level of emissions, particularly if plants must be shut down for temporary periods, the maximization of net benefits does not call for instantaneous adjustment of emissions to every change in external costs.

The uniformity issue is particularly severe for automobiles in view of the variability in their emission damages with weather and traffic. While the costs of instantaneous adjustments appear prohibitive, the current legislative approach aiming to require the same control at all times everywhere maximizes uniformity losses. Preliminary results by Richard Zerbe suggest promise for varying standards by area and other intermediate strategies.

Location of Industry in Different Parts of the Country

Applying the same emission standard everywhere in the nation carries uniformity losses to a maximum. Costs for areas where emissions are reduced below $\overline{s}$ could exceed gains for areas of positive net benefits. If an industry's production costs are not very different among locations, shifts toward areas of low damage may be cheaper than relying solely on control in place. One of the least costly ways to reduce pollution losses could be to induce a different locational pattern. At odds with this idea is current legislation, which aims to freeze location. Standards for new plants are the same everywhere and are more stringent than for existing plants, impeding even normal locational adjustment.

While prominent in classic externality discussions (Coase, Pigou) spatial considerations have not been fully resolved partly because of need to more adequately consider land. The usual conclusion that a quantitative restriction, tax or salable right have the same effects on a producer is valid only if he stays in business at the same location. Consider the bid for an industrial site by a polluter. He will have higher profits with a quantitative restriction on emissions requiring no payments, than he will with necessity to pay an emissions tax, even though the two forms of control induce the same emissions reduction if he produces at the site. With the quantitative restriction he can bid more for the site and is more likely to bid out competing users. To foster optimum allocation, it appears his tax payments should equal total damages imposed making his bid correctly reflect x-benefits less y-losses. In deciding where to locate, externalities will then be internalized.

One pitfall to avoid is levying only a marginal tax rate. If an emission tax rate is made equal to marginal damages caused by a pound of pollutant, the total tax paid will equal total damages only if marginal and average damages are equal. If the idea is correct that marginal tends to be above average damages, the tax collected would be greater than total damages calling for a lump sum rebate. Under any policy, there may be a difference between marginal payments for emissions and the damages caused. As another example, under an emission standard marginal payments are zero. A general rule is to make lump sum rebate (or tax) equal to the difference between the sum of marginal payments and the total damages caused.

Payments by polluters should be coupled either with no compensation to those damaged (Baumol) or more equitably one payment for all compensated. As applied to land values, with no compensation, external costs will cause negative windfalls to owners of land on which external costs are borne, and taxpayers will gain the proceeds of the tax. With compensation, there need be no such wealth transfer. The present value of the damages is transferred from polluters to owners of affected land exactly offsetting the loss in sale value of their land due to the damages. Note that compensation is not related marginally to damages and is to be paid to land owners, not land users. If frequent compensation is expected and if land owners can manage to have excessive losses incurred on which to base the compensations, the expectation of compensations might conceivably induce nonoptimal use of land on which damages occur. However, the idea of deciding land use for expected compensation ignoring market revenues seems somewhat far fetched, and the practice would be limited by abilities of outside adjudicators to verify gross cases of excessive losses.

With agreement that the foregoing norms would induce optimal adjustment to pollution, estimates of costs of departures from them could be made. This requires comparing locational differences in costs of polluting production with differences in damages, a job in which little progress has yet been made.

Conclusion

Any environmental policy instrument reduces pollution damages through some combination of reducing emission producing inputs, installing pollution control devices, reducing product output or changing the location of activities. This paper has examined benefits and costs for these dimensions. Instead of seeking a single best instrument the approach has been to compare alternatives, recognizing that resource allocation is not the sole consideration in environmental policy. Yet the very reason for concern with the environment is to correct resource allocation failures. Resource allocation is more important in environmental policy than in many other policy areas.

References

1. Baumol, W. "On Taxation and the Control of Externalities," *American Economic Review* (1972).
2. Becker, G. *Economic Theory*. New York: Knopf, 1969.
3. Bohm, P. "Pollution, Purification and the Theory of External Effects," *Swedish Journal of Economics* (1970).
4. Buchanan, J. *The Demand and Supply of Public Goods.* Chicago: Rand-McNally and Company, 1968.
5. Ciriacy-Wantrup, S. V. "The Economics of Environmental Policy," *Land Economics* (1971).
6. Clarke, E. "Introduction to Theory for Optimum Public Goods Pricing," *Public Choice* (1971).
7. Coase, R. "The Problem of Social Choice," *Journal of Law and Economics* (1960).
8. Crocker, T. and R. Anderson, "Property Values and Air Pollution," *Committee on Urban Economics Conference Proceedings,* Chicago, 1970.
9. Dales, J. *Pollution, Property and Prices.* Toronto: University of Toronto Press, 1968.
10. Harberger, A. "The Economics of Waste," *American Economic Review* (1964).
11. Kneese, A. and B. Bower. *Managing Water Quality: Economics Technology, Institutions.* Baltimore: Johns Hopkins, 1968.
12. Kamien, M., W. Schwartz, and F. Dolbear. "Asymmetry Between Bribes and Charges," *Water Resources Research* (1966).

13. Mishan, E. "The Postwar Literature on Externalities," *Journal of Economic Literature* (1971).
14. Pigou, A. *The Economics of Welfare*. New York: Macmillan, 1946.
15. Ridker, R. *Economic Costs of Air Pollution*. New York: Praeger, 1967.
16. Tideman, N. "The Efficient Provision of Public Goods," *Public Prices for Public Goods,* edited by S. Mushkin. Washington, D.C.: The Urban Institute, 1972.
17. Tolley, G. "Water Resources," *International Encyclopedia of Social Sciences* Vol. 16. New York: Macmillan, 1968.
18. Turvey, R. "On Divergences Between Social Cost and Private Cost," *Economica,* 1963.
19. U.S. Environmental Protection Agency. *The Costs of Clean Air*. Washington, D.C.: Government Printing Office, 1970.
20. U.S. Environmental Protection Agency. *The Costs of Clean Water*. Washington, D.C.: Government Printing Office, 1969.
21. Upton, C. "The Allocation of Pollution Rights," *National Tax Journal* (forthcoming).
22. Wolozin, H. *The Economics of Air Pollution*. New York: Norton, 1966.
23. Wright, C. "Some Aspects of the Use of Corrective Taxes for Controlling Air Pollution Emissions," *Natural Resources Journal* (1969).
24. Zerbe, R. "Theoretical Efficiency in Pollution Control," *Western Economic Journal* (1970).

COMMENT

Gardner Brown, Jr., University of Washington

It is common knowledge that the best level of pollution occurs when the marginal damage of pollution offsets the marginal cost of an action which reduces pollution. This rule can be characterized either graphically or mathematically. The first sections of George Tolley's paper provide us with an original treatment of some thorny conceptual issues involved in identifying these functions.

It is also true that goods or bads, as the case may be, have a space and time dimension. Carrots from the San Joaquin valley differ in a fundamental economic sense from carrots from Chicago, and the economic value of a given environmental quality level in New York City only fortuitously is the same as the level in the desert region of Utah. Therefore, most would agree that a uniform air or water quality standard is not likely to make economic sense. The fourth and fifth sections of the paper discuss the effects of uniformity, those heavy-handed environmental policies which fail to recognize the spatial and temporal characteristics of bads.

The necessary ingredients for evaluating alternative environmental policies are benefit and cost functions. In too many publications, empirical benefit functions capture the value of changes in environmental quality by looking only at changes in the total expenditures of variable factors. Such estimates are conceptually biased, since changes in environmental quality impinge on the value of producers' and consumers' surpluses, necessary elements in the true estimate of net benefits. For plausible values of the elasticities of demand and supply the bias is shown to be significant. Tolley's main contribution here is to treat households as firms after the fashion of Becker.

Turning next to loss functions, the analysis begins with single product profit maximization in an environment where pollution occurs but voluntary regulation is ruled out. The first point of departure assumes a control policy on emissions. Since the control policy is restrictive—otherwise why have it?—benefits are reduced compared to the unregulated market outcome, reduced by the marginal cost of control. Labels may be misleading. The focus is on the production of conventional goods in combination with bads. Forcing firms to produce less bads increases their costs, referred to in the text as change in benefits. It should be emphasized that the author is not here discussing what is commonly referred to in the literature as the benefit or damage reduction function.

The regulator's second policy alternative involves the choice of a level of control inputs. It is exemplified by the prescription: Use secondary treatment in the case of water quality management. For any given level of control input, the optimal level of the pollution intensive input will be greater in the second policy relative to the first unless it is in a world of fixed factor proportions. As long as factor substitution is technically feasible, policy two is second best, involving greater private costs for any given initial level.

A third policy requires regulating the level of the pollution intensive input, while the fourth policy entails controlling emissions by using a final output level.

Emission regulation is superior to all other policies because the entrepreneur is left with more choice. He can select best values for x, f, c, and u whereas with the remaining policies, one of the first three variables is exogenously determined. Tolley proves that policy three, restricting pollution intensive inputs, is superior to the output restriction policy four after assuming a constant marginal benefit (loss) function and a specific form for the emission relation (11). The same conclusion holds much more generally. The policy which has the flattest marginal benefit function (3) is best. Therefore, if the fuel restriction policy is better than restricting output,

$$b(x)_{ss}^{\bar{f}}/b(x)_{ss}^{\bar{x}} < 1,$$

where $b(x)_{ss}$ is the slope of the marginal benefit function. Since

$$b(x)_{ss}^{\bar{f}} = (1/S_f^2)(M/M_{ff}), \tag{15f}$$

$$b(x)_{ss}^{\bar{x}} = (1/S_f^2)(M_{xx}/M_{xf})(M/M_{xf}), \tag{15x}$$

using the original notation, then

$$b(x)_{ss}^{\bar{f}}/b(x)_{ss}^{\bar{x}} = \frac{M_{xf}^2}{M_{ff}M_{xx}} < 1$$

or

$$M_{ff}M_{xx} - M_{xf}^2 > 0.$$

One can show that this expression indeed is positive by computing its value from the determinant exhibited above (11) in the original paper. The only requirement is that the production function is concave.

Regrettably, a complete general ranking does not seem possible since the virtues of the fuel policy (3) relative to the fixed control expenditures policy (2) depend on the choice of elasticities and parameter values.

The last pages of Tolley's paper discuss some of the problems which arise when a uniform policy is applied to an area or time period in which the net benefit function varies. Uniformity of pollution policy is expensive because no longer are there incentives to make intertemporal production adjustments or substitutions from a region of high opportunity cost to one of low opportunity cost of pollution.

Zoning and salable rights are additional tools available to policy makers. As the author rightly emphasizes, if the rights are not well-defined in space, the rights will be distributed inappropriately unless exchange involving spatial transfer is accompanied by a charge which reflects spatial opportunity cost differences.

This section offers fertile ground for other investigators to till. To cite one example, suppose meteorology was a fine-tuned science enabling us to predict weather with certainty. Suppose weather changes affect the net benefit pollution function and policy-makers had chosen a policy of charging for emissions. What is the optimal rate of price change? Surely economists would not recommend changing price with each change in the weather. Nor would they likely recommend no price change. Where on this broad spectrum does the correct rule, in fact, lie?

In the first section, benefits due to a change in pollution level, correctly measured, are the changes in the variable factor cost plus changes in profit plus changes in consumers' surplus. In the second section, the cost to producers of changes in the level of pollution is equal to the change in profits. It appears that there is double counting with changes in profit showing up in both the benefit and cost function. This is false. Note that the arguments of the benefit function are pollution level and commodity y referred to in one place as a thinness of dust cover. In contrast, the costs are due to changes in profit from producing commodity x and pollution (more accurately, emissions). Since one component in the objective function in the second section is the total consumers' surplus of x which is distinct from good y, there is no double counting. Tolley finds it helpful to structure his analysis of the net benefits of environmental policies in a unique fashion. For this effort he deserves credit, but if he would provide travelers with a few more signposts along the way, he would reduce the cost of the trip and gain more adherents to his preferred route.

In section five, Tolley cautions policymakers to avoid using *marginal* tax rates. When marginal damages are above average damages, he reasons, total tax take will be greater than total damages, calling for a lump sum rebate. There may be arguments, stemming from distributive concerns, why total taxes ought to equal total damages, but efficient resource allocation calls for marginal rules and air resources are not an exception. If I wish to buy an additional unit of air quality, surely I should pay its "owner" his opportunity cost. If he is a wealth maximizer, surely he will not sell that unit on the basis of the average value of all units in his stock of wealth.

What types of problems are amenable to Tolley's formal analysis and in what problem setting do his general results hold? His story is cast within the framework of the short run and there is no uncertainty. Only one pollutant exists thus disposing of the problem of synergism in production and consumption. Consumers have access to the best available knowledge as do producers and they respond to changes instantaneously. Producers are profit maximizers where the latter notion is given a simple straightforward textbook interpretation. The cost of actualizing and maintaining each policy is the same. Policies do not differ with respect to political feasibility, administrative ease, difficulties of enforcement and other factors discussed in the companion piece by Baumol and Oates.

Some may wish to argue that Tolley cast his net in too modest an arc. I would argue that even artificially simple but sound approaches are meritorious because they place in bold perspective crucial relationships which may change only in degree when a more encompassing model is developed. Understanding of the complex generally proceeds from a deep appreciation of the less complex to which this paper is a contribution.

Tolley is engaged in a substantial research project on environmental quality. This paper probably can be regarded as an interim report in which the author spells out the analytical framework of the larger study. Some of the topics are better developed than others, leaving the reader with an impression of rugged terrain rather than a polished surface. Nevertheless, the paper is a provocative and original contribution to the literature on environmental quality and bears the imprint of an insightful innovative mind. With this paper as openers, I'd bet heavily on the final study.

EMPIRICAL ANALYSIS
OF DOMESTIC PROBLEMS

Operational Problems in Large Scale Residuals Management Models

Walter O. Spofford, Jr., Resources for the Future, Inc., Clifford S. Russell, Resources for the Future, Inc., and Robert A. Kelly, Resources for the Future, Inc.

Introduction

Over the past three years, we at Resources for the Future, Inc. have been working on the development of a regional residuals management model which in general form is in the classical mold, but which includes certain departures in detail that we consider important.[1] Like the classical models, it is designed to find the least-cost way of meeting ambient environmental quality standards given knowledge of the costs facing residuals dischargers and of the natural systems intervening between these dischargers and the points throughout the region at which quality is constrained. Unlike the earlier models, however, it is designed to deal with air and water quality and solid waste problems simultaneously because of the tradeoffs among airborne, waterborne and solid residuals implied by the conservation of mass and energy. In addition, we have developed industrial models, which are included as modules in the overall regional model, which

NOTE: We are grateful for the many helpful comments received from our colleagues Blair T. Bower and Allen V. Kneese, and from the conference discussant, J. Hayden Boyd.

1. A pathbreaking effort in this field was the work of the Delaware Estuary Comprehensive Study funded by the federal government to provide the basis for choosing stream standards and setting effluent standards (load allocations) in the Delaware Estuary. For a description of their model see Federal Water Pollution Control Administration, *Delaware Estuary Comprehensive Study* (Philadelphia, Pa.: U.S. Department of the Interior, July 1966).

can reflect the impact on residuals generation of changes in product mix, raw material quality, etc., and which include methods other than end-of-pipe treatment for altering residuals discharges.[2] Finally, the model and its method of optimum seeking are designed to be flexible with regard to the kinds of models of the natural environment which can be used. Thus, in particular, we do not limit ourselves to the linear transformation functions which traditionally have been used to connect discharges and ambient concentrations, but allow for inclusion of more complex formulations, including nonlinear simulation models.

Our approach thus far has been to construct small, "didactic" versions of this framework in order to test and develop our ideas without running up tremendous computer bills or getting buried in mountains of data. Two didactic applications have been constructed and are reported elsewhere.[3] In the first (see footnote 3, Russell and Spofford, 1972), appropriate demand functions and economic damage functions associated with ambient residuals concentrations at various locations throughout the region were assumed to exist, the environmental models—air dispersion and water quality—were assumed to be linear, and the objective function was one of net regional benefits. The institutional framework envisioned for this case was a regional management authority with powers to set effluent charges or standards.

In a follow-up, but still didactic application (see footnote 3, Russell, Spofford and Haefele, 1972), the model was expanded to provide information on the sociogeographic distribution of costs and benefits associated with meeting different levels of environmental quality. It was applied to an hypothetical region similar to the first one, and linear environmental models were again employed. The institutional framework envisioned for selecting levels of environmental quality, and for subsequent implementation of policy, was a legislative body.

Both these applications are reported elsewhere, hence, there is no need to go into further detail here. By way of introduction, though, we do show the overall model framework schematically in figure 1.

2. See C. S. Russell, "Models for the Investigation of Industrial Response to Residuals Management Action," *Swedish Journal of Economics* Vol. 73, No. 1 (1971): 134–156.

3. See C. S. Russell and W. O. Spofford, Jr., "A Quantitative Framework for Residuals Management Decisions," in *Environmental Quality Analysis: Theory and Method in the Social Sciences,* Kneese and Bower, eds. (Baltimore: Johns Hopkins Press, 1972). For a discussion of the framework as modified for use in a legislative setting, see C. S. Russell, W. O. Spofford, Jr., and E. T. Haefele, "Residuals Management in Metropolitan Areas" (paper delivered at the International Economics Association Conference on Urbanization and the Environment, 19–24 June, 1972, Copenhagen).

Figure 1

Schematic Diagram of the Regional Residuals Management Model

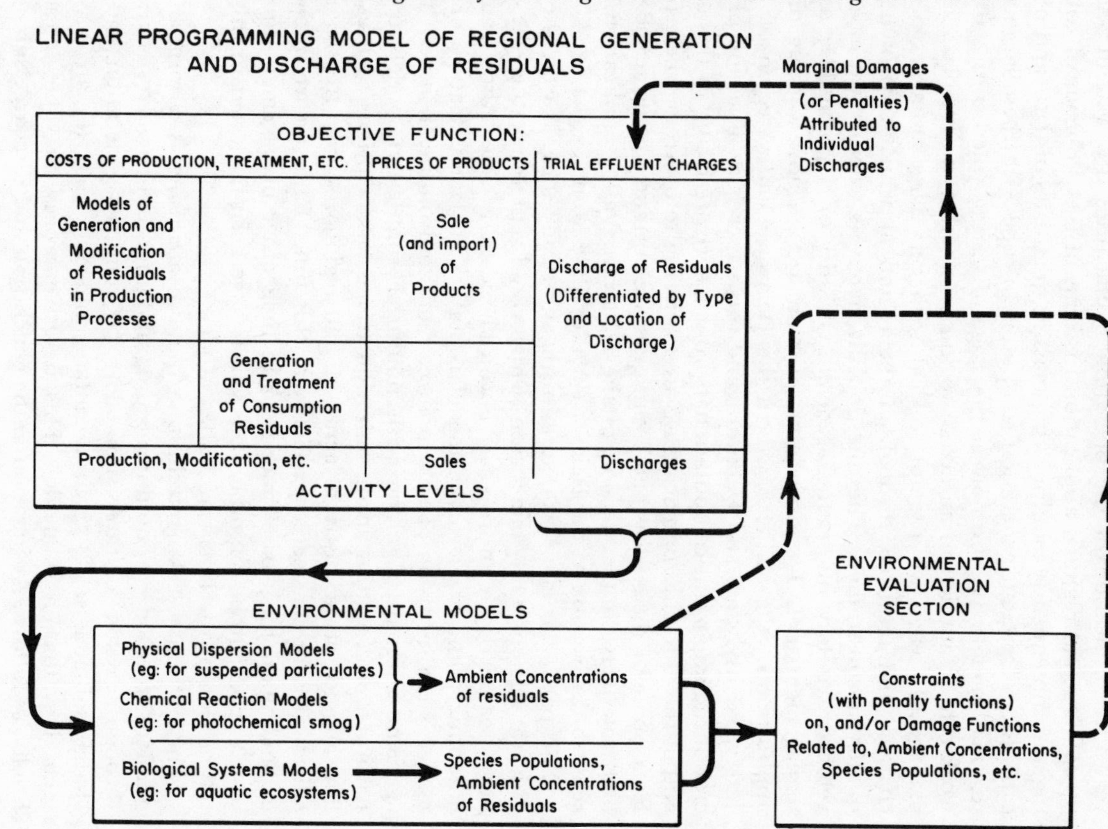

We have learned from our experience with didactic models that this approach is operationally feasible, at least for small scale applications. However, small scale applications to hypothetical regions provide us with very little indication of the operational difficulties involved in scaling up to an actual regional application in terms of the problems of collecting and subsequently manipulating massive quantities of data, and of the capability of present generation computers to cope with these large scale regional models. We are now at the stage of testing whether this framework can be applied to an actual region or whether it will become unmanageable when we attempt to deal with very large numbers of dischargers and locations throughout the region at which environmental quality is constrained. The question ultimately is whether we have developed a mildly interesting academic curiosity or a potentially useful management tool. To answer this question, we are now working on an application of the model based on the Delaware Valley region of New Jersey, Pennsylvania and Delaware, and this paper is a discussion of several of the important computational problems we are facing in this effort and of the different approaches we are exploring to overcome these problems. Some of these ideas and techniques are currently being tested in a first, relatively simple, version of the Delaware Valley application. We refer to this version as the Delaware Valley Base Model.

This particular model is deterministic and steady state, as were the two didactic versions. Only one season (which could represent either the "low flow" season or an entire year) is considered at a time. Also, from an economic point of view, the model is static. The main feature of this model is the inclusion of both nontreatment and on-site treatment management alternatives, along with nonlinear simulation models of the natural world, within an optimization framework. Options such as low flow augmentation, instream aeration, and regional sewage treatment facilities are not considered explicitly at this time. Later on we intend to expand upon this "base" model to include other management options which appear to be important but which have been neglected in this initial version of the Delaware Valley Model.

The optimum seeking technique that we are using is a form of the gradient method of nonlinear programming and involves iterating through a system of three submodels: (1) residuals generation and discharge submodels; (2) environmental submodels; and (3) an environmental evaluation submodel. This iterative process may be described, briefly, as follows. At iteration k, the generation and discharge submodel, which is structured as a linear programming problem, is solved using a set of effluent charges which is based on the state of the natural world on the $(k-1)$th iteration. The resulting discharges are passed to the environmental models which transform them into information on ambient concentrations and species populations. These data on the resulting state of the natural world are then compared to exogenously specified standards of environmental quality. "Penalty" functions are used to reflect the solution's failure in meeting these standards; marginal penalties associated with each discharge of each type of residual are computed and returned to the generation and discharge model as prices on residuals discharges for the $(k+1)$st iteration. When all the constraints are met (within some predetermined tolerance) and no further improvement in the objective function is possible, successive sets of both discharges and effluent charges will be the same, and the algorithm has found an optimum.

Ultimately, we would hope that such a management model might be useful either to an executive agency, such as a regional environmental quality management authority, or to a legislative body. The model is purposely designed to be flexible enough to deal with environmental quality damage functions (if and when they are available) or sets of standards on ambient environmental quality. With nonlinear environmental models, meeting environmental quality standards is, as we shall see, more difficult computationally than employing economic damage functions. In this initial model, as a test for our optimization algorithm, we assume ambient standards must be met.

We shall report here on what we are learning from use of the Delaware Valley Base Model, and on some of the specific programming techniques we are using. It is hoped that these details will be of interest to others engaged in large-scale modeling projects.

Some Operational Problems of Large Scale Modeling Efforts

Models of residuals generation and discharge

Over the past decade, Resources for the Future, Inc. has conducted considerable research in the areas of industrial water use and residuals generation and discharge.[4] A number of linear programming models of industrial plants has been one of the outgrowths of this research program. These models include beet sugar plant, thermal electric plant, petroleum

4. See the paper in this volume by Blair T. Bower for a discussion of this research.

refinery, and integrated iron and steel production.[5] It has been our intention all along to include these models in the residuals generation and discharge portion of our Delaware Valley residuals management model. But the question of how best to do so has raised a number of practical problems. The major problem is model size as related both to round-off error in matrix inversion and to computer time required for solution. In this section, we shall discuss the pros and cons of two approaches for coping with the problem of size—decomposition, and construction of condensed models of the industrial plants.

Condensed Models of Industrial Plants. The full-scale industry models, which were developed for the individual industry studies, have the significant advantage of incorporating a large range of alternative responses open to the plant in the face of effluent charges or discharge standards. In addition, they make it possible to show how residuals generation and discharge, and response to management actions, change with such exogenous (to the regional residuals problem) influences as factor input costs, product mix, and available production—materials recovery—by-product technology. The problem is, of course, that the more the model incorporates, the larger it becomes. For example, the full-scale models developed for petroleum refining and steel production have between 300 and 500 rows. If we combined a number of these models into a single LP matrix by arraying the individual plant models along the diagonal, the resulting regional management model would exceed the computational reliability of the LP routines now available before even a fraction of a large, complex region's industries had been included. For the LP algorithm we are using (IBM's MPSX package), the upper limit on solution reliability is probably between 2,000 and 3,000 rows even though some have reported success with as many as 4,000 rows. As a general rule, though, for problems any larger than about 1,500 rows, care should be taken in checking and interpreting results.[6]

5. See Appendix II to *Future Water Demands: The Impacts on the Water Use Patterns of Selected Sectors of the United States Economy: 1970–1990,* a Study for the National Water Commission by C. W. Howe, C. S. Russell, and R. A. Young, assisted by W. J. Vaughan, all of Resources for the Future, Inc., June 1970; and *Residuals Management in Industry: A Case Study of Petroleum Refining,* C. S. Russell (Baltimore: Johns Hopkins Press, 1973).

6. These statements are based, in part, on the experiences of D. P. Loucks and D. H. Marks. There seems to be no agreement on the upper limit of the number of rows as it relates to solution reliability. Some have had trouble getting a reliable solution with as few as 1,500 rows. Others claim to have been successful with as many as 4,000 rows. The upper limit on row size depends, among others, on the condition of the matrix of coefficients which can differ tremendously among problems. The condition of a

One possible way around this problem is the construction of condensed, or collapsed, versions of the full-scale plant models. The condensation process consists of the following:

1. a choice of a limited number of important inputs, outputs (products), and residuals which would determine the number of rows in the new model;

2. a repeated solution of the larger, full-scale model for different residuals discharge constraint sets, as well as for different constraints on inputs or outputs;

3. a characterization of each solution as a vector with entries in the rows determined in 1; (these entries would be reduced in proportion to some standard unit of input or output, i.e., a natural unit for the petroleum refinery is a barrel of crude oil charged.)

4. an expression of the objective function value from the full-scale model's solution in terms of the same standard unit chosen in 3;

5. an addition to the set of summary vectors just derived, the necessary explicit discharge activities to which trial effluent charges may be attached.

Care must, however, be taken in the developmental stage of a condensed model to anticipate the subsequent price stimuli to be used in actual operation of the overall regional management model. The unit costs used as objective function entries for the summary vectors, and the additional stimuli to be applied in the regional model, are intimately related. Objective function entries for the summary vectors should comprise only those costs (and prices) which will not be accounted for explicitly when the condensed models are included as modules within the overall regional management model. Residuals discharges, for example, are priced separately in the regional model. Hence, in developing the condensed models, zero prices are used on these activities in the full-scale industry models. This insures that the objective function entries for the summary vectors do not include any charges for residuals discharges.

We have investigated this technique for making use of full-scale plant models; the report on the Delaware Valley Base Model detailed later in this paper includes collapsed models of two petroleum refineries. However, we have found problems with this approach. The most important one is that in order to duplicate even a fraction of the flexibility of the full-scale model, we must include a very large number of columns (i.e.,

---

matrix can usually be improved by proper scaling. For a more extensive discussion of this point, see W. Orchard-Hayes, *Advanced Linear-Programming Computing Techniques* (New York: McGraw-Hill, 1968), Chapter 6.

alternative solutions) in the condensed version. Considering, for a moment, only a single residual, the response of the condensed model to an effluent charge will more closely approximate the full-scale version the finer the grid of discharge constraints on which the condensation is based. But, it is important to note that this same statement applies to the multidimensional space containing the vector of all residuals of interest. If we have five residuals of interest, and if we confine ourselves to a very rough grid (e.g., high, medium and low levels of discharge), there are still 243 ($3^5 = 243$) alternative solutions of the full-scale model to be obtained and expressed in the appropriate vector form. If we increase the grid fineness to four levels of discharge, we increase the number of solutions and, hence, summary vectors to 1,024. It is clear that the expense and bookkeeping involved in constructing collapsed models is considerable and that their column size can become very large.

Whether or not the condensed model approach would be a solution to the row-size problem depends, of course, on the number of rows in the resulting condensed models, and on the number of significant residuals dischargers in the region. If the average size of the condensed models could be kept to ten rows, and if we ignore the requirements for artificial bounds for the step-size selection part of the overall solution method (to be discussed later), we could construct a single LP model for a region of between 150 and 200 dischargers. But keeping these condensed models to ten rows is not easy. Thus, if we wish to include only one input, one output (product), six primary residuals and two secondary residuals (sewage sludge and solids from particulate removal, for example), we are up to ten rows. Every refinement on the product or residuals side reduces the number of individual sources we can include. And we cannot, of course, neglect the necessity for artificial bounds, so that our "capacity" is very much lower than 150–200 plants; a guess would be 40–50. Now, for many regions this would be sufficient, but in a large industrialized region, such as the Delaware Valley, this would not begin to cover the significant sources of air and waterborne residuals, particularly when we realize that at least the largest municipal incinerators and sewage treatment plants have also to be included and provided with discharge reduction alternatives.[7]

Decomposition. Another alternative for dealing with model size is to

7. Metropolitan Philadelphia Interstate Air Quality Control Region, "Inventory of Emission Sources," Office of Air Programs, Environmental Protection Agency lists about 300 individual industrial plants, plus another 30 or so municipal incinerators and large institutional heating plants.

subdivide the large regional LP problem, which would be created by lumping all sources of residuals for which discharge reduction alternatives are available, into a series of smaller linear programs. This amounts to recasting the problem in a standard decomposed form:[8]

$$\min \quad \{c_1X_1 + \cdots + c_nX_n\}; \tag{1}$$

$$\begin{aligned} \text{s.t.}\quad A_{11}X_1 & \geq b_1 \\ \vdots & \\ A_{n1}X_n & \geq b_n \end{aligned} \tag{2}$$

$$A_{12}X_1 + \cdots + A_{n2}X_n \leq b_{n+1}. \tag{3}$$

When there are no shared constraints (equation 3) among individual "decomposed" components, each of the smaller LP's may be solved, in turn, as a separate subproblem prior to entering the environmental model subroutines for determination of resulting ambient environmental quality. This is, in fact, the case with the Delaware Valley Base Model which was purposely divided into two LP's to be solved separately. Dividing the optimization problem up this way is conceptually straightforward and certainly appealing from a computational point of view, even though there are certain practical difficulties in using the MPSX routine in this manner. But these difficulties are primarily matters of keeping MPSX outputs and inputs straight when there are many discharges, artificial bounds (step sizes), and trial effluent charges to be passed back and forth among various LP's and FORTRAN subroutines.

In a more complex model of a region, it may be impossible to provide individual, unconnected LP subproblems. This, of course, depends upon the interconnections—both market and nonmarket—among activities in the region. For example, if the environmental models were linear, and if they were dealt with in the regional model as part of the constraint set, it would be virtually impossible to subdivide the regional model into separate, unconnected submodels. However, even the elimination of environmental models as part of the constraint set does not guarantee us that we will be able to subdivide the regional model into a series of sepa-

8. For a discussion of the decomposition principle for linear programming, see G. B. Dantzig, *Linear Programming and Extensions* (Princeton: Princeton University Press, 1963) or G. H. Hadley, *Linear Programming* (Reading, Mass.: Addison-Wesley Publishing Co., Inc., 1962).

rate submodels. There are other types of relationships that inherently link activities together. For example, in our simple models, such as the Delaware Valley Base Model, we have had a market link between the petroleum refineries and home heating through the purchase of various grades of distillate fuel oil. The form of these constraints has been a simple one: production plus imports must be greater than or equal to regional use. (Heating for domestic purposes has been assumed price inelastic.)

The obvious approach to this problem is to treat the set of linear models with shared linear constraints as a classical decomposed linear programming problem and to solve it as such before entering the environmental models. Since decomposition algorithms are themselves iterative, we would be building in a set of iterations within each iteration of the overall management model, and this may involve us in a significant increase in computation time. A major drawback of this approach is that the LP algorithm does not have a decomposition algorithm built into it. Hence, to take advantage of decomposition, we would have to improvise. We are presently exploring this possibility.

In summary, the main concern we have with models of regional generation and discharge of residuals, when included as part of a regional residuals management model of a large, complex region, is with sheer model size. Our concern relates not only to the problems associated with round-off errors, but also to computational time and expense. We have proposed various approaches to the size problem and have investigated many of them using the Delaware Valley Base Model. Currently, it appears that a combination of collapsed versions of the full-scale industry models, standard decomposition, and sequential solution of a set of LP submodels is a feasible approach to the problem of attaining solutions to large scale, regional residuals management models.[9]

9. There is, at least in principle, a third possibility for reducing model size. All of the industry models which we intend to include in our regional residuals management model are linear, but their constraint sets contain a significant number of equality constraints. Each equality constraint could be used to eliminate a variable (column) in the LP. But once an industry model is built, this is a time consuming procedure and errors are likely to result. In addition, some of the eliminated variables are likely to provide useful information for management decisions and, hence, would have to be computed anyway after the optimization phase of the analysis were complete. Thus, the choice of variables to be retained is most important. For example, it would not be desirable to eliminate the residuals discharge vectors because both discharges (which are input to the environmental models) and prices on these discharges change from iteration to iteration. Although we recognize elimination of equality constraints as a possibility for dealing with model size, up to this point in our research, we have not given it serious consideration. In the future, however, we may.

Environmental models

Environmental models—air and water dispersion, chemical reaction, and biological systems—are used to describe the impact on the natural environment of energy and material residuals discharged from the production and consumption activities of man. We use these models to predict steady state concentrations of residuals and related substances (e.g. algae, oxygen in the estuary) at various points in the regional environment, given: (a) a set of residuals discharge levels from the linear programming submodel of regional generation and discharge of residuals; and (b) a set of values for the environmental parameters such as stream flow and velocity, wind speed and direction, atmospheric stability, and atmospheric mixing depth.

Some environmental models are easier to deal with than others within an optimization framework. It depends, in general, upon the mathematical structure of the model. In terms of the complexity involved, we find it useful to distinguish among four broad categories: (1) linear, explicit functions; (2) linear, implicit functions; (3) nonlinear, explicit functions; and (4) nonlinear, implicit functions.

We are currently using two environmental submodels in conjunction with our regional residuals management model. The first, a linear atmospheric dispersion model, is used to predict ambient concentration levels throughout the region of sulfur dioxide and airborne particulates. It was provided to us by the Environmental Protection Agency.[10] The second, a nonlinear aquatic ecosystem model, used to predict various ambient concentrations in the estuary, was developed at Resources for the Future specifically for our Delaware Valley residuals management study.[11] The inclusion of these environmental models has hopefully increased the usefulness of the overall management model for purposes of better informing public policy, but has raised several computational problems also. These two environmental models represent the extremes of complexity for inclusion within a management framework. A discussion of each model will raise some of the important issues and will reveal some of the problems involved.

Atmospheric Dispersion Model. Of the various atmospheric quality models which are available now, physical dispersion models are the most advanced. Chemical reaction models, such as for photochemical smog, are

10. Division of Applied Technology, Office of Air Programs, Environmental Protection Agency, Durham, N.C.

11. R. A. Kelly, "Conceptual Ecological Model of the Delaware Estuary," *Systems Analysis and Simulation in Ecology* Volume IV (New York: Academic Press, 1974).

being developed, and carbon monoxide models for urban areas are appearing. The most successful modeling efforts to date have been associated with predicting both steady and nonsteady state concentration distributions of sulfur dioxide and suspended particulates of 20 microns or less in diameter. Because of the availability of an existing air dispersion model, we selected ambient levels of sulfur dioxide and suspended particulates to represent the air quality of our region.[12]

The atmospheric model which we are using is the air dispersion model from the federal government's Air Quality Implementation Planning Program (IPP).[13] This model uses a dispersion model developed by Martin and Tikvart which evaluates concentrations downwind from a set of point and area sources on the basis of the Pasquill point source, Gaussian plume formulation. The Gaussian plume formulation may be used to estimate ambient concentrations under deterministic, steady state conditions. For any given source-receptor pair, production process and abatement device, specified meteorologic conditions, and discharge rate of unity, this nonlinear equation reduces to a linear coefficient relating ambient concentrations with residuals discharge rates.

The necessary inputs to this model are: *x-y* coordinates of all sources and receptors in the region; emission rates for each source—point and area; physical stack height, stack diameter, stack exit temperature, and stack exit velocity for each point source; a seasonal joint probability distribution for wind speed, wind direction, and atmospheric stability; a mean seasonal temperature and pressure; and a mean atmospheric mixing depth for the period of interest.

The output of this air dispersion model represents arithmetic mean seasonal concentrations of sulfur dioxide and airborne particulates based on the probabilities of occurrence of 480 discrete meteorological situations. For this computation, 16 wind directions, 6 wind speed classes, and 5 atmospheric stability classes are considered for each source-receptor pair with the occurrence of all combinations possible (hence, $16 \times 5 \times 6 = 480$ total possibilities). The joint probabilities of occurrence for each of these 480 combinations are determined from actual meteorological data.

12. We should point out that the selection of sulfur dioxide and suspended particulates as measures of air quality in our model coincides with real world considerations. These are, in fact, the first two airborne residuals for which environmental quality standards have been set in the United States.

13. See TRW, Inc., *Air Quality Implementation Planning Program* Vols. I and II (Washington, D.C.: Environmental Protection Agency, 1970), also available from National Technical Information Service, Springfield, Virginia, 22151, accession numbers PB 198 299 and PB 198 300 respectively.

For a given set of meteorological conditions and physical parameters, the vector of mean seasonal concentrations of sulfur dioxide and particulates, R, may be expressed linearly, in matrix notation, as;

$$R = AX + B, \tag{4}$$

where X is a vector of sulfur dioxide and particulate discharge rates; A is a matrix of transfer coefficients which specify, for each source-receptor pair in the region, the contribution to ambient concentrations associated with a residuals discharge rate of unity; and B is a vector of background concentration levels. The matrix of transfer coefficients, A, is the output of the dispersion model.

The important thing to note from equation 4 is that the state of the natural world (R) is expressible directly in terms of linear, explicit algebraic functions. This particular mathematical form is relatively easy to deal with in an optimization framework. In fact, equation 4 in its present form may be incorporated directly within the constraint set of a standard linear program when one of the management objectives is to constrain ambient concentrations of residuals.

As we shall see in the next section, one of the requirements of our optimization scheme is the availability of an environmental response matrix, $\partial R_i/\partial x_j$, $i = 1, \ldots, m$; $j = 1, \ldots, n$; where m is the total number of environmental quality indicators at all the designated receptor locations in the region and n is the total number of residuals discharges in the region. This matrix may be obtained by differentiating equation 4 with respect to all the residuals discharges in the region. That is,

$$\left(\frac{\partial R}{\partial X}\right) = A. \tag{5}$$

Before we leave this section, we should point out that not all atmospheric quality models are as easy to deal with as the physical dispersion models which are expressed in linear, explicit analytical form. Chemical reaction models, such as for photochemical smog, for example, would be significantly more difficult to handle within our optimization framework. The kinds of problems we would face with them are revealed in the discussion of a nonlinear aquatic ecosystem model which follows.

Aquatic Ecosystem Model. There are a variety of indicators which are commonly used for describing the quality of a body of water. Among them are pathogenic bacterial counts (or counts of an indicator thereof), algal densities, taste, odor, color, pH, turbidity, suspended and dissolved solids, dissolved oxygen, temperature, and population sizes of certain plant and animal species. Because of the importance of dissolved oxygen to virtually all species of higher animals, and the relative ease with which it can be measured and modeled for a river or estuary, its concentration has been, and still is, one of the most frequently used criteria for setting general water quality standards.

Streeter-Phelps type dissolved oxygen models have been used for many years to predict water quality as a result of discharges of organic material (most notably, sanitary sewage).[14] Given certain assumptions about the natural environment, these DO models can be expressed as a set of linear algebraic relationships analogous to the linear air dispersion models discussed previously. From a computational point of view, they are very easy to deal with. This is, in fact, one reason for their continued popularity.[15]

However, these models have three deficiencies which we feel warrant the exploration of more sophisticated aquatic ecosystem models. First, we are really interested in the dissolved oxygen level only insofar as it is an accurate indicator of such things as algal densities and the population sizes of certain species of fish. To the extent that these densities and populations can vary independently of dissolved oxygen concentrations, we need information about them if policies on water quality are to be established intelligently. Second, materials other than organics (for example, nutrients and toxics) are known to have significant effects on aquatic ecosystems. Consequently, these inputs should be included along with the organics in order to evaluate more fully the impact on the environment of residuals discharges. Finally, systems ecologists feel that aquatic ecosystem models based on at least some biological (or ecological) theory, which includes the mechanisms of feeding, growth, predation, excretion, death, and so on, are more reliable for predicting dissolved oxygen levels than the more empirically based models of the Streeter-Phelps variety.

The aquatic ecosystem model we have developed is based on a trophic level approach.[16] The components of the ecosystem are grouped in classes

14. H. W. Streeter and E. B. Phelps, "A Study of the Pollution and Natural Purification of the Ohio River," *Public Health Bulletin No. 146* (Washington, D.C.: U.S. Public Health Service, 1925).

15. Models of the BOD-DO type are in widespread use. A typical example is given for the Delaware Estuary by R. V. Thomann in *Systems Analysis and Water Quality Management* (New York: Environmental Science Services Division of Environmental Research and Applications, Inc., 1972), pp. 160–81.

16. For examples of this approach, see R. B. Williams, "Computer Simulation of Energy Flow in Cedar Bog Lake, Minnesota, based on the classical studies of Linde-

("compartments") according to their function, and each class is represented in the model by an endogenous, or state, variable. Eleven compartments are designated in our model. The endogenous variables representing these eleven compartments are nitrogen, phosphorus, turbidity (suspended solids), organic material, algae, bacteria, fish, zooplankton, dissolved oxygen, toxics, and heat (temperature). In addition, the following exogenous variables (parameters) are considered: turnover rate (or advective estuary flow), and inputs (of the eleven chemical and biological materials above). Carbon is assumed not to be limiting and, hence, is not considered as either an endogenous or exogenous variable. Material flows among compartments within a given reach of the river (or estuary) are depicted in figure 2.

Figure 2

Diagram of Materials Flows Among Compartments Within a Single Reach

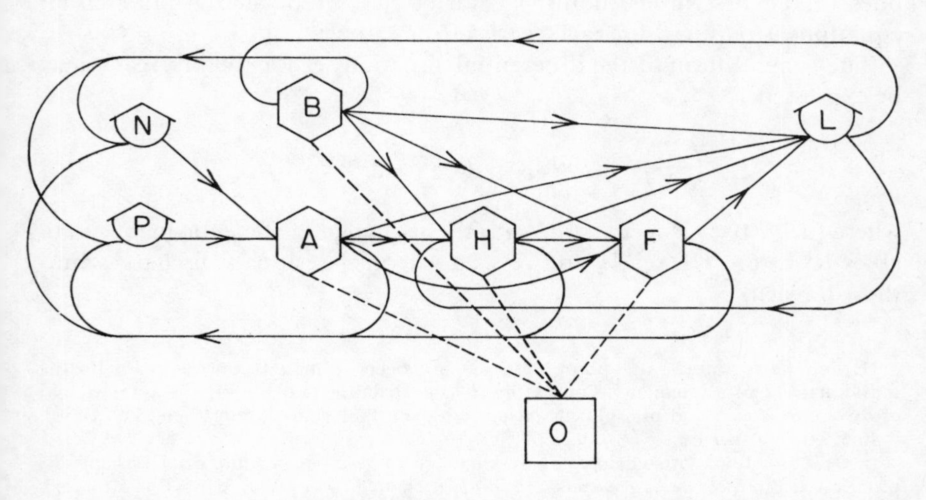

The inputs to the estuary model from the residuals discharges in the region are organic material measured by its BOD, total nitrogen, phosphorus, phenols (toxics), and heat. The outputs of this model of concern

---

man," *Systems Analysis and Simulation in Ecology* Vol. I, B. C. Patten, ed. (New York: Academic Press, 1971), pp. 543–82, and H. T. Odum, *Environment, Power and Society* (New York: Wiley-Interscience, 1971).

to us are densities of fish biomass, algal densities, and dissolved oxygen levels. The levels of these outputs are constrained; that is, environmental standards are imposed. In addition, concentrations of nitrogen, phosphorus, suspended solids, and organic material; temperature; and mass of bacteria and zooplankton are also available as by-product outputs of this model.

The time rate of change of material in each compartment is expressed in terms of the sum of the transfers among other compartments, and between adjacent sections of the estuary (since the material is distributed spatially as well as temporally). To insure mass continuity of the materials considered, material entering and leaving a compartment is explicitly accounted for.[17] The mathematical description of material transfers among compartments is based on the theoretical-empirical formulations given by Odum.[18]

Each compartment requires a separate differential equation to describe mass continuity, and in general, these equations must be solved simultaneously. In this particular case, the differential equations are ordinary ones of the first order, nonlinear variety. A set of similar differential equations is required for each reach of the estuary.[19]

The general form of the differential equation set for the kth reach may be expressed as,

$$\left(\frac{dR}{dt}\right)^k = f[R(t)^{k-1}, R(t)^k, R(t)^{k+1}, X(t)^k], \tag{6}$$

where $(dR/dt)^k$ is a vector of time rates of change of the endogenous variables R^k in the kth reach, and X^k is a vector of residuals discharges into the kth reach.

17. For the two nutrients—phosphorus and nitrogen—a mass balance is made on the individual chemical elements. For species, a mass balance is made on the total weight of carbon, nitrogen and phosphorus (assuming a constant ratio among them, i.e., C:N:P = 40:5:1 for all species).

18. H. T. Odum, "An energy circuit language for ecological and social systems: its physical basis," in *Systems Analysis and Simulation in Ecology* Vol. II, B. C. Patten, ed. (New York: Academic Press, 1972), pp. 139–211.

19. Finite difference forms of the more general partial differential equation set for describing mass continuity are used for the distance (space) variable. This is why we are able to write a separate set of differential equations for each reach (section) of the estuary. However, within each reach, time is expressed continuously (thus, the set of total differential equations rather than algebraic equations). When these differential equations are solved with analog computers, the concentrations of materials are continuous in time. When these equations are solved using digital computers, time must also be expressed in the finite difference form, and in the process, they reduce to a simultaneous set of nonlinear algebraic equations.

There are two problems associated with this ecosystem model formulation which we wish to discuss in more detail: (1) that of obtaining a steady state solution, and (2) that of obtaining an environmental response matrix. The first relates to models of this type in general whereas the second relates only to those situations where ecosystem models are to be included within an optimization framework.

Solution Methods. In its present form, equation 6 represents a set of ordinary nonlinear differential equations—one equation for each compartment, and one set of compartmental equations for each estuary reach—which must be solved simultaneously. If we were interested in the transient (or nonsteady) states of the system, simulation techniques, i.e., numerical integration (simulating first over space and then time) provide us with a readily available means of solution. However, we are interested only in the steady state solution.

For determining steady state solutions, there are two possibilities (or a combination thereof), neither of which guarantees finding a stable point equilibrium: (1) simultaneous simulation of a nonlinear differential equation set, and (2) simultaneous solution of a set of nonlinear algebraic equations. If we neglect inputs to, and outflows from, each reach due to longitudinal dispersion, the system can be dealt with first over time, and then space, starting with the uppermost reach and progressing systematically down the estuary.[20] In this case, equation 6 for the kth reach would reduce to,

$$\left(\frac{dR}{dt}\right)^k = f[R(t)^{k-1}, R(t)^k, X(t)^k]. \tag{7}$$

Now, only the eleven compartmental equations within each reach must be solved simultaneously. The state of the system within a particular reach depends only upon the inputs from upstream, $R(t)^{k-1}$ and the residuals discharges to the kth reach, $X(t)^k$, both of which may now be treated as exogenous inputs. In addition, if the resulting steady state

20. Neglecting longitudinal dispersion, even in an estuary, is not as unreasonable as it first appears. Finite difference techniques for solving these differential equations introduce a numerical diffusion effect into the model. Inputs are immediately mixed in the volume, not because of any physical effects, but solely because of the numerical procedure. See D. J. O'Connor and R. V. Thomann, "Water Quality Models: Chemical, Physical, and Biological Constituents," *Estuarine Modeling: An Assessment* (Washington, D.C.: Environmental Protection Agency Stock No. 5501-0129, February 1971), Chapter III, p. 138.

solution, $(R^*)^k$, is independent of the time paths of rates of inputs, $R(t)^{k-1}$ and $X(t)^k$, equation 7 reduces to,

$$\left(\frac{dR}{dt}\right)^k = f[(R^*)^{k-1}, (X)^k, R(t)^k]. \tag{8}$$

Usually, ecological models are solved by simulation. Simulation of the differential equation set (a set of equations similar to equations 6 through 8 poses no particular problem, but the steady state solution, if one exists at all, may take considerable time. Oscillations can, and do, occur, and solutions may be otherwise unstable; they may become infinitely large. However, May[21] has demonstrated for a set of reasonable assumptions and a similar predator-prey nonlinear model, that these systems possess either a stable point equilibrium or a stable limit cycle.

Even when a steady state solution can be found, an additional problem is that there may be more than one stable point equilibrium. To investigate this problem, we ran an experiment with our ecosystem model. We used a random number generator to provide us with a set of random starting points. Twenty-five random starts resulted in the same steady state solution which indicates that our model is probably well behaved in this respect. However, another model may not be.

At steady state, $dR/dt = 0$, and thus the differential equation set above, equation 8, reduces to a set of nonlinear algebraic equations of the following form.

$$0 = f[(R^*)^{k-1}, (X)^k, (R^*)^k]. \tag{9}$$

The endogenous variables, $(R^*)^k$, are implicitly expressed in this formulation.

Various numerical methods, such as Gauss-Seidel and Newton's, have been used with success for solving simultaneous nonlinear algebraic equation sets, but each has its faults. Gauss-Seidel (also known as "Successive Approximation") has slow convergence properties, but it is relatively stable. Newton's method has more rapid convergence properties, but it is sensitive to initial conditions and it is often unstable.

Determination of steady state values for the endogenous variables in a nonlinear ecosystem model is difficult due to the nonuniqueness and

21. R. M. May, "Limit Cycles in Predator-Prey Communities," *Science* Vol. 177 (September 1972), pp. 900–902.

complexity of the solution. Even the stability characteristics of the steady state solution cannot be determined prior to its solution. With linear models, we can solve for the eigen (or characteristic) values of the differential equation set. These will tell us whether or not the time independent solution converges to a finite set of values, or diverges to infinity, or even if oscillations are involved—stable, diverging, or converging. For the nonlinear differential equation set, the best we can do is linearize the system at some point, and examine the eigen values of the resulting linear form. But this only tells us what is happening locally.

At this time, we are using a combination of Newton's method and simulation. These techniques are being used in the following way. Starting with the first reach, a solution is attempted by Newton's method using an estimate of the steady state values of the endogenous variables as an initial point. If a steady state solution is obtained, a solution for the second reach is attempted, using the steady state values of the endogenous variables from the first reach as a starting point. This procedure is repeated until the steady state solution is obtained for the last reach, or a reach is encountered which cannot be solved by Newton's method. When a solution cannot be obtained, an approximate solution is generated by numerically integrating the equation set over a one hundred day period. Empirical observation of the solution behavior indicates this is a fairly decent steady state solution. The simulation solution is then used to solve the next reach by Newton's method, and so on.

The Ecosystem Response Matrix. To include this nonlinear aquatic ecosystem model within the residuals management model, in addition to determining a set of steady-state values, it is also necessary to evaluate the response throughout the ecosystem to changes in the rates of the residuals discharges. That is, it is necessary to know, for example, the effect on algae in reach 17 of an additional BOD load discharged into reach 8, and so on. This requirement results in a considerable number of additional computations, but this knowledge of the system response, in conjunction with the penalty functions to be discussed in the next section, is the key to being able to use these complex ecosystem models within the optimization framework.

The response matrix we wish to compute may be expressed in matrix notation as, $\partial R/\partial X$, where R is a vector describing the state of the system throughout the entire length of the estuary, and X is a vector of residuals discharges throughout the region. Using equation 9 for each reach of the estuary, and the relationship,

$$Z^k = q^k R^{k-1} + X^k, \tag{10}$$

where Z^k is a vector of inputs to the kth reach, R^{k-1} is a vector of concentrations of materials in reach $k - 1$, q^k is the estuary advective flow rate into the kth reach, and X^k is a vector of residuals discharges to the kth reach, a section of the system response matrix may be computed accordingly,

$$\frac{\partial R^i}{\partial X^j} = \frac{\partial R^i}{\partial Z^i} \cdot \frac{\partial Z^i}{\partial R^{i-1}} \cdot \frac{\partial R^{i-1}}{\partial Z^{i-1}} \cdots \frac{\partial R^{j+1}}{\partial Z^{j+1}} \cdot \frac{\partial Z^{j+1}}{\partial R^j} \cdot \frac{\partial R^j}{\partial Z^j} \cdot \frac{\partial Z^j}{\partial X^j}. \quad (11)$$

From equation 10 we note that,

$$\frac{\partial Z^k}{\partial R^{k-1}} = q^k I, \quad (12)$$

and,

$$\frac{\partial Z^k}{\partial X^k} = I, \quad (13)$$

where I is the identity matrix. Thus, the $(Z^k/\partial R^{k-1})$ terms are known a priori and are exogenous parameters in the ecosystem model.

The other terms, $(\partial R^k/\partial Z^k)$, are evaluated from equation 9 according to the rules for differentiating implicit functions.[22] That is,

$$\frac{\partial f}{\partial R} \cdot \frac{\partial R}{\partial Z} = -\frac{\partial f}{\partial Z}, \quad (14)$$

or,

$$\frac{\partial R}{\partial Z} = -\left(\frac{\partial f}{\partial R}\right)^{-1} \frac{\partial f}{\partial Z}. \quad (15)$$

This operation involves the inversion of the Jacobian matrix $(\partial f/\partial R)$. In addition, because the system of equations is nonlinear, the Jacobian matrix $(\partial f/\partial R)$ must be recomputed for each resulting state of the natural world.

It should be clear, then, from the above discussion, that the major problem associated with including environmental models within our management framework is one of computer time. Nonlinear representations of the natural world increase the complexity and the number of

22. See, for example, I. S. Sokolnikoff and R. M. Redheffer, *Mathematics of Physics and Modern Engineering* (New York: McGraw-Hill Book Co., Inc., 1958), pp. 237–41.

calculations necessary for each iteration, but hoping that they will also increase both the realism and predictive capability of the model.

Management model formulation and optimization scheme

In this section, we (a) present a formal mathematical description of our regional residuals management model; (b) indicate the method of handling certain kinds of constraints which are difficult, in fact in some cases impossible, to deal with in the traditional manner; and finally (c) we discuss the optimization procedure we are using.

Model Formulation. The objective function we are currently using is expressed in the form of a net benefit function. Hence, the objective is to maximize. The positive elements in this function include gross revenues from the sale of various products. The negative elements include: all the opportunity costs of traditional production inputs; all liquid and gaseous residuals modification (treatment) costs; and all collection, transport, and landfill costs associated with the disposal of solid residuals.

There are, basically, three types of constraints in the management model: traditional resource availability (inequality) constraints; continuity relationships (equality constraints); and residuals management (inequality) constraints. The latter, which involve the use of environmental models, are employed to constrain the levels of ambient environmental quality. The nature of all three types of constraints has been discussed in detail elsewhere.[23] We do not elaborate again on the first two types here. The third type is discussed in a slightly different context, as we are now treating these constraints a little differently than we did before.

Before proceeding, let us state the residuals management problem formally.[24]

$$\max \{F = f(X\ R)\}; \tag{16}$$

$$\text{s.t.} \quad g_i(X) = 0 \qquad i = 1, \ldots, m < n - q, \tag{17}$$

$$g_i(X) \geq 0 \qquad i = m + 1, \ldots, p, \tag{18}$$

23. C. S. Russell and W. O. Spofford, Jr., "A Quantitative Framework," and C. S. Russell, W. O. Spofford, Jr., and E. T. Haefele, "Residuals Management."

24. Note that the environmental relationships could have been written directly as,

$$h_i(X) \leq S_i, i = 1, \ldots, q.$$

However, we choose to deal explicitly with the variables R_i, $i = 1, \ldots, q$, here as they will be useful to us in a later development.

$$h_i(X) = R_i \qquad i = 1, \ldots, q, \tag{19}$$

$$R_i \leq S_i \qquad i = 1, \ldots, q, \tag{20}$$

$$R_i \geq 0 \qquad i = 1, \ldots, q, \tag{21}$$

$$X_i \geq 0 \qquad i = 1, \ldots, n, \tag{22}$$

where $f(X\ R)$ is, in general, a nonlinear objective function; $g_i(X) = 0$, $i = 1, \ldots, m$, is a set of linear equality constraints; $g_i(X) \geq 0$, $i = m + 1, \ldots, p$, is a set of linear inequality constraints; $h_i(X) = R_i$, $i = 1, \ldots, q$, represents a set of environmental functions which relate ambient concentrations of residuals to residuals discharges; X_i, $i = 1, \ldots, n$, is a vector of decision variables, including residuals discharges; R_i, $i = 1, \ldots, q$, is a vector of ambient levels of residuals concentrations and population sizes of species; and S_i, $i = 1, \ldots, q$, is a vector of ambient environmental quality standards (e.g., sulfur dioxide and particulates in the atmosphere, and algae, fish, and dissolved oxygen in the water).

As we have pointed out previously, some of the necessary environmental functions $h_i(X) = R_i$ are available in linear form (e.g., the air dispersion relationships and the Streeter-Phelps type dissolved oxygen models).[25] Others are only available in nonlinear analytical form, while still others are available in various other forms. As we pointed out in our discussion of nonlinear aquatic ecosystem models, no analytical expressions for them —either linear or nonlinear—of the form $h(X) = R$ are available. The variables R_i, $i = 1, \ldots, q$ are expressible only as a set of implicit nonlinear functions and, hence, simulation and other iterative techniques must be used to compute their values. From this discussion, we note that, in general, the environmental constraint set, equation 19, represents a variety of functional forms, many of which are difficult, or even impossible, to deal with using traditional mathematical programming techniques.

Because our optimization scheme, to be described below, requires that all the constraints be linear, we remove the environmental relationships from the constraint set and deal with them in the objective function. This modification of the problem requires the use of the penalty function concept which we shall discuss below.[26]

25. When environmental functions are expressible in this particular linear analytical form, their coefficients are known in the literature as transfer coefficients.

26. The use of "penalty functions" for eliminating constraints is not a new idea. It is a well-known technique and is in frequent use in one form or another under a variety

The new optimization problem may be stated formally as,

$$\max \{F = f(X) - P(X)\}; \tag{23}$$

$$\text{s.t.} \quad g_i(X) = 0 \qquad i = 1, \ldots, m, \tag{17}$$

$$g_i(X) \geq 0 \qquad i = m + 1, \ldots, p, \tag{18}$$

$$x_i \geq 0 \qquad i = 1, \ldots, n, \tag{22}$$

where,

$$P(X) \equiv \sum_{i=1}^{q} p_i[S_i\ R_i = h_i(X)], \tag{24}$$

and where $p_i\ (S_i\ R_i)$, $i = 1, \ldots, q$ are the penalty functions associated with exceeding the environmental standards, S_i, $i = 1, \ldots, q$.

Although our optimization scheme requires only that we remove those constraints (environmental relationships) which are not of the linear form $R = AX$, we note from the formulation of the new problem, equations 23, 17, 18 and 22, that even the linear environmental models have apparently been removed (as constraints). This is optional and depends upon the model formulation and its size. If model size, the number of rows and columns, is of no consequence and if the entire management model is contained within a single linear program (LP), it is more efficient to keep the linear environmental relationships as part of the constraint set.

If, on the other hand, model size is a problem and it is desirable, as discussed above, to divide the management model up into a number of smaller LP's, disposition of the linear environmental models is not as straightforward. No matter how the larger LP is subdivided, the environmental relationships, which involve all the liquid and gaseous residuals discharges throughout the region, invariably link the smaller LP's. In this case, if the linear environmental models are retained as part of the constraint set, one of the available decomposition techniques must be employed.

---

of names. For example, Zangwill refers to this technique as "penalty" and "barrier" methods depending upon whether an optimum is approached from outside or within the feasible region. See W. I. Zangwill, *Nonlinear Programming: A Unified Approach* (Englewood Cliffs, N.J.: Prentice-Hall, Inc., 1969), Chapter 12. Fiacco and McCormick, on the other hand, refer to this as "exterior" and "interior" point methods, respectively. See A. V. Fiacco and G. P. McCormick, *Nonlinear Programming: Sequential Unconstrained Minimization Techniques* (New York: John Wiley and Sons, Inc., 1968).

The Penalty Function. The scheme which we are using to eliminate environmental relationships from the constraint set and still meet the environmental quality standards, S_i, $i = 1, \ldots, q$, is known as a penalty or exterior point method (as opposed to a barrier or interior point method). The name derives from the fact that throughout the optimization procedure we allow the vector of standards, S, equation 20, to be violated, but only at some "penalty" to the value of the objective function. The objective of the approach is to make this penalty severe enough such that at the optimum the standards will be satisfied, within some tolerance.

Because in our optimization scheme we evaluate the gradient, ∇F, at each step in the procedure, we require that the objective function, equation 23, be continuous and have continuous first derivatives. A quadratic penalty function of the following form satisfies these requirements.[27]

$$p_i(X) = \max \left[(h_i(X) - S_i)\ 0 \right]^2 \qquad i = 1, \ldots, q. \tag{25}$$

For computer applications, equation 25 may be written more conveniently as,

$$p_i(X) = \left\{ \frac{[h(X) - S] + |[h(X) - S]|}{2} \right\}^2 \qquad i = 1, \ldots, q, \tag{26}$$

a form which gives $p_i(X) = 0$ when $h_i(X) < S_i$.

The major difficulty with the penalty function expressed above as equation 25 is that, in general, it is not steep enough in the vicinity of the boundary (that is, the standard) and, consequently, the "unconstrained" optimum is apt to lie substantially outside the original feasible region. However, a slight modification to the P function remedies this situation. If $r > 0$, and $p_i(X) \neq 0$, the new penalty function,

$$\frac{1}{r_i} p_i(X), \tag{27}$$

approaches infinity as $r_i \rightarrow 0$. Specifying a sequence of decreasing values for r has the effect of moving the unconstrained optimum closer and closer to the boundary of the feasible region. From a computational standpoint, it is sufficient that r only be made small enough to ensure

27. Note that the second derivative of this function is also defined and that it is positive for $h(X) > S$.

that the unconstrained optimum is within a preselected distance of the boundary.[28]

This situation adds substantially to the computational requirements of our iterative optimization scheme. Not only do we have to find an optimum for the management problem given a set of penalty functions and associated parameter values, but now we have to find a new optimum for a sequence of values of r. Obviously, the fewer r's we need to use during the ascent procedure, the better off we will be. The reason for the small values of r, as noted before, is to ensure that the optimum is sufficiently close to the boundary of the feasible region. The reason for a relatively large value of r in the beginning of the ascent procedure is strictly a computational one. It is related to the efficiency of the optimization scheme employed. Rapid changes in the response surface are difficult, in general, to deal with except when the optimum is being approached and the step size is relatively short. It is difficult to know a priori what a good starting value of r would be.[29]

From an operational point of view, selection of an appropriate set of penalty functions and sequence of values for the penalty function parameter r is a real concern to us. The efficiency of the optimization scheme is directly dependent on how this is handled. We hope that with some experience with the operational behavior of a specific model, it will be possible to specify a range of values for r from which a reasonably small subset could be selected. We will investigate this question using the Delaware Valley Base Model.

The Optimization Procedure. A formal presentation of the nonlinear programming algorithm we are using to optimize a nonlinear objective function subject to a set of linear constraints has been presented elsewhere.[30] Only the essence of the scheme is repeated here. Relevant equations and expressions used for this procedure are restated, and the objective function, equation 23 is modified accordingly.[31]

28. Because at the optimum the standards are met only within some tolerance, it should be noted that $h_i X > S_i$ for some i and, hence neither the penalty, $P(X)$, nor the vector of the marginal penalties, $\partial P(X)/\partial X$, reduces to zero.

29. In addition, it should be pointed out that a relatively large value of r in the beginning of the procedure ensures that neither the value, nor the slope, of the penalty function exceeds the largest value that the computer can deal with.

30. C. S. Russell and W. O. Spofford, Jr., "A Quantitative Framework," pp. 126–37.

31. Before we proceed, it should be pointed out that other nonlinear programming algorithms do exist. Considerable progress has been made in the last decade in the development of general, nonlinear algorithms that can handle nonlinear objective functions and nonlinear constraints of both the equality and inequality type. See, in particular, F. A. Fiacco and G. P. McCormick, *Nonlinear Programming*; J. B. Rosen, "The

The optimization scheme we are using is analogous to the gradient method of nonlinear programming. The technique consists of linearizing the response surface in the vicinity of a feasible point, X^k. To do this, we construct a tangent plane at this point by employing the first two terms of a Taylor's series expansion (up to first partial derivatives). This linear approximation to the nonlinear response surface will, in general, be most accurate in the vicinity of the point X^k and less accurate as one moves farther away from this point. Because of this, a set of "artificial" bounds (constraints) is imposed on the system to restrict the selection of the next position along the response surface to that portion of the surface most closely approximated by the newly created linear surface. The selection of the appropriate set of artificial bounds is analogous to choosing a step size in other gradient methods of nonlinear programming.

Because the newly created subproblem is in a linear form, we are able to make use of standard linear programming techniques for finding a new optimal point, X^{k+1}. This point locates the maximum value of the linearized objective function within the artificially confined area of the response surface. Because, in general, the linearized surface will not match the original nonlinear surface, the original nonlinear objective function must be evaluated at this point to determine whether or not this new point, X^{k+1}, is, in fact, a better position than the previously determined one, X^k. That is, the following condition must be satisfied:

$$F(X^{k+1}) > F(X^k). \tag{28}$$

If this condition is satisfied, a new tangent plane is constructed at the point X^{k+1} and a new set of artificial bounds is placed around this point. As before, a linear programming code is employed to find a new position, X^{k+2}, which maximizes the linearized objective function, and so on until a local optimum is reached. This procedure, like all gradient methods, finds only the local optimum. If the response surface contains more than one optimum, the problem becomes one of finding the global optimum. One way of approaching this is to start the procedure at different points within the feasible region, R, where the starting points may be chosen at random.[32]

Gradient Projection Method for Nonlinear Programming, Part I: Linear Constraints," *S.I.A.M. Journal on Applied Mathematics* 8, no. 1 (1960): 181–217; and J. B. Rosen, "The Gradient Projection Method of Nonlinear Programming, Part II: Nonlinear Constraints," *S.I.A.M. Journal on Applied Mathematics* 9, no. 4 (1961): 514–32.

32. For techniques on random starts within a feasible region defined by a linear constraint set, see, P. P. Rogers, "Random Methods for Non-Convex Programming" (Ph.D. diss., Harvard University, 1966).

As we have just seen, this optimization procedure requires that we linearize the objective function at a point X^k. We do this according to the following formulation:

$$F(X^{k+1}) = \nabla F(X^k) \cdot X^{k+1} + \gamma, \tag{29}$$

where γ is a constant. Expressing our revised objective function, equation 23, along with the modification suggested by expression 27, in terms of equation 29 results in,

$$\begin{aligned} F &= \nabla f(X) \cdot X - \nabla P(X) \cdot X + \gamma \\ &= \frac{\partial f(X)}{\partial X} \cdot X - \frac{\partial P(X)}{\partial X} \cdot X + \gamma. \end{aligned} \tag{30}$$

In our residuals management problem, $\partial f(X)/\partial X$ is a vector of linear cost coefficients associated with traditional production inputs, and residuals handling, modification, and disposal activities; and $\partial P(X)/\partial X$ is a vector of marginal penalties associated with the discharge of each residual.

Given that $R = h(X)$, equation 19, we see from equation 25 that,

$$\frac{\partial P(X)}{\partial X_j} = \sum_{i=1}^{q} \frac{2}{r_i} \{\max [(R_i - S_i), 0]\} \frac{\partial R_i}{\partial X_j} \qquad j = 1, \ldots, n. \tag{31}$$

The term

$$\left(\frac{2}{r_i} \{\max [(R_i - S_i), 0]\}\right),$$

represents the slope, dp_i/dR_i, of the ith penalty function evaluated at the point R_i. The term $\partial R_i/\partial x_j$ represents the marginal response of the ith descriptor of the natural world (or ecosystem) to changes in the discharge of the jth residual. Equation 31 may be expressed more generally as,

$$\frac{\partial P(X)}{\partial X_j} = \sum_{i=1}^{q} \frac{dp_i}{dR_i} \cdot \frac{\partial R_i}{\partial X_j} \qquad j = 1, \ldots, n, \tag{32}$$

or in matrix notation as,

$$\frac{\partial P(X)}{\partial X} = \left(\frac{\partial R}{\partial X}\right)^T \cdot \frac{dp}{dR}. \tag{33}$$

For linear environmental systems, $\partial R_i/\partial X_j$ is an element of the matrix of transfer coefficients, A, when the environmental functions are expressed, linearly, as,

$$R = h(X) = A \, . \, X. \tag{34}$$

Hence, for the case of linear environmental systems, the marginal penalties (equation 33) may be expressed in matrix notation as,

$$\frac{\partial P}{\partial X} = A^T \cdot \frac{dp}{dR}, \tag{35}$$

where $\partial P/\partial X$ is a vector of marginal penalties, A is a matrix of environmental transfer coefficients, and dp/dR is a vector of slopes of the penalty functions evaluated at R.

For the case of nonlinear environmental models, the situation is similar except that evaluation of the environmental response matrix, $\partial R/\partial X$, is somewhat more involved and in addition, because the response is nonlinear, it must be recomputed for each state of the natural world.

The Linearized Subproblem. Now that we have presented the essence of the optimization scheme that we are using, including a discussion of the LP subproblem which is necessary for us to both construct and solve at each step along the ascent procedure, we can restate our management problem in these terms.

$$\max \left\{ F[X^{k+1}] = \Big[\nabla F[X^k] - \nabla P[X^k] \Big] \cdot X^{k+1} + \gamma \right\}; \tag{36}$$

$$\text{s.t.} \qquad g_i(X) = 0 \qquad i = 1, \ldots, m, \tag{17}$$

$$g_i(X) \geq 0 \qquad i = m + 1, \ldots, p, \tag{18}$$

$$x_i \geq 0 \qquad i = 1, \ldots, n, \tag{22}$$

$$x_j \leq \beta_j \qquad j = 1, \ldots, s, \tag{37}$$

$$x_j \geq \alpha_j \qquad j = 1, \ldots, s, \tag{38}$$

where β_j and α_j are, respectively, upper and lower bounds on the s discharge variables at the $(k + 1)$th iteration. The efficiency of the optimization scheme depends directly on how these bounds are selected. Investigation of various procedures for selecting bounds is, perhaps, the most important use of the Delaware Valley Base Model. The techniques we are currently using will be presented later with a discussion of some results of the Base Model.

The Delaware Valley Base Model: An Illustration

In this section, we address ourselves to the computational problems discussed in the second section. To explore various solution methods and programming techniques, we constructed what we call the Delaware Valley Base Model. It is based on the Delaware Valley region in terms of geographic characteristics, economic activities, and residuals dischargers. It employs the same atmospheric dispersion and aquatic ecosystem models that the future full-scale model of this region will use. The major differences between this model and the full-scale residuals management model are the number of residuals generation and discharge activities provided with residuals management options, and the areal extent of the region considered.

The primary objective of this modeling effort is to build a computer model with all the features (hardware and software) of the proposed large scale regional model; one which can be expanded easily, but which remains small enough to experiment with programming techniques and ideas. Specifically, our aims include:

1. demonstrating the feasibility of solving a number of individual linear programs in sequence prior to entering the FORTRAN coded environmental models;
2. gaining experience with the use of complex, nonlinear ecosystem models as an integral part of the residuals management framework;
3. experimenting with the penalty function concept for meeting standards (or constraints) on ambient concentrations;
4. experimenting with various step size selectors on a reasonable size problem (136 discharge variables);
5. providing reality in terms of the Delaware Valley region.

The Delaware Valley region

The eleven county Delaware Valley region we ultimately intend to model is shown in figure 3.[33] This region consists of Bucks, Montgomery, Ches-

33. Much has been written about this particular region, especially in the water resources area. For a general discussion of water quality modeling efforts in the Delaware Estuary, see A. V. Kneese and B. T. Bower, *Managing Water Quality: Economics, Technology, Institutions* (Baltimore: Johns Hopkins Press, 1968).

ter, Delaware, and Philadelphia counties in Pennsylvania; Mercer, Burlington, Camden, Gloucester, and Salem counties in New Jersey; and New Castle county in Delaware. The rivers of interest in this region are the Schuylkill, which enters the Delaware at Philadelphia; the Delaware Estuary which runs approximately 85 miles from the head of Delaware Bay to the head of tide at Trenton, New Jersey; and a short reach of the Delaware River above Trenton. For modeling purposes, the estuary is divided into 30 sections, 16 of which are shown in figure 4.[34] For purposes of air quality management, a 10 kilometer grid is superimposed on the eleven county region. This is shown in figure 3.

The Delaware Valley region, with a 1970 population of approximately 5.5 million people,[35] is one of the most industrialized areas in the United States. For example, this region contains 7 major oil refineries, 7 steel plants, 13 paper (or pulp and paper) mills, 14 important thermal power generating facilities, numerous chemical and petrochemical plants, and 6 large municipal sewage treatment plants.[36]

The portion of the eleven county Philadelphia region which we include in the Base Model is outlined in figure 3 and shown in detail in figure 4. As depicted in these figures, the area of interest runs from Wilmington to Philadelphia. It is a rectangular area, 45 by 55 kilometers, upon which a 9 by 11 grid with equal spacings of 5 kilometers is superimposed. The river bounded by this area includes sixteen sections of the Delaware Estuary Model—sections 6 through 21.

34. These are the same 30 sections which were originally established by the Delaware Estuary Comprehensive Study (DECS) and the ones which are still being used by the Delaware River Basin Commission (DRBC) in their modeling efforts of the estuary. See Federal Water Pollution Control Administration, Delaware Estuary Comprehensive Study, and Delaware River Basin Commission, "Final Progress Report: Delaware Estuary and Bay Water Quality Sampling and Mathematical Modeling Project," May 1970.

35. This figure represents the 1970 population for the eleven county area described above; Delaware PC(VI-9), New Jersey PC(VI-32), Pennsylvania PC(VI-40). U.S. Department of Commerce, *1970 Census of Population* (Washington, D.C.: Government Printing Office, 1970).

36. Information about major dischargers in the region can be obtained from three major sources: (1) the Delaware River Basin Commission (DRBC), Trenton, N.J., for information on liquid residuals discharged to the estuary (see, for example, Delaware River Basin Commission, "Final Progress Report"; (2) Metropolitan Philadelphia Interstate Air Quality Control Region, "Inventory of Emission Sources," for information on gaseous residuals discharged within the Delaware Valley region; and (3) Greater Philadelphia Chamber of Commerce, *Business Firms Directory of Greater Philadelphia* (Philadelphia: Greater Philadelphia Chamber of Commerce, 15th edition, 1971), for addresses and employment data on plants.

Figure 3

Delaware Valley Region

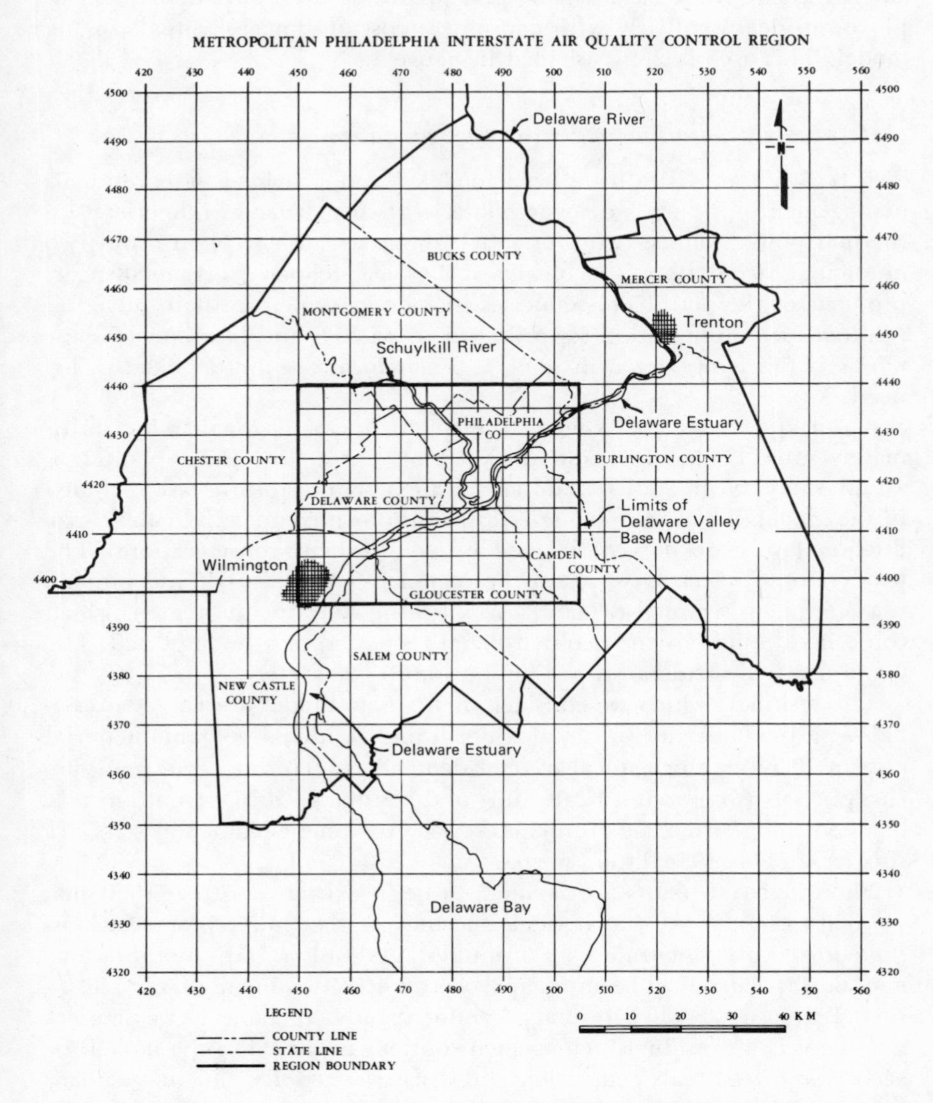

Note: The grid is in kilometers and is based on the Universal Transverse Mercator Grid System (UTM).

The area represented by the 99 (5 kilometer) grid squares has been divided into two subareas—a 50 grid square urban-suburban area, and a 49 grid square rural area. The 50 grid square urban-suburban area is employed to deal explicitly with consumer (postconsumer) residuals in the model. This area is shown shaded in figure 4.

Residuals generation and discharge activities

Ten regional activities have been modeled and provided with residuals management options; two sugar refineries, two petroleum refineries, two thermal power plants, two municipal sewage treatment plants, and two municipal incinerators. The output of the sugar refineries is refined sugar; products of the petroleum refineries include gasoline, distillate fuel, and residual fuel. Information regarding the capacities and locations of these ten activities is presented in table 1. Their locations are depicted in figure 4.[37]

As an alternative in reducing the quantities of residuals which ultimately must be handled and disposed of, outputs of these plants are allowed to vary. In addition, import and export possibilities are included in the model. The sugar refineries can shut down completely if their production levels are not constrained by employment considerations. The production of electricity within the region can be reduced, and imports used to fill up regional demand. Also, heating fuel for the region, which could be supplied by the two petroleum refineries, may be imported. The two municipal incinerators can be shut down completely.

The residuals which we consider in the base model are gaseous residuals—particulates and sulfur dioxide; liquid residuals—organic material measured by its biochemical oxygen demand (BOD), nitrogen, phosphorus, phenols (toxics) and heat; and solid residuals—furnace bottom ash, digested sludge from the municipal sewage treatment plant, wet scrubber slurries, and municipal solid wastes.

The industrial plants are assumed to be in existence so that their major features (such as the thermal efficiency of the power plants) are assumed fixed over the time span of interest, and only certain modifications may be carried out. The municipal wastewater treatment plants are assumed to have installed primary treatment only. In table 2, we summarize the various residuals management options available in the model for each type of residuals generation and discharge activity and the primary

37. It is intended that these activities be representative of industries and residuals modification facilities at these locations, but they are not necessarily accurate.

Figure 4

Map of Delaware Valley Base Model Region

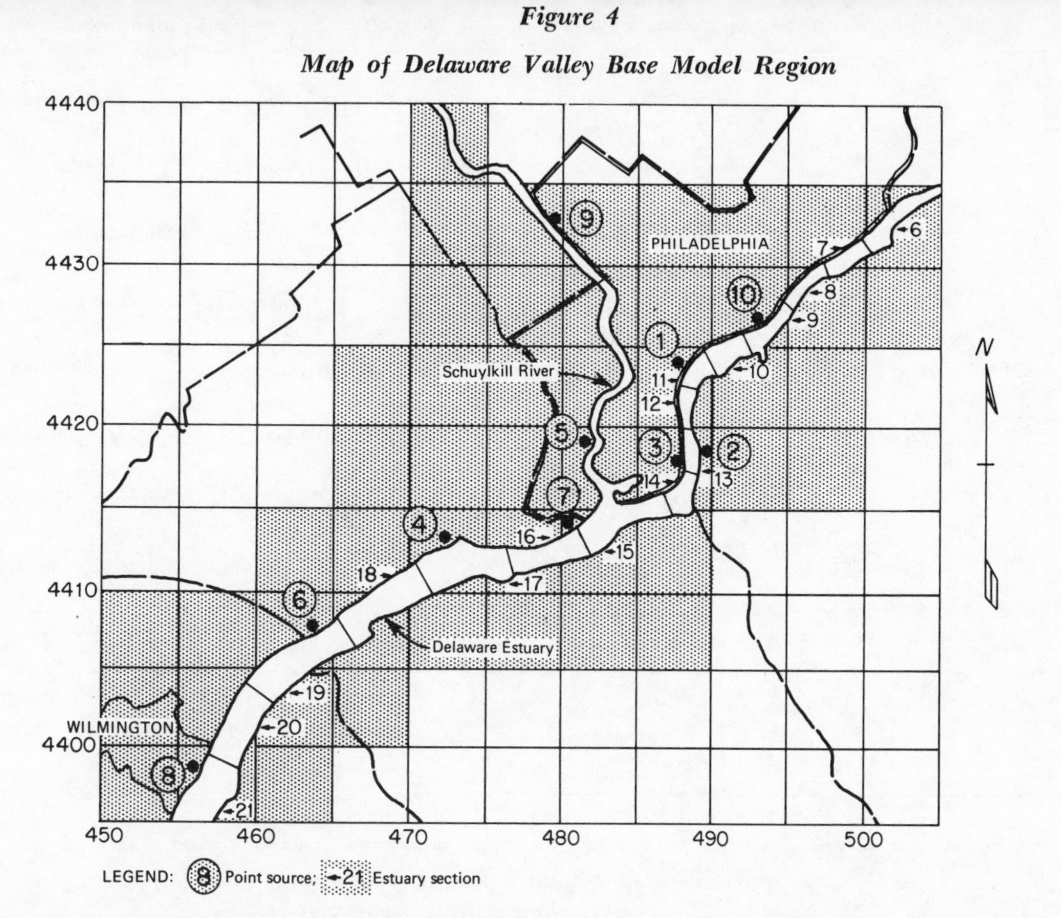

Note: The grid squares are numbered consecutively from bottom to top starting with the lower left corner and ending with the upper right corner.

TABLE 1

Delaware Valley Base Model: Regional Residuals Generation and Discharge Activities

Source Number	Type Activity	Capacity	Location: City	River Reach	River Mile	UTM Coordinates x	UTM Coordinates y
1	Sugar refinery	400 tons of refined sugar per day	Philadelphia, Pa.	11	101.8	488.6	4423.7
2	Sugar refinery	400 tons of refined sugar per day	Camden, N.J.	13	97.85	489.7	4418.7
3	Thermal power generating facility	345 MW	Philadelphia, Pa.	13	97.5±	488.1	4418.0
4	Thermal power generating facility	707 MW	Eddystone Borough, Pa.	17	85±	472.4	4412.2
5	Petroleum refinery	162,000 barrels of crude oil per day	Philadelphia, Pa.	15[a]	92.47[a]	482.7	4419.1
6	Petroleum refinery	172,000 barrels of crude oil per day	Marcus Hook, Pa.	19	78.95	463.8	4406.9
7	Municipal sewage treatment plant	160 mgd.	Philadelphia, SW Philadelphia, Pa.	16	90.7	481±	4413±
8	Municipal sewage treatment plant	60 mgd.	Wilmington, Delaware	21	71.3	456±	4398±
9	Municipal incinerator	565 tons per day	Philadelphia, NW Philadelphia, Pa.			479.5	4431.9
10	Municipal incinerator	298 tons per day	Philadelphia, NE Philadelphia, Pa.			493.2	4425.9

[a] This oil refinery is 2.5 miles up the Schuylkill River. The Schuylkill River enters Delaware Estuary at river mile 92.47.

TABLE 2

Residuals Management Options Available to the Various Types of Dischargers in the Region

Management Option Available	*Primary Residuals Reduced*	*Secondary Residual Generated*
	Sugar Refineries	
Partial or full reuse of flume water	BOD	Sludge
Secondary and tertiary wastewater treatment	BOD	Sludge
Cooling tower(s)	Heat	Heat is rejected to the atmosphere along with water vapor
Burn lower sulfur coal	SO_2	None
Electrostatic precipitators (3 alternative efficiencies; 90, 95, 98 per cent)	Particulates (fly ash)	Bottom ash
Sludge digestion and landfill	Sludge	The secondary residual here is digested sludge at a different location
Sludge dewatering and incineration	Sludge	Particulates
Dry cyclone; 90 per cent efficiency	Particulates	Bottom ash
	Petroleum Refineries	
Secondary and tertiary treatment, and various reuse alternatives (cooling tower water makeup, desalinate water, boiler feedwater)	Nitrogen Phenols BOD	Sludge
Cooling tower(s)	Heat	Heat is rejected to the atmosphere along with water vapor
Burn lower sulfur fuel	SO_2	None
Refine lower sulfur crude	SO_2	None
Sell, rather than burn, certain high sulfur products (e.g., refinery coke)	SO_2	None
Cyclone collectors on catcracker catalyst regenerator (2 efficiencies; 70, 85 per cent)	Particulates	Bottom ash
Electrostatic precipitator; 95 per cent efficiency	Particulates	Bottom ash

(continued)

TABLE 2 (*Concluded*)

Management Option Available	*Primary Residuals Reduced*	*Secondary Residual Generated*
Sell, rather than burn, high sulfur refinery coke	SO_2 Particulates	None
Sludge digestion and landfill	Sludge	The secondary residual here is digested sludge at a different location
Thermal Power Generating Plants		
Cooling tower(s)	Heat	Heat is rejected to the atmosphere along with water vapor
Burn lower sulfur coal	SO_2	None
Limestone injection-wet scrubber; 90 per cent efficiency	SO_2	Slurry
Electrostatic precipitators; 90, 95, 98 per cent efficiency	Particulates	Bottom ash
Settling pond; 90 per cent efficiency	Slurry	Solid ash
Municipal Sewage Treatment Plants		
Secondary or tertiary wastewater treatment	BOD Nitrogen Phosphorus	Sludge
Sludge digestion, drying and landfill	Sludge	The secondary residual here is digested sludge at a different location
Sludge dewatering and incineration	Sludge	Particulates, bottom ash
Dry cyclone; 80, 95 per cent efficiency	Particulates	Bottom ash
Municipal Incinerators		
Electrostatic precipitators (2 alternative efficiencies; 80, 95 per cent efficiency)	Particulates	Bottom ash

residuals which are reduced and the secondary residuals generated as a result of each of the management alternatives.[38]

38. The sugar refinery alternatives are based on information in G. O. G. Löf and A. V. Kneese, *The Economics of Water Utilization in the Beet Sugar Industry* (Washington, D.C.: Resources for the Future, 1968). The petroleum refineries are condensed versions of a model developed at RFF and described in C. S. Russell, "Residuals

As shown in table 2, there are some trade-offs among forms of residuals generated and subsequently discharged to the environment. Slurries, resulting from the removal of sulfur dioxide using limestone injection and wet scrubbing, can cause water quality problems. Particulates, resulting from the incineration of sludge, can cause air quality problems, and so on. Options do exist in the model, however, for converting the major portion of both gaseous and liquid residuals to solid residuals. Sludges are either digested, dried, and landfilled, or dewatered and incinerated, with the residue going to landfill. Furnace bottom ash is trucked directly to landfill. The slurries produced by the limestone injection-wet scrubber at the two thermal power generating facilities are lagooned on-site.

Consumption residuals

The three types of postconsumer (or consumption) residuals considered in the model are municipal sewage; particulates and sulfur dioxide from household space heating activities; and municipal solid residuals. With regard to management options in the model, we assume these residuals are generated only within the 50 grid square urban-suburban area. The quantities of sewage input to the two municipal sewage treatment plants are based on average daily BOD loadings at the Wilmington, Delaware and Philadelphia, SW wastewater treatment facilities during 1968.

Household heating requirements for the 50 grid square urban-suburban area were estimated on the basis of the Environmental Protection Agency's (EPA) inventory of gaseous emissions for this region. Sulfur dioxide emissions for each grid square, assuming the use of high sulfur distillate fuel for all household heating needs, were matched with the area source emissions.[39] This enabled us to compute the daily quantity of distillate

---

Management." The information for the electric power plants and their associated alternatives came largely from J. K. Delson and R. Frankel, "Residuals Management in the Coal-Energy Industry," unpublished manuscript; and Paul H. Cootner and G. O. G. Löf, *Water Demand for Steam Electric Generation* (Washington, D.C.: Resources for the Future, 1965). The municipal treatment plant vectors were constructed using the data compiled by Robert Smith and reported in "Costs of Conventional and Advanced Treatment of Waste Water," *Journal of the Water Pollution Control Federation* (September 1968), pp. 1546–74. The municipal incinerator characteristics are based on information contained in Combustion Engineering, Inc., *Technical Economic Study of Solid Waste Disposal Needs and Practices* Vol. IV (Rockville, Md.: United States Department of Health, Education and Welfare, Bureau of Solid Waste Management, 1969).

39. EPA's inventory of emission sources for this region consists of two kinds of sources, point and area. A point source consists of a single stack. Large sources, in terms of emissions, are usually dealt with as point sources. An area source consists of the aggregation of many smaller sources over a grid square of designated size—usually 2.5, 5.0, 10.0 or 20.0 kilometers on a side.

fuel oil required for each grid, and then assuming a heat content of 5.84 $\times 10^6$ BTU per barrel for distillate fuel oil, we were able to compute the home heating requirements (in BTU's) for each grid.

These heating requirements can be met in the model using any one of three sulfur grades of fuel oil. Sulfur dioxide emissions are assumed to be two times the sulfur weight in the fuel burned. Particulate emissions are assumed to be 0.504 pounds per barrel burned, independent of sulfur content.[40] The model can then select a policy for the sulfur content of domestic fuel oil in the urban-suburban area.

Inhabitants of the urban-suburban area have three options for dealing with their solid residuals: (1) incineration, with the residue (bottom ash) being disposed of at one of the available landfill sites; (2) disposal directly to a local landfill site with low and high compaction alternatives; and (3) disposal at a more distant, rural site with low and high compaction alternatives. Two of Philadelphia's municipal incinerators are considered in the model: Northeast, with a maximum process rate of 298 tons per day; and Northwest, with a maximum process rate of 565 tons per day.[41] We provide inputs to these incinerators by assuming an area served by each, a population for each area, and a per capita municipal solid residuals collection rate of 5.72 pounds per day.[42]

Background residuals

In order to keep the Base Model down to what we consider a reasonable size for our purposes here, only a small portion of all the residuals generation and discharge activities within the region were modeled and provided with residuals management options. Unfortunately, the resulting ambient concentrations due only to the ten point and 50 area (household space heating) sources were unreasonably low. To make the problem more interesting in terms of reality for the region, all of the remaining sources of residuals—liquid and gaseous, point and area—were then added to the model. In terms of the management framework, the newly added sources are treated as "background" inputs to the environmental models.

40. Emissions for fuel oil combustion are based on data in Environmental Protection Agency, *Compilation of Air Pollution Emission Factors* (Washington, D.C.: Environmental Protection Agency, Office of Air Programs, February 1972), pp. 1–4 to 1–7.

41. These capacities are based on EPA's inventory of emission sources for this area.

42. This is the average quantity of solid residuals collected per person in the urban United States. R. J. Black et al., *The National Solid Waste Survey: An Interim Report* (Rockville, Maryland: Department of Health, Education and Welfare, 1968), table 2.

Background inputs of residuals to the Delaware Estuary—both direct discharges and tributary loads—were estimated from data provided by the Delaware River Basin Commission (DRBC) and from surface water records published by the United States Geological Survey (USGS).[43] From the residuals discharge data on individual industrial and municipal dischargers collected by the DRBC, total inputs of nitrogen, phosphorus, and BOD were estimated for each source not already included as one of the ten activities in the Base Model. These data were averaged over the last three or four years, depending on the availability of data and the variability exhibited by the discharge stream.

Tributary loads were estimated by extrapolating gagged stream flow to the total area of each major watershed (or where tributaries were not gagged, by multiplying per area surface water runoff of an adjacent tributary by the total drainage area of the ungagged tributary), and multiplying this flow times the concentration of nitrogen, phosphorus, and BOD as reported in EPA's STORET data bank. Assumed inputs to the Delaware Estuary of BOD, nitrogen, and phosphorus have been aggregated by reach and are presented in table 3.

Background generation of sulfur dioxide and particulates was estimated as follows: All point sources in the emission source inventory, except one very large source in Grid Square 21 (which was maintained as a "background" point source) and all those treated explicitly in the Base Model as individual dischargers with management options, were aggregated over the 5 kilometer grid in which they are located. We call these "aggregated" point sources and include 44 (out of a total of 99) in the Base Model. (The remaining 55 sources were either zero or relatively small. In the latter case, they were added to the area sources of corresponding grid squares.) We assume each aggregated point source discharges through a single stack at the center of its respective grid. Stack characteristics—stack height, diameter, exit velocity, and stack exit temperature—of each aggregated point source have been specified such that the locations (heights) above the ground of the centers of mass of the

43. United States Geological Survey, *Water Resources Data for Pennsylvania: Part 1, Surface Water Records* (Washington, D.C.: United States Department of Interior, published annually); United States Geological Survey, *Water Resources Data for Maryland and Delaware: Part 1, Surface Water Records* (Washington, D.C.: United States Department of Interior, published annually); United States Geological Survey, *Water Resources Data for New Jersey: Part 1, Surface Water Records* (Washington, D.C.: United States Department of Interior, published annually). These reports are based on the water year beginning October 1 and ending September 30.

TABLE 3

Inputs of Organic Material, Nitrogen, and Phosphorus to the Delaware Estuary (pounds per day)

Estuary Sections (DRBC)	Point Sources								
	Industrial[a]			Municipal[b]			Total, Point Sources		
	BOD_5	*N*	*P*	BOD_5	*N*	*P*	BOD_5	*N*	*P*
1	0	0	0	8,065	1,210	504	8,065	1,210	504
2	2,758	2,940	807	1,668	250	104	4,426	3,190	911
3	1,111	3,169	14	2,282	342	143	3,393	3,511	157
4	1,857	328	30	0	0	0	1,857	328	30
5	2,167	202	39	644	97	40	2,811	299	79
6	0	0	0	825	124	52	825	124	52
7, 8	0	0	0	897	135	56	897	135	56
9, 10	5,530	155	31	115,525	17,329	7,220	121,055	17,484	7,251
11, 12	9,450	32	164	0	0	0	9,450	32	164
13, 14	23,659	4,380	552	141,925	21,289	8,870	165,584	25,669	9,422
15	39,973	20,385	319	4,280	642	268	44,253	21,027	587
16	25,340	5,121	176	100,564	15,085	6,285	125,904	20,206	6,461
17	63,070	17,539	455	7,093	1,064	443	70,163	18,603	898
18	16,914	5,359	277	1,803	270	113	18,717	5,629	390
19	57,036	6,800	1,311	0	0	0	57,036	6,800	1,311
20	0	0	0	673	101	42	673	101	42
21, 22	123,807	38,367	2,670	37,279	5,592	2,330	161,086	43,959	5,000
23, 24	0	0	0	335	50	21	335	50	21
25, 26	12,000	12,400	1,087	0	0	0	12,000	12,400	1,087
27, 28	0	0	0	1,365	205	85	1,365	205	85
29	0	0	0	0	0	0	0	0	0
30	0	0	0	0	0	0	0	0	0
Total	384,672	117,177	7,932	425,223	63,783	26,576	809,895	180,960	34,508

(continued)

TABLE 3 (Concluded)

Estuary Sections (DRBC)	Area Sources						Totals, All Sources		
	Tributaries[e]			Storm Water[d]					
	BOD_5	N	P	BOD_5	N	P	BOD_5	N	P
1	1,720[c]	407	514	1,360	367	103	11,100	1,980	1,120
2	2,090	931	483	0	0	0	6,520	4,120	1,390
3	0	0	0	0	0	0	3,390	3,510	157
4	182	206	238	0	0	0	2,040	534	268
5	2,000	950	425	0	0	0	4,810	1,250	504
6	4,110	1,810	724	0	0	0	4,940	1,930	776
7, 8	211	159	150	1,810	489	137	2,920	782	343
9, 10	1,880	564	439	12,960	3,499	982	136,000	21,500	8,670
11, 12	1,500	625	179	21,260	5,740	1,612	32,200	6,400	1,950
13, 14	2,790	1,785	431	9,490	2,562	719	178,000	30,000	10,600
15	17,181	4,455	5,746	18,860	5,092	1,430	80,300	30,600	7,760
16	1,430	563	299	0	0	0	127,000	20,800	6,760
17	2,470	1,737	1,000	1,950	526	148	74,600	20,900	2,050
18	879	1,012	1,037	0	0	0	19,600	6,640	1,430
19	0	0	0	0	0	0	57,000	6,800	1,310
20	0	0	0	0	0	0	673	101	42
21, 22	11,900	3,940	1,319	8,320	2,246	631	181,000	50,100	6,950
23, 24	0	0	0	0	0	0	335	50	21
25, 26	0	0	0	0	0	0	12,000	12,400	1,090
27, 28	0	0	0	0	0	0	1,370	205	85
29	0	0	0	0	0	0	0	0	0
30	0	0	0	0	0	0	0	0	0
Total	50,343	19,144	12,984	76,010	20,523	5,762	936,248	220,627	53,254

(Footnotes for Table 3)

Note: Organic material is reported in pounds of five-day biochemical oxygen demand (BOD_5) per day; nitrogen, in pounds of nitrogen (N) per day; and phosphorus, in pounds of phosphorus (P) per day.

[a] Industrial loads were estimated from unpublished data supplied by the Delaware River Basin Commission.

[b] Based on 1968 sewage treatment plant discharges, in Delaware River Basin Commission, "Final Progress Report: Delaware Estuary and Bay Water Quality Sampling and Mathematical Modeling Project," May 1970. Nitrogen and phosphorus loads were estimated from the BOD loads according to data presented in G. A. Rohlich and P. D. Uttormark, "Wastewater Treatment and Eutrophication," *Nutrients and Eutrophication* Special Symposium Volume 1 (Ann Arbor: American Society of Limnology and Oceanography, Inc., 1972).

[c] This quantity does not include inputs from the Delaware River above Trenton.

[d] BOD loads are based on 1964 stormwater overflow in R. V. Thomann, *Water Quality Management*. Nitrogen and phosphorus loads have been estimated from BOD loads according to data given in Environmental Protection Agency, *Storm Water Management Model* (Washington, D.C.: U.S. Government Printing Office, 1971), p. 180.

[e] Tributary loads are based on the low flow season averaged over a three year period.

resulting plumes of both the aggregated point source and its corresponding individual point sources (within the same grid) are the same.[44]

The area sources from the source inventory were aggregated (or disaggregated) to the 5 kilometer grid size shown in figures 3 and 4. However, we have made one modification to these sources. From the fifty area sources within our urban-suburban area, we have subtracted the gaseous emissions resulting from household space heating. Particulate and sulfur dioxide emissions from this activity are included in the residuals generation and discharge submodel, and are thus eliminated from background residuals.

We felt it was necessary to separate aggregated point sources from the area sources in the region because of the difference in effective stack heights (sum of the physical stack height plus plume rise) of these two types of sources.

The estuary and its quality

For the period October 1912 to September 1965, the mean annual flow of the Delaware River at Trenton, New Jersey, was 11,550 cubic feet per

44. This treatment does not, however, insure the same distribution of ground level concentrations. The latter is a complex nonlinear function of many arguments including effective stack height. Emissions close to the ground contribute, proportionally, more to local ambient concentration levels than do the emissions discharged higher up. The higher emissions tend to spread out over the region more. This technique was used here as an expedient measure. Consideration of each point source as a separate entity in the air dispersion model would have been an expensive proposition, and the additional accuracy would not have been warranted for the kinds of investigations we are making with our Base Model.

second (cfs). The Delaware is similar to most rivers in the northeastern United States with respect to its seasonal variation of flow: maximum mean monthly flows are experienced in March and April (20,000 to 25,-000 cfs), and minimum mean monthly flows in the late summer months of August and September (about 2,500 cfs).[45]

Although a Streeter-Phelps type dissolved oxygen computer model for the Delaware Estuary is available from the DRBC, we are using the aquatic ecosystem model described in an earlier section. The DO model of this estuary used 30 reaches between Trenton, New Jersey and Liston Point, Delaware. By combining some of the shorter reaches, we have reduced this number to 22 in order to save computer time. The Delaware Valley Base Model employs 11 of the new reaches: those which lie within the area outlined in figures 3 and 4.

The atmosphere and its quality

In order to relate ambient concentrations throughout the region with gaseous residuals discharges, we needed both an air dispersion model and data on the meteorology of the Delaware Valley region. The atmospheric dispersion model which we are using in the Base Model was presented in an earlier section. Necessary data inputs to the model are a joint probability distribution for wind speed, wind direction, and atmospheric stability based on meteorological data for the Delaware Valley region. We assumed, based on EPA data, that the mean monthly maximum atmospheric mixing depth for the season of interest is 1,000 meters, and that the mean seasonal temperature and pressure for this region are 68°F (20°C) and 1,017 millibars (30.03 inches of mercury) respectively.

For purposes of the air dispersion model, each of the ten activities presented in table 1 is treated as a single point source even though most have, in fact, several stacks—one, for example, has over 40 stacks listed in the gaseous emission inventory. Where more than one stack was listed for a particular plant, a "virtual" stack was calculated on the basis of the center of mass of the individual plumes. This method of aggregation was discussed previously in this section. Stack characteristics which are required by the IPP model for computing effective stack height include physical stack height, stack diameter, and stack exit velocity and temperature of the gas.

45. United States Geological Survey, *Compilation of Records of Surface Waters of the United States, October 1961 to September 1965: Part I–B North Atlantic Slope Basins, New York to York River* U.S. Geological Survey Water-Supply Paper 1902 (Washington, D.C.: U.S. Government Printing Office, 1970).

The computer program

The optimization algorithm we are using consists of a standard linear program (IBM's MPSX–360) and four FORTRAN coded subprograms.[46] The algorithm is set up in such a way that two, or more, linear programming problems can be solved sequentially prior to entering the four FORTRAN coded routines (to be described below). Our experience to date indicates that this is a workable arrangement as far as the MPSX software is concerned.

The ten sources in the region for which residuals management options have been provided are divided equally between two linear programs, MPSX–1 and MPSX–2. Home heating and municipal solid residuals management options for the 50 grid squares comprising the urban-suburban area have also been divided evenly between the two linear programs.

Given prices (marginal penalties) on discharges, evaluated in the environmental and constraint evaluation submodels to be discussed below, these two linear programs are solved, subject to the appropriate constraints, and the resulting residuals discharges are passed as inputs to the environmental models. See figure 5 for overall program flow.

The first of the FORTRAN subprograms, DATA SORT, is used to call the remaining three, in turn, to aggregate liquid residuals discharges entering the same reach or section of the estuary, and to assign marginal penalties to all the liquid residuals dischargers prior to returning to MPSX control for another iteration of the linear programming models.

The next FORTRAN routine, AQUATIC ECOSYSTEM, is the nonlinear aquatic ecosystem model. This model computes the steady state values for the eleven endogenous variables in each of the eleven sections of the estuary; this will be expanded to 22 reaches with the full-scale Delaware Valley Model. In addition, given penalty functions for exceeding standards on various quality measures of the natural world, this routine evaluates total penalties associated with water quality, as well as the marginal (water quality) penalties attributable to all liquid residuals dischargers in the region.

The third FORTRAN routine, ATMOSPHERIC DISPERSION, is the linear atmospheric dispersion model. In this routine, steady state levels of suspended particulates and sulfur dioxide at all ninety-nine grid

46. IBM's linear programming code was used because of its capability for handling large problems, our previous experience with it, and the availability of the IBM 360 system for our use.

Figure 5

Schematic Diagram of Program Flow: Optimization Algorithm

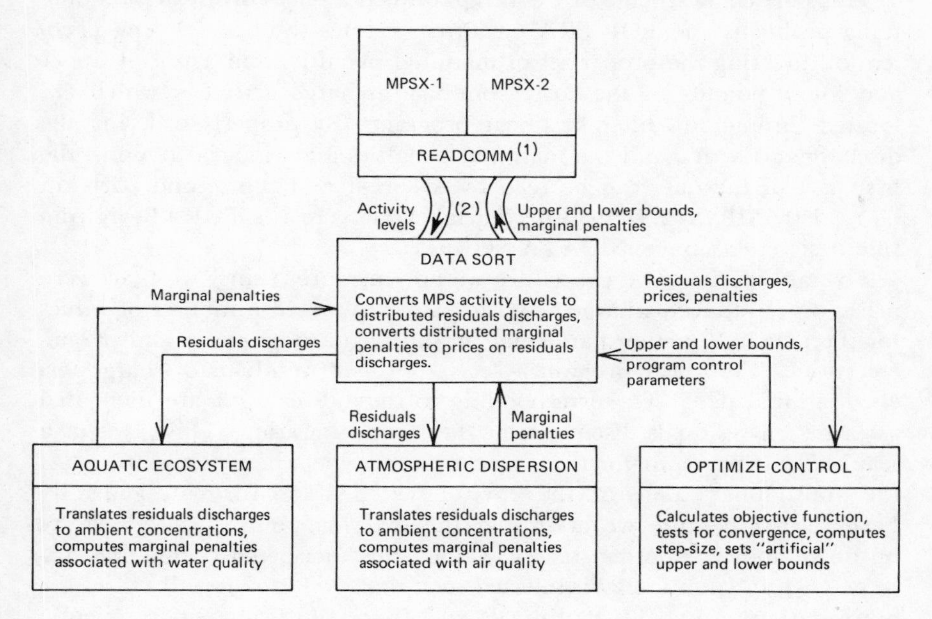

1. READCOMM is an IBM supplied FORTRAN coded subroutine which allows interaction between MPSX and other FORTRAN routines. For a description of its use see IBM Manual SH20-0960-0, *Mathematical Programming System Extended (MPSX) Read Communications Format (READCOMM) Program Description Manual* (White Plains, New York: IBM, 1971).
2. Indicates an input-output link.

points throughout the region are computed. In addition, total penalties associated with exceeding air quality standards are evaluated, and marginal (air quality) penalties are computed for each discharger in the region.

The fourth and last FORTRAN routine, OPTIMIZE CONTROL controls the allowable changes in levels of residuals discharges between successive iterations, ensures that the new solution, in terms of the objective function value, is better than the previous one, and determines when the program should terminate. Specifically, it selects the step sizes at each iteration for the 136 residuals discharge variables in the Base Model, it sets the artificial bounds (constraints) associated with each discharge variable, and it determines when the procedure has found a local optimum.

This particular scheme of iterating between a series of linear programming problems and FORTRAN routines creates the bookkeeping problem of directing the proper set of marginal penalties and artificial upper and lower bounds to the corresponding discharge activities which are located throughout different linear programs. By properly ordering the discharge activities, and the marginal penalties and artificial bounds, the first part of the data can be read by MPSX–1 and the second part can be read by MPSX–2. This ordering takes place in the FORTRAN routine designated above as DATA SORT.

To start the iterative procedure we are currently using, we place zero prices on residuals discharge activities, and assume an initial set of values for the penalty function parameter, r. Subject to these prices and parameter values, the linear program selects "optimal" residuals discharge levels. Marginal penalties, corresponding to these discharges, are then used as prices on residuals discharges at the second iteration. This gives us a feasible starting point for the steepest ascent type scheme we employ. At the "optimum," some of the environmental standards will generally be exceeded because we are using the exterior point penalty scheme outlined above. Another set of values for the penalty function parameter, r, in equation 27 is selected such that $r^{k+1} < r^k$ for all r, and the optimization process is continued, and so on until no environmental standard, S_i, is exceeded by more than a previously specified amount. Because the response surface may be multipeaked, this procedure yields, at best, a local optimum.

One of the step size selectors which we are currently experimenting with is described in our earlier work.[47] Another, which is based on the former, will be presented in the next section where we present some results of the model. Both step size selectors require a set of ranges for the discharge variables. The feasible range for all 136 discharge variables has been estimated from a knowledge of the processes and activities employed.

For any given set of penalty function parameter, r, the program is coded to end computations when any one of the following conditions is met:

(i) the value of the objective function increases by no more than a specified amount;

(ii) the number of iterations equals the specified maximum number;

47. C. S. Russell and W. O. Spofford, Jr., "A Quantitative Framework," equation 2.1–18, p. 135.

(iii) the number of different step size sets equals the specified maximum number.

Some results

Our interest in the computational results lies primarily in the evidence they afford on the following questions:

Does the heuristic algorithm we have described appear to converge to an optimum when faced with a moderately large problem?

Is the solution obtained characterized by only small violations of the ambient environmental quality standards?

Is there evidence that the solution represents a global optimum?

What can we say about the effect of various choices for penalty-function parameters and step-size selector schemes in relation to the above questions?

Finally, is the cost of the computational exercise so high as to promise that a full-scale regional model would be simply a white elephant?

In this section, we discuss the light shed on these important questions by five separate runs of the Base Model. These runs are distinguished by the ways in which the penalty function parameters are handled, and by the step-size selector employed.

Penalty Function Parameters. Recalling that the penalty function for a particular constraint may be written $1/r \cdot p(X)$, we distinguish the following sets of values for $1/r$ in table 4:

TABLE 4

Sets of Values for 1/r

	Penalty Function Parameter Set			
Indicator	*a*	*b*	*c*	*d*
Sulfur dioxide	100	1000	100	10
Suspended particulates	10	100	10	1
Algae	10^6	10^6	10^5	10^4
Fish	10^6	10^6	10^5	10^4
Dissolved oxygen	10^6	10^6	10^5	10^4

Step-Size Selector Scheme. Three step-size selector schemes were used for the five runs of the Base Model presented in this section. For all three step-size selectors, the initial step-size vector is the same. However, subsequent step-size vectors differ.

Given the vector of ranges, R, for all 136 discharge variables, the initial step-size vector, δ^1, is computed as follows.

$$\delta_i^1 = \frac{R_i}{10} \qquad i = 1, \ldots, 136. \tag{39}$$

For the *first* step-size selector scheme, subsequent step-size vectors are computed using the following formulation.

$$\delta^{k+1} = \frac{\delta^k}{10} \qquad k = 1, \ldots, n, \tag{40}$$

where n is the number of step-size sets.

For the *second* step-size selector scheme, step-size vectors, after the first, are computed as follows.

$$\delta^{k+1} = \frac{\delta^k}{2} \qquad k = 1, \ldots, n. \tag{41}$$

The *third* step-size selector scheme is a little more involved. Rather than shortening all elements of the step-size vector proportionately at a given step-size change (as in the case with the first two schemes), this selector reduces elements of the vector selectively. According to this scheme, at any given step-size change, an element of the step-size vector is either reduced according to equation 40, or left unchanged.

We decide whether to reduce the step-size of a particular discharge variable as follows. First, we try to assess whether the objective function decrease, which originally signaled the step-size change, is a result (on aggregate) of discharging too much or too little to the environment. We do this by comparing consecutive net benefits (the sum of the net benefits of MPSX–1 and MPSX–2 net of effluent charges), and consecutive total penalties, for the last two iterations of the management model.

If the total penalties decrease *and* at the same time the net benefits decrease, we assume that the step-size change was caused by discharging (on aggregate) too little to the environment. On the other hand, if the total penalties increase, regardless of what the net benefits do, we assume the step-size change was caused by discharging (on aggregate) too much to the environment.

Once we have established this, we must then compare, element by element, the last two sets of discharges, and the last two sets of marginal penalties. Disposition of each element of the step-size vector is summarized as follows.

Assume discharge too much. If the discharge increases, and the marginal penalty increases, change step-size according to equation 41. For other combinations, leave step-size unchanged.

Assume discharge too little. If the discharge decreases, and the marginal penalty decreases, change step-size according to equation 41. For other combinations, leave step-size unchanged.

The Five Computer Runs. The five runs may be described as follows:

Run 1: the first step-size selector with $1/r$ set at level "a" initially and kept there for 40 iterations;[48]

Run 2: the first step-size selector with $1/r$ set at level "d" initially, kept there for 20 iterations, reset to level "a" and maintained for a further 20 iterations;

Run 3: the first step-size selector with $1/r$ set initially at level "d," changed to level "c" at the first step-size change and to level "b" with the second step-size change; the program is allowed to run out to the 40th iteration;

Run 4: the second step-size selector with $1/r$ set at level "a" initially and kept there for 40 iterations;

Run 5: the third step-size selector with $1/r$ set at level "a" initially and kept there for 40 iterations.

Ambient Standards. For each run, the ambient environmental quality standards were set as follows in table 5. The standards on the airborne residuals are much less stringent than current federal standards. They were chosen this way only because the Base Model has so few controllable sources relative to the discharges we have included as "background."[49] Standards approaching the current federal limits are infeasible in this test

48. That these runs all involve stopping after 40 iterations is simply a feature of our initially cautious experimentation. Ordinarily the algorithm itself would determine the stopping point.

49. It should also be pointed out that the air dispersion model was not calibrated prior to using it. Because of this, the computed ambient concentrations were somewhat higher than those actually observed in the Philadelphia region. If we use the calibration relationships for this model from an EPA air quality study of this region (1967–1968 data) and modify the air quality penalty function parameter sets accordingly, comparable results for our regional residuals management model could be obtained by reducing the sulfur dioxide standard in our model to $175\mu g/m^3$ ($0.416 \times 420 = 174.7$) and the suspended particulates standard to $88\mu g/m^3$ ($35 + 0.532 \times 100 = 88.2$). For these calibration relationships, see EPA, "Application of Implementation Planning Program Modeling Analysis: Metropolitan Philadelphia Interstate AQRC," Air Quality Management Branch, Applied Technology Division, Office of Air Programs, EPA, Durham, N.C., February 1972, (mimeo), figures 2 and 3.

TABLE 5

Ambient Environmental Quality Standards

Indicator	*Location*	*Standard*
Sulfur dioxide	center of each grid square	$\leq 420\ \mu g/m^3$ [b]
Suspended particulates	center of each grid square	$\leq 100\ \mu g/m^3$ [b]
Algae[a]	each reach	≤ 2.0 mg/L
Fish[a]	each reach	≥ 0.29 mg/L
Dissolved oxygen	each reach	≥ 3.0 mg/L

[a] Based on biomass concentrations in terms of the total weight of carbon, nitrogen, and phosphorus.
[b] Standards used with uncalibrated air dispersal model.

situation even under the uninteresting policy choice of shutting down all six industrial emitters and the two municipal incinerators.

Results of the Five Runs. The results of the five computer runs are presented in tables 6, 7, and 8, and in figures 6 through 10. In table 6, we show, for the five computer runs, production levels at the 40th iteration for all six industrial plants and for the two municipal incinerators. In addition, we show production levels for the first iteration (zero prices on residuals discharges) and the second iteration (maximum prices on residuals discharges). The results of the first two iterations are the same for all five runs. Also depicted in table 6 are the net benefits (sum of the individual MPSX objective function values corrected or plus for the effluent charges "paid"), the total penalties, and the objection function values at the 40th iteration for the five computer runs.

In table 7, we present for the five computer runs the discharge levels of air-borne, liquid, and solid residuals at the 40th iteration for the ten activities in the region for which residuals management options have been provided. Also shown are the residuals discharge levels for the first and second iterations (which are the same for all five runs).

Ambient concentrations which exceed the standards on the 40th iteration are displayed in table 8 for all 5 runs. There we show the standard which is violated, the amount of the violation, the location of the violation (estuary reach or air quality grid square), and the per cent violation based on the level of the standard.

Figures 6 through 10 depict a plot of the objective function value vs. iteration for the 5 runs. Also shown in these figures are the iterations

TABLE 6

Production Levels and Objective Function Values for the Five Computer Runs

Source Number	*Production Outputs and/or Inputs*	*Units*	*Iteration*		*Results at 40th Iteration for Runs*				
			1	*2*	*1*	*2*	*3*	*4*	*5*
1	Refined sugar	tons/day	400	0	58	0	0	50	0
2	Refined sugar	tons/day	400	0	107	0	41	85	5
3	Electricity	MW hrs./day	6,040	0	6,040	6,040	6,040	6,040	6,040
4	Electricity	MW hrs./day	11,550	11,550	11,550	11,550	11,550	11,550	11,550
5	Crude oil processed	bbls./day	162,000	0	129,300	147,200	145,300	129,900	144,300
	Gasoline	bbls./day	116,200	0	57,200	70,200	69,200	57,600	67,400
6	Crude oil processed	bbls./day	172,000	130,800	172,000	172,000	172,000	172,000	172,000
	Gasoline	bbls./day	123,400	53,200	99,500	123,400	123,400	105,800	121,000
9	Solid residuals processed	tons/day	495	0	506	512	516	507	505
10	Solid residuals processed	tons/day	298	0	298	298	298	298	298
Net benefits (MPSX–1 + MPSX–2)*		$ per day	492,300	69,700	297,600	335,700	317,300	306,100	329,700
Total penalties		$ per day	2,277,500	0	1,600	22,600	48,400	800	600
Objective function value		$ per day	(1,785,200)	69,700	296,000	313,100	268,900	305,300	329,100

* Effluent charges not included.

TABLE 7

Residuals Discharge Levels for the Five Computer Runs

Source Number	Residual	Units	Iteration		Results at 40th Iteration for Runs				
			1	*2*	*1*	*2*	*3*	*4*	*5*
1	Sulfur dioxide	tons/day	6.9	0	0.8	0	0	0.9	0
	Particulates	tons/day	22.1	0	1	0	0	1	0
	BOD	lbs./day	33,400	0	134	0	0	117	0
	Nitrogen	lbs./day	2,790	0	78	0	0	68	0
	Heat	10^6 BTU/day	2,380	0	0	0	0	0	0
	Total solids	tons/day	6.4	0	25.4	0	0	22.4	0
2	Sulfur dioxide	tons/day	6.9	0	1.2	0	0.7	1	0.1
	Particulates	tons/day	22	0	2.3	0	0.7	2.2	0.1
	BOD	lbs./day	33,400	0	327	0	95	198	12
	Nitrogen	lbs./day	2,790	0	180	0	56	116	7
	Heat	10^6 BTU/day	2,380	0	0	0	0	0	0
	Total solids	tons/day	6.4	0	47.2	0	18.2	37.1	2.3
3	Sulfur dioxide	tons/day	54.4	0	12.4	17.6	16.2	14.3	14.3
	Particulates	tons/day	111.7	0	23.2	22.5	21.3	23	23
	Heat	10^6 BTU/day	32,400	0	5,400	21,400	19,500	1,700	1,900
	Solids	tons/day	50.7	0	66.3	64.4	64.9	65.6	65.6
	Slurry	tons/day	0	0	173.4	152.2	157.8	165.8	165.8
4	Sulfur dioxide	tons/day	104	7.6	58.6	104	81.7	66.1	63.5
	Particulates	tons/day	309.8	9.7	140.5	275.9	226.7	168.4	184.6
	Heat	10^6 BTU/day	61,900	0	15,700	42,100	37,100	14,900	11,000
	Solids	tons/day	97	70.2	113.8	127.4	108.2	111.2	112
	Slurry	tons/day	0	285.2	187.6	0	92.2	156.6	167.2

5	Sulfur dioxide	tons/day	148.2	0	33.9	35.9	33.4	34.2	34.4
	Particulates	tons/day	13.6	0	2.4	3.5	3.5	2.4	3.2
	BOD	lbs./day	12,500	0	4,500	6,500	6,300	4,500	6,200
	Nitrogen	lbs./day	568	0	290	369	365	291	361
	Phenols	lbs./day	2,023	0	626	963	943	632	916
	Heat	10^6 BTU/day	53,200	0	740	28,400	28,000	500	0
6	Sulfur dioxide	tons/day	157	10.5	88.2	157	157	101	96
	Particulates	tons/day	14.4	1.3	7.7	14.4	14.4	9.3	12.4
	BOD	lbs./day	13,300	4,700	9,800	13,300	13,300	10,600	12,600
	Nitrogen	lbs./day	601	327	451	601	601	460	571
	Phenols	lbs./day	2,150	670	1,540	2,150	2,150	1,680	2,030
	Heat	10^6 BTU/day	56,500	0	27,200	52,700	44,900	31,700	52,900
7	Particulates	tons/day	49.4	2.1	25.4	21.3	23.2	28.6	27.3
	BOD	lbs./day	272,500	19,500	33,900	27,200	31,200	34,000	32,300
	Nitrogen	lbs./day	32,400	11,400	14,200	13,400	15,600	14,200	13,900
	Solids	tons/day	82.4	108.8	93.4	134.4	148.9	97.7	107.6
8	Particulates	tons/day	19.1	1.1	10.8	21.8	19.2	12.4	19.1
	BOD	lbs./day	105,000	7,500	27,500	72,200	58,200	31,800	105,000
	Nitrogen	lbs./day	12,500	4,400	8,300	10,800	10,200	8,900	12,500
	Solids	tons/day	31.8	55.9	58.4	36.4	42.1	54.9	31.8
9	Particulates	tons/day	9.9	0	4.9	7.4	6.6	5.7	9.5
	Solids	tons/day	123.9	0	131.9	130.8	132.8	131.3	126.8
10	Particulates	tons/day	5.9	0	2.9	4.8	4.5	3.4	5.7
	Solids	tons/day	74.5	0	77.5	75.6	76.0	77.1	74.8

TABLE 8

Violation of Ambient Standards on the 40th Iteration for the Five Computer Runs

Run	*Number of Violations*	*Indicator*	*Units*	*Location (Grid/Reach)*	*Violation, Actual*	*Violation, Per Cent of Standard*
1	4	Sulfur dioxide	$\mu g/m^3$	59	1.4	0.33
		Fish	mg/L	15	0.01	3.45
		Dissolved oxygen	mg/L	17	0.034	1.13
		Dissolved oxygen	mg/L	15	0.009	0.3
2	5	Sulfur dioxide	$\mu g/m^3$	59	5.1	1.21
		Sulfur dioxide	$\mu g/m^3$	20	0.5	0.12
		Fish	mg/L	15	0.012	4.14
		Dissolved oxygen	mg/L	15	0.13	4.33
		Dissolved oxygen	mg/L	17	0.053	1.77
3	5	Sulfur dioxide	$\mu g/m^3$	59	0.5	0.12
		Sulfur dioxide	$\mu g/m^3$	20	0.1	0.02
		Suspended particulates	$\mu g/m^3$	59	0.1	0.1
		Dissolved oxygen	mg/L	17	0.176	5.87
		Dissolved oxygen	mg/L	15	0.131	4.37
4	4	Sulfur dioxide	$\mu g/m^3$	59	1.9	0.45
		Suspended particulates	$\mu g/m^3$	58	0.2	0.2
		Fish	mg/L	15	0.001	0.34
		Dissolved oxygen	mg/L	17	0.022	0.73
5	6	Sulfur dioxide	$\mu g/m^3$	59	1.3	0.31
		Suspended particulates	$\mu g/m^3$	59	0.9	0.9
		Suspended particulates	$\mu g/m^3$	58	0.3	0.3
		Fish	mg/L	15	0.001	0.34
		Dissolved oxygen	mg/L	15	0.014	0.47
		Dissolved oxygen	mg/L	17	0.014	0.47

Figure 6

Objective Function Value vs. Iteration, Run 1

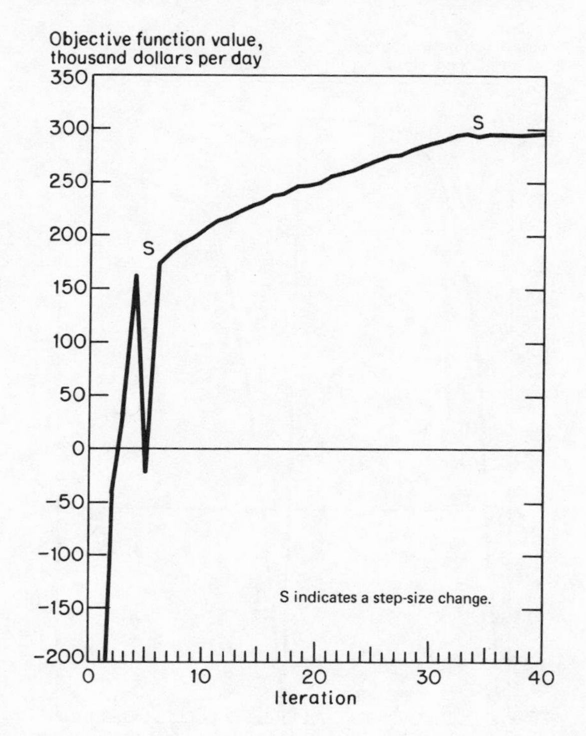

where step-size changes occurred (indicated by an "S"), as well as those where penalty function parameter sets were changed (indicated by a "P").

Discussion of the Results. The iterative optimization scheme which we are using for the Delaware Valley Base Model can find, or at least come very close to finding, a local optimum. It may take a long time, but it will eventually get there. The question we are exploring here is one of computational efficiency—which step-size scheme in combination with which sets of penalty function parameters permits climbing to the optimum the fastest. We attempt to answer this question with 5 runs of the Base Model. Caution should be exercised in comparing the results of the 5 computer runs, however, because each of the runs meets a different mix of levels of environmental quality (see table 8).

The most significant conclusions to be derived from the results of the runs follow.

Figure 7

Objective Function Value vs. Iteration, Run 2

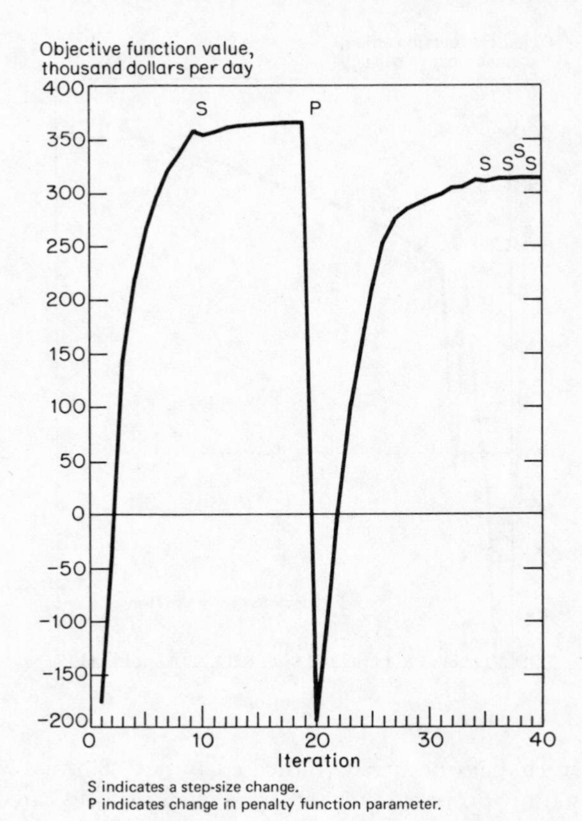

S indicates a step-size change.
P indicates change in penalty function parameter.

1. Technically, none of the runs have reached an optimum by the 40th iteration, although evidence suggests that Runs 2, 4, and 5 all may be quite close. Plots of typical residuals discharges throughout the iterative optimization procedure (not shown here) indicate that Runs 1 and 3 are quite far from the optimum. We might also point out that of the 5 runs, Runs 1 and 3 obtained, by the 40th iteration, the lowest objective function values (see table 6).

2. From the 5 runs presented, it is not possible to ascertain whether they are approaching the same optimum or different optima. The question of local vs. global optimum cannot be answered, but we suspect that we are dealing with a multi-peaked response surface.

Figure 8

Objective Function Value vs. Iteration, Run 3

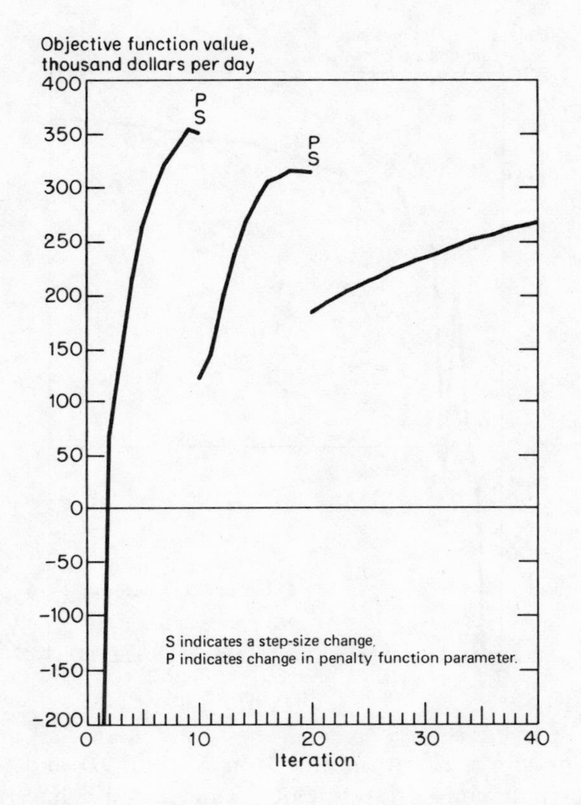

3. Of the 5 runs presented, Run 5 not only appears to be the most efficient in terms of its hill-climbing ability (see figure 10), but also it appears to have achieved the best solution by the 40th iteration. This conclusion is based on the fact that: on aggregate, Run 5 did one of the best jobs of meeting the environmental quality standards (see table 8); its objective function value is the highest of the five; its penalties are the lowest of the 5; and finally, its net benefits are the second highest of the 5 runs (see table 6).

In addition, there is evidence, albeit weak, to suggest that neither Runs 2 nor 4 are as close to the optimum as Run 5. This is shown in the liquid discharge data (BOD and nitrogen) for source 8, and the sewage treatment plant for Wilmington (see table 7). There are no standards violated

Figure 9

Objective Function Value vs. Iteration, Run 4

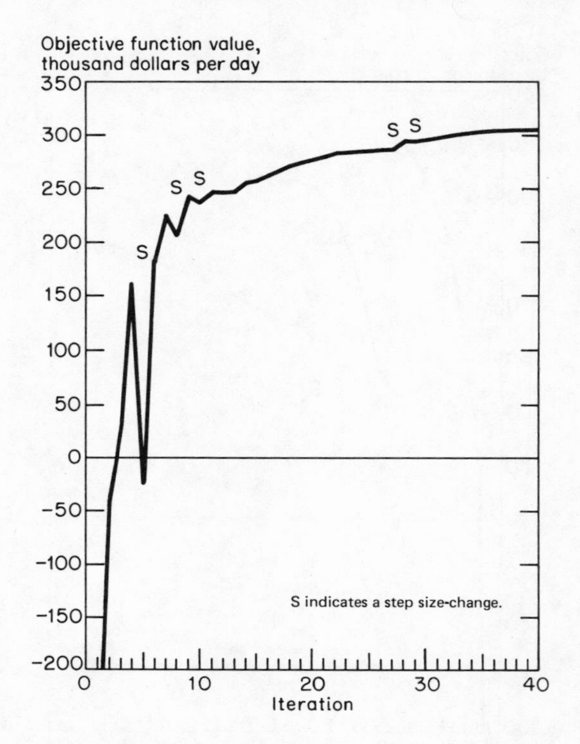

downstream of source 8, yet only in Run 5 are BOD and nitrogen discharges at their maximum levels. Both Runs 2 and 4 have quite a way to go before reaching this upper limit.

4. The penalty function scheme described in this paper was quite successful in meeting environmental quality standards (see table 8). A comparison of the sulfur dioxide violations for Run 3 (see table 8) with those of Runs 1 and 2 indicate the effect of increasing the penalty function parameter. Recall that the penalty function parameter for Run 3, on the 40th iteration, is ten times that of either Runs 1 or 2.

5. The third step-size scheme with no change in penalty function parameter, Run 5, appears to be superior to the other schemes which were tested. However, the step-size choice continues to remain the weakest part of our optimization technique and there is much room for improvement.

Figure 10

Objective Function Value vs. Iteration, Run 5

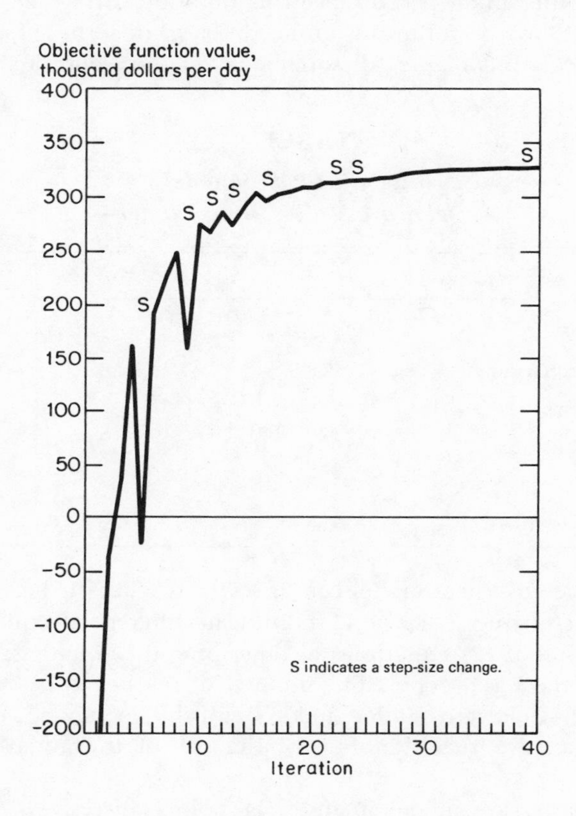

6. Examination of residuals discharge data throughout the iterative optimization scheme (not shown here) suggests that the response surface in the vicinity of the optimum is relatively flat, at least inside the feasible region, and that substantial tradeoffs among different residuals and different dischargers, in many cases, are taking place. Examination of the net benefit function values on the 40th iteration (see table 6) together with the rather significant variations in discharge data for this iteration (see table 7) indicates the same thing. That is, there are wide ranges of discharges among the 5 runs, yet the net benefits vary only a few per cent (maximum of between nine and ten per cent for Runs 2, 4, and 5). This difference is even more pronounced when net benefits and discharge data for Runs 2 and 5 are compared.

Costs of a Computer Run. What can we say about costs, and the prospects for running a much larger version of this model within a reasonable computer budget? To place this question into some perspective, we summarize, in the following table, the size, in terms of the number of rows and columns of the LP submodels, of the Delaware Valley Base Model.

TABLE 9

Size of LP Submodel
—Delaware Valley Base Model—

	MPSX–1	*MPSX–2*	*Total*
Variables (columns)			
Total	320	320	640
Residuals discharges	68	68	136
Constraints (rows)			
Total	178	178	356
Equality	135	135	270
Inequality	43	43	86
Step-size bounds (rows)	136	136	272

The aquatic ecosystem model comprises 11 reaches of the estuary, with each reach requiring a set of 11 nonlinear differential equations. This results in a total of 121 equations for describing the aquatic system within the limits of the Base Model. The number of reaches to be considered in the full-scale Delaware Valley model will be expanded to 22. This amounts to an eventual doubling of the size of the aquatic ecosystem model.

The air dispersion model includes 11 point sources, 44 "aggregated" point sources, and 99 area sources for a total of 154 sources. Ambient concentrations are computed for 99 receptor locations. In order to predict the ambient concentrations of sulfur dioxide and suspended particulates, one 154 by 99 element matrix of transfer coefficients is required for each residual. It is difficult to estimate at this time how many sources will eventually be considered in the full-scale Delaware Valley model, but the receptor locations probably will not be increased by more than 50 per cent.

The cost per iteration of the Base Model is about \$8.70 [49] for the cur-

50. This is based on a CPU running time of 0.42 minutes per iteration on an IBM 360–65 with a core requirement of 300 K.

rent computer program. A large fraction of this cost is for internal input-output time (65 per cent); a smaller fraction for central processing unit (CPU) time (26 per cent); and the smallest part for computer printout (9 per cent). It goes without saying that there are large savings possible if the number of iterations required to reach, or at least come close to, an optimum can be reduced through more efficient means of selecting step-sizes and sequences of penalty function parameter sets. It is our plan to continue to search for more efficient combinations. But the cost information above also implies that there are potentially large savings from reducing the quantity of information which must be stored and transferred in and out of core with the various submodels each iteration.

One possibility for saving internal input-output time would be to include the linear air dispersion model within the constraint set of the residuals generation and discharge LP submodel. This would involve adding 198 rows (99 for ambient concentrations of sulfur dioxide, and another 99 for suspended particulates) to the Base Model, but could reduce the internal I–0 time by, perhaps, ten per cent. In addition, it would probably improve the efficiency of the iterative optimization scheme since in this case penalty functions would be needed only for standards on algae, fish, and, dissolved oxygen. However, a major drawback of this approach is that individual (unlinked) LP submodels, in general, would not be possible with this scheme and some form of decomposition would be necessary to deal with model size.

We are currently searching for other ways to reduce internal input-output time, even at the expense of increasing CPU time, and plan to investigate some of them with the Delaware Valley Base Model.

As for cost increases with model size, the problem appears serious, but certainly not serious enough to abandon the project at this point. It appears to us that, currently, most of the CPU time is used for sorting both discharges and marginal penalties, for controling the program flow, and for the solution (including penalties and marginal penalties) of the environmental submodels. CPU costs, however, should not increase in proportion to the increase in the number of discharges because many of the computations necessary to set marginal penalties at each iteration are already being done in the Base Model.

On the other hand, the input-output time required for the transfer of residuals discharges, marginal penalties, and the changing upper and lower bounds on each discharge will increase linearly with the number of discharges. This is a problem, and one which lends little hope for improvement as long as nonlinear simulation models of the natural world are included within the optimization framework. But as noted above, this time requirement for the Delaware Valley model could be reduced substantially as long as we remain with linear atmospheric dispersion models and include them as part of the constraint set of the LP submodel.

The costs of solving the MPSX packages themselves appear to be a relatively small part of the problem, at least at this time, but it also appears that it will be more efficient to have a few large MPSX LP's rather than many smaller ones. The primary reason for this is that each MPSX module has certain minimum internal input-output requirements independent of its size.

On balance, we are hopeful that the model size can be increased considerably with a substantially less than proportional increase in the cost of a computer run.

Summary and Conclusions

In the Introduction, we presented, very briefly, our regional residuals management framework and indicated that two didactic applications to hypothetical regions had previously been made and are reported elsewhere. In the first application, both demand functions and economic damage functions were assumed, and the institutional framework envisioned was a regional management authority. In the second application, the model was expanded to provide information on the socio-geographic distribution of costs and benefits associated with meeting *different levels* of environmental quality. The institutional framework envisioned for selecting levels of environmental quality was a legislative body. We further indicated in the Introduction that the main purpose of this paper is to explore the computational problems associated with scaling up from small didactic models to a large scale regional application.

We have discussed, in some detail, the problem of model size and ways to cope with it, and also presented a way to include nonlinear models of the natural world within an optimization framework. As a means for testing our ideas and techniques for dealing with some of the computational problems presented we constructed a relatively simple residuals management model of the Delaware Valley region. This application is referred to by us as the Delaware Valley Base Model. Results of five computer runs indicate that our optimization scheme, which allows for the inclusion of nonlinear models of the natural world, is operationally feasible, but that more work is necessary to improve the computational efficiency of the procedure so that computer costs can be reduced.

In conclusion, there is no question that the management model we are experimenting with is more complex than previous residuals management models, and that it requires a substantial amount of computer resources. Three questions we face and hope to answer as a result of our research efforts on large scale regional residuals management models are as follows.

1. Is it necessary to include all forms of residuals within a single computational framework? In principle, it is necessary, but in practice, does the additional simultaneous information on air and water quality, and on the generation of solid residuals, warrant the added effort and expense?

2. Nontreatment management options such as changes in production processes, raw material input mixes, and product specifications, and by-product production and recycling, have all been shown (under some circumstances) to be less costly alternatives to reducing, or modifying, residuals than waste treatment alternatives, but can we consider even a portion of these alternatives for a large complex region consisting of many different types of economic activities without (a) exceeding the computational reliability of present day computer hardware and software; and (b) exceeding the computer budget of a waste management agency given that the first condition could even be satisfied?

3. Models of aquatic ecosystems are able to provide additional useful information for making public policy on environmental resources, but (a) can we incorporate these models within an optimizing framework without completely expending the computer budget of a regional residuals management agency; and (b) is it necessary to go to nonlinear ecosystem models given all the other uncertainties, in both data and model components, in the other parts of the regional residuals management model?

Our research to date indicates that the inclusion of nonlinear models of the natural world within an optimization framework is expensive, but not unreasonably so. Whether it is necessary to include these nonlinear models, or whether it would suffice to employ linear models of the natural world, we cannot say at this time. We are currently in the process of exploring this question.

Regarding the second question, given that the intent of our regional residuals management modeling effort is to be able to generate distributional information on costs, benefits, and environmental quality for a wide range of alternative management strategies for meeting ambient environmental quality standards, a priori elimination of management options, in many cases, would be a difficult, and at best arbitrary, task. Our research thus far has shown that nontreatment alternatives are frequently less costly than the more traditional abatement alternatives, but even more important to us, it has shown that, in most cases, a priori selection of alternatives for least-cost solutions is not possible because of all the links—both market and nonmarket—which exist for any complex situation. In order to be able to provide as many management options in the regional model as seems desirable for the Delaware Valley region, we plan to continue to search for ways of coping with model size.

Finally, as far as the first question is concerned, none of us will know the answer to this until someone or some group tries an integrated approach to residuals management modeling and compares its output with that of other kinds of residuals management models.

COMMENT

J. Hayden Boyd, The Ohio State University

The paper by Spofford, Russell and Kelly is a progress report on the Delaware Base Model, an ongoing, large scale modeling effort at Resources for the Future. An earlier paper by Russell and Spofford also reported on the progress of the model, and gives much useful background on its conception and planning.[1] Two sorts of questions naturally arise in examining a paper such as this. First, what are the technical goals of the modeling effort which it describes, and how well have these technical goals been achieved? Second, what are the nontechnical goals of the model? That is, how do the authors plan to use the model to increase the value of our environment, as we consume its waste disposal and environmental quality services?

The authors wish to advance the state of the art in several ways. First, several waterborne, airborne, and solid wastes, and several aspects of water and air quality, are considered simultaneously. For waterborne wastes and water quality, the model contains 11 endogenous variables, including 5 residuals (N, Ph, BOD, phenols, heat), and 3 quality parameters (algae, fish, DO). For air, sulfur dioxide and particulates are modeled. Solid wastes such as furnace bottom ash, digested sludge from municipal sewage plants, wet scrubber slurries, and municipal solid wastes are also included. Second, the environment is modeled using a nonlinear

1. Clifford S. Russell and Walter O. Spofford, Jr., "A Quantitative Framework for Residuals Management Decisions," *Environmental Quality Analysis,* Allen V. Kneese and Blair T. Bower (Baltimore: The Johns Hopkins Press, 1972).

aquatic model and a linear atmospheric model. The linear atmospheric model is simple and straightforward, while the aquatic model is quite complex and expensive to compute. The authors claim that "nonlinear representations of the natural world increase the complexity and number of calculations ncessary for each iteration, but also they increase both the realism and predictive capability of the model." They present no evidence to support this conclusion.

Third, and significantly, the authors are modeling production modifications in ten waste dischargers. They are allowing the amount of waste generated to be modified as a pollution control strategy, and they are also looking at trade-offs among wastes. For example, sewage treatment produces sludge which when burned produces particulates which can be precipitated and carried to a landfill.

This is an impressive and ambitious list of technical features, and represents an advance in the state of the art. The talents of competent systems analysts and the capacities of a large digital computer are challenged by an effort of this scale. It is important, however, to keep perspective by considering what the authors were not able to do. They did not consider alternatives such as regional treatment plants or bypass piping to move discharge points elsewhere in the estuary. These kinds of alternatives typically reduce the cost of achieving a given level of water quality. They did not consider expansion and contraction of the scale of current waste generators, not to mention the options of entry and exit from the region. Stochastic elements or cyclical variations in waste generation or in environmental reaction were not considered. Environmental modifications (for example, low flow augmentation) are not studied in the present paper, although the earlier paper by Russell and Spofford indicated that at one time there were plans to include environmental modifications. Explicit demands for environmental quality (the so-called "damage functions"), also included in the earlier plans, have apparently been dropped.

How well have the limited, albeit ambitious technical goals been achieved? As the authors are the first to acknowledge, the results are not yet in. The model is complex enough that analytical solution is impossible, and gradient methods must be used. There is little theory about which hill climber is fastest or how to speed up convergence. I have no doubt, however, that the authors will succeed in solving these knotty technical problems.

In its present form, the model is far short of fulfilling the authors' technical goals. Only ten waste dischargers are modeled explicitly. "Background" residuals seem to dominate environmental quality. Shutting down the ten modeled sources would not meet federal standards, but adding more sources explicitly will only add to computational difficulties.

There were also some expositional difficulties. I had difficulty following the description of the industry submodels. Several alternative techniques for shoehorning the industry models into the overall model were described, but it was not indicated which one was actually being used. The inference is that the outputs of plants are allowed to vary as a pollution control strategy, since imports seem to be a substitute for the plants' outputs in some cases. But, I was not able to find an explicit statement about how plant outputs do, in fact, vary as part of the overall strategy of pollution abatement. It is not clear exactly how the choice between low sulfur fuel and high sulfur fuel for household heating is being handled. Finally, the explanation of why penalty payments on discharges ought not to appear in the overall objective function was hard to follow. The reason is of course that while a payment for this service is a cost to the firm, it is not a cost to society, as it represents a rent on the scarce environmental resource.

The particular gradient method which the authors have selected deserves further comment. The "penalty functions" may appear to be surrogate prices on effluents, but they are not. The marginal penalties go to zero as the environmental quality constraints are met. In fact, it is this very feature which leads to computational difficulties, because the algorithm gets progressively "lazier" in the vicinity of the feasible region. The $1/r$ parameters are designed to give an extra push as the feasible region is approached. The behavior of the algorithm reflects the implicit economic assumption made by this and other such programming models: that the demand for the constraint is zero elastic. In other words, improvements in environmental quality beyond the standard are valueless, while violations of the standard cost infinite consumer surplus, so that the standards will be met (if at all possible), whatever the cost. The penalty or exterior point method of programming temporarily relaxes the constraint as a computational expedient, but the end result is a maximum which lies in the "feasible" region.

Environmental quality constraints are always somewhat arbitrary, particularly here where background residuals are so high that the federal standards can't be met in any case. It would be better to retain in the model an explicit demand function for environmental quality, even if it has to be assumed at this stage of our knowledge. The aquatic and atmospheric models could then compute the marginal opportunity cost of the pollutant disposal services of the environment and pass them to the industry submodels as effluent charges. These effluent charges would not go to zero as the feasible region (satisfying a set of arbitrary environmental quality constraints) is approached. An explicit quality demand function may help to speed up convergence of the model. Even if not, it would certainly make easier the economic interpretation of the model's solutions.

The overall goal of the modeling effort is to help rationalize the management of the environment. Neither in this paper, nor in the earlier paper by Russell and Spofford, is the institutional framework within which this model is to be used discussed in detail. But the authors do envision a regional management authority with power to set effluent charges or standards. In other words, the authority has ownership rights in the environment, and allocates its capacity among various outputs, including the absorption of various kinds of wastes and the provision of various aspects of environmental quality. The model is to be a staff tool for such an authority.

Regional environmental management models are not new, but as yet they have been little used to guide the actual allocation of the environment's scarce services. Any attempt by a regional authority to ration the environment's waste absorption services affects the wealth of dischargers, giving incentives to combine with others to use the political mechanism to influence the effluent charges or standards. If the regional management authority uses effluent charges, raising charges to reduce effluents results in consumer surplus losses to the dischargers. If discharge standards are used, pollution reductions cause changes in both consumer surplus and the implicit expenditure rectangle associated with the consumption of the environment's waste disposal service. If a systems analysis technique such as the Delaware Base Model is used to get a "least cost" solution, there may be severe transfers of relative wealth among dischargers, as some are cut back more than others. Next year's model, which considers a different set of alternatives, may lead to quite a different distribution of relative wealth. The political system seems to resist procedures which lead to capricious wealth transfers. In the absence of institutions for side payments, it deems to serve up uniform treatment standards (e.g., secondary treatment for all, best available technology), even when lower (total) cost solutions have already been documented.

According to the authors, the purpose of this paper is to answer the question, "Have we developed a mildly interesting academic curiosity or a potentially useful management tool?" One infers that the authors believe that if the computer program should converge fairly rapidly then the Delaware Base Model is a useful management tool. Yet, in the absence of innovative institutions to cope with the wealth transfer problem, it seems doubtful that this model will actually lower the social cost of a clean environment.

One object of this modeling effort is to "find a 'best' set of policy instruments (charges and limits) for imposition by . . . the authority." [2] The authors wish to shed light on the choice among alternative allocating institutions: standards (allocation by fiat) vs. use of the price mechanism; effluent charges vs. discharge rights (perhaps auctioned, perhaps transferable among polluters).

I wonder if a super systems analysis, modeling simultaneously as many relevant aspects of the universe as can be put into the computer, is really the vehicle to answer such questions. Disaggregated, piecemeal research would seem to be indicated, rather than a large scale computer model, to inform the choice among alternative allocating institutions. We need evidence on information requirements and administrative costs. What about the ability of alternative institutions to cope with stochastic and cyclical elements in waste generation and in the environment? What about the effects of the choice of institutions on the growth and decay of various industries? What about the ability of institutions to cope with the entry and exit problem?

The Delaware Base Model surely has social utility even if, because of the expense of its computation, it turns out to be a "mildly interesting academic curiosity." But it would seem that much information beyond that capable of being generated by such large scale models is required to aid in our search for a better environment at least social cost.

2. Russell and Spofford, "A Quantitative Framework," p. 119.

Input-Output Analysis and Air Pollution Control

Robert E. Kohn, Southern Illinois University, Edwardsville

Introduction

Some significant research on economics of the environment is taking place in the field of input-output analysis. Leontief has published an important article relating pollution abatement to the economic structure.[1] Isard and his associates are expanding the input-output scope to include ecologic commodities and environmental processes.[2] Miernyk and others have presented papers on the impact of pollution and pollution control on regional economies.[3]

The present paper is, essentially, an empirical sequel of Leontief's. In that article, we are presented with the hypothetical model of a region, in which 30 grams of air pollution is a maximum allowable flow. Because 60 grams are being generated by economic activities, an antipollution

Note: This research was supported by National Science Foundation Grant No. GS-2892. The writer is grateful to Ben-chieh Liu, William H. Miernyk, Edwin S. Mills, and Hugh O. Nourse for a critical reading of an earlier version of this paper.

1. Wassily Leontief, "Environmental Repercussions and the Economic Structure: An Input-Output Approach," *The Review of Economics and Statistics* (August 1970): 262–271.

2. Walter Isard, C. Choguill, J. Kissin, et al., *Ecologic and Economic Analysis for Regional Planning* (New York: The Free Press, 1971).

3. William H. Miernyk, "Environmental Management and Regional Economic Development," *Annual Meetings of the Southern Economic Association and the Southern Regional Science Association,* Miami Beach, Florida, November 6, 1971. J. R. Norsworthy and Azriel A. Teller, "Estimation of the Regional Interindustry Effects of Pollution Control," *Winter Meetings of the Econometric Society,* New Orleans, Louisiana, December 26–29, 1971. David L. Raphael and Ernest E. Enscore, Jr., "The Direct and Indirect Impact of Regional Air Pollution," *Annual Meeting of the Air Pollution Control Association,* Cleveland, Ohio, June 1967.

sector is created to eliminate the excess emissions. The cost of abatement is $3.00 per gram; thus it might appear that the total cost of pollution control would be $90.00. This, however, is not the case. The inputs required by the antipollution sector generate increases in the production levels of other sectors which are themselves sources of pollution. When the various multiplier effects are taken into account, it is found that 33.93 grams of air pollution must be eliminated. The total cost of abatement is therefore $101.80.[4]

Let us define some terms suggested by Leontief's example. The cost of reducing air pollution to some predetermined allowable level, ignoring the economic impact of abatement itself, is Z. The cost of achieving the allowable levels, taking into account the additional flow of pollution resulting from abatement activities, is Z'. For purposes of analysis we then define *the abatement multiplier* as the ratio, Z'/Z. In Leontief's hypothetical model, the abatement multiplier is $101.80/$90.00, or 1.131.

The Leontief model assumes a single pollution control process or a fixed combination of processes. In reality, there are many processes for controlling pollution and it is unlikely that they are combined in fixed proportions. The extent to which the abatement multiplier necessitates an expanded level of control and alters the optimal mix of control methods will be investigated in the context of a specific model.

The writer has developed a linear programming model in which maximum allowable flows of five separate pollutants in the St. Louis airshed are specified.[5] A set of air pollution control method activity levels is selected which achieves the allowable flows at the least cost, Z. The abatement multiplier, Z'/Z, for the St. Louis airshed will be calculated, using this same model in conjunction with an input-output study of the St. Louis region.[6]

The value of Z' is determined by augmenting the cost-effectiveness model with appropriate feedbacks. The feedback steps are as follows (the symbols in parentheses are matrix products which will be defined subsequently):

(Hx), the value of economic inputs associated with any set of control method activity levels, x;

4. Wassily Leontief, "Environmental Repercussions," p. 268.

5. Robert E. Kohn, "A Linear Programming Model for Air Pollution Control in the St. Louis Airshed," Ph.D. dissertation, Washington University, 1969.

6. Ben-chieh Liu, *Interindustrial Structure of the St. Louis Region, 1967* (St. Louis: St. Louis Regional Industrial Development Corporation, 1968).

(GHx), the changes in regional production levels necessary to sustain any set of control method activity levels;

($FGHx$), the changes in the levels of polluting activities associated with changes in production levels;

(E^*FGHx), the changes in emission flows (assuming the base year level of control) corresponding to changes in the levels of polluting activities.

The efficient set of control methods, x_o, is now that set which eliminates excess pollution, including the incremental pollution associated (directly and indirectly) with pollution control itself, at the least cost, Z'.

It is presumed that a cost-effectiveness model can be a useful guide for regulatory agencies in selecting an efficient strategy for achieving a predetermined set of air quality goals. In this paper, we investigate the significance of the abatement multiplier for such policy making. This research should provide insight as to how individual cost models and input-output studies can be made more useful for environmental planning. It may also enable us to evaluate the feasibility of incorporating a pollution control sector in the structure of an input-output model.

The Linear Programming Model for Determining Z

In matrix notation, the linear programming model for air pollution control is:

$$\begin{aligned} \text{minimize} \quad & Z = cx; \\ \text{subject to} \quad & Ux = s; \\ & Ex \leq a, \\ & x \geq 0, \end{aligned} \tag{1}$$

where x is an $(N \times 1)$ vector of air pollution control method activity levels and c is a $(1 \times N)$ vector of unit control costs. The vector s is an $(M \times 1)$ vector of pollution source levels, such as the number of tons of coal burned per year in a particular power plant, and the $(M \times N)$ matrix U is a distributive matrix which equates control method activity levels to pollution source levels. The element u_{ij} is 1 when the jth control method is defined for the ith pollution source and zero otherwise. Thus, there are as many 1's in any row of the U matrix as there are control methods defined for the pollution source which corresponds to that row. For the sum of control method activity levels to equal the corresponding pollution source level, it is necessary that both are measured in the same units, that the control method set include M control methods representing the activity levels of the M existing types of control (or noncontrol), and that each control method be uniquely defined for a single pollution source. The element, e_{ij}, of the $(P \times N)$ matrix E is the quantity of pollutant i emitted per activity unit of control method j, and the $(P \times 1)$ vector a is the allowable annual emissions for the P pollutants in the airshed. A simple example which illustrates the model is presented in the appendix to this paper.[7]

Model 1 was implemented as follows. The vector s contains projected levels of 94 polluting sources for the St. Louis airshed in 1975. These levels were projected from observed growth rates or from estimates by industrial representatives. The control method vector x, which includes over 215 alternative control methods in addition to the 94 existing control methods, and the emissions matrix E were developed from available engineering data, with control method costs, c, based on 1968 prices. The allowable annual emission flows, a, were derived from official maximum allowable concentrations for five pollutants in the St. Louis airshed.[8] The annual cost of achieving the air quality goals in 1975 was found to be an estimated $35.3 million over and above the cost of abatement that would be expended given the existing, preregulatory level of control.[9]

The Linear Programming Model for Determining Z'

The increase in polluting activities associated with pollution abatement is Δs. The cost of eliminating excess pollution, including the incremental emissions associated with Δs, is the solution of:

7. The model is described with more detail in Robert E. Kohn, "Optimal Air Quality Standards," *Econometrica* Volume 39 (November 1971): 983–995.

8. The relationship of the annual average concentration of a pollutant to the total annual emissions of that pollutant in the airshed is based on a *proportional model* which is defined on page 15,490 of "Requirements for Preparation, Adoption and Submittal of Implementation Plans," *Federal Register* Volume 36, Number 158 (August 1971). An alternative approach, in which annual emissions are related to ambient air concentrations by means of Gaussian diffusion formulas is used in Robert E. Kohn, "Industrial Location and Air Pollution Abatement," *Journal of Regional Science* Volume 14, Number 1 (April 1974): 55–63.

9. It is the economic impact of incremental costs of abatement that are examined in this paper. For simplicity, the costs of existing control methods are taken as zero and the costs of alternative control methods for any pollution source are incremental costs over and above that of the existing control method. The *total* annual cost of abatement would be in excess of $45 million.

$$\begin{aligned} \text{minimize} \quad & Z' = cx; \\ \text{subject to} \quad & Ux = s + \Delta s; \\ & Ex \leq a, \\ & x \geq 0. \end{aligned} \tag{2}$$

The increase in polluting activities, Δs, is assumed to be a linear function of abatement,[10]

$$\Delta s = FGHx, \tag{3}$$

where F is an $(M \times L)$ matrix, whose element f_{ik} is the change in polluting level i per dollar change in sales of sector k; G is an $(L \times L)$ matrix of intersectoral multipliers, where g_{kK} is the increase in sales of sector k per dollar increase in final demand sales of sector K; and H is an $(L \times N)$ matrix whose element h_{Kj} is the value of inputs from sector K per activity unit of control method j. Substituting for Δs in model 2 and moving $FGHx$ to the left-hand side gives:

$$\begin{aligned} \text{minimize} \quad & Z' = cx; \\ \text{subject to} \quad & (U - FGH)x = s; \\ & Ex \leq a, \\ & x \geq 0. \end{aligned} \tag{4}$$

There may be some incompatibility here in combining an open input-output model, where all production sectors can expand simultaneously, and a cost-effectiveness model in which full employment is implicitly assumed. No attempt will be made here to formulate additional assumptions which might ensure the internal consistency of the model. Most likely such assumptions would allow for an inflow of resources into the airshed. Note that while the costs, cx, are the value of national resources allocated to abatement, it is only the impact on local economic activities, s, that is examined. The effect of pollution control in the St. Louis air-

10. Ayres and Kneese have called attention to the increase in water and soil polluting activities which may be a consequence of air pollution abatement. See R. U. Ayres and A. V. Kneese, "Production, Consumption, and Externalities," *American Economic Review* Volume 59 (June 1969): 282–297. This aspect of air pollution control was examined empirically in Robert E. Kohn, "Joint-Outputs of Land and Water Wastes in a Linear Programming Model for Air Pollution Control," *1970 Social Statistics Section, Proceeding of the American Statistical Association,* Washington, D.C., 1971, pp. 207–214. In the present paper it is the increase in *air* polluting activities as a consequence of the control of air pollution itself which is being examined.

shed on pollution levels and pollution control costs in *other* airsheds is ignored.

The procedures used to calculate the *H*, *G*, and *F* matrices are now described.

The H matrix

The *H* matrix was implemented to conform with the sector classification in Liu's *Interindustrial Structure of the St. Louis Region, 1967* (see footnote 6). Although Liu distinguished 23 sectors, it was found impractical to allocate each unit control cost among 23 component inputs. Accordingly, six direct input sectors were selected, resulting in an *H* matrix with 6 nonzero and 17 zero rows. The six sectors are listed in rows 1 through 6 of Table 1. The remaining inputs, which were assumed to have no intersectoral impact, are noted in rows 7 through 10. This table includes selected control methods from the model and the inputs associated with these control methods. The entries in rows 1 through 6 typify elements h_{Kj} of the *H* matrix and denote the requirements from sector *K* (rows 1 through 6) per unit of activity of control method *j* (columns 1 through 8).

The inputs which would be purchased from the chemical, petroleum, and rubber sector include, for example, the increased cost of low sulfur fuel oil (see column 2), dolomite for wet scrubbing stack gases (see column 5), gasoline and diesel fuel for refuse hauling and landfilling equipment (see column 7), etc.

The value of inputs from the machinery sector represents purchases of scrubbers, dust collectors, afterburners, etc. (for convenience all machinery is assigned to the nonelectric machinery sector), while the purchase of automotive control devices and refuse hauling vehicles are attributed to the transportation equipment sector. Although these capital expenditures may be made within the space of five to seven years, the 1975 sector purchases are assumed to be equal to the annual depreciation of the equipment. Since the latter is, in general, based on a longer equipment life than seven years, the purchases from these two sectors may be understated for the year 1975.[11]

A negative purchase from the mining sector represents the value of coal replaced by natural gas (see column 3) while a positive value is the incremental cost of low sulfur coal (see column 4).

The transportation, communication, and utilities sector includes the

11. The assumption that equipment expenditures in any year are equal to depreciation would be more appropriate for a steady state economy.

purchase of natural gas for air pollution control, both as a substitute for coal and a fuel for afterburners. In the case of the dolomite wet scrubbing control method (see column 5), purchases from this sector represent the value of scrubbing water and of electricity to power pumps and fans.

Purchases from the household sector are for labor to maintain and operate control equipment. A negative purchase from the household sector (see column 3) indicates a saving of labor associated with a particular control method.[12]

Miscellaneous unallocated inputs include such items as the equipment and facilities for maintaining control equipment, the nonlabor costs for disposing of nonrecyclable by-products such as dolomite waste (see column 5), the cost of outside soil for landfilling (see column 7), the value of recovered steam in the operation of a carbon monoxide waste heat boiler, etc. While such costs should be allocated to primary sectors, there are other unallocated costs which should not. For example, the saving in household labor when domestic furnaces are converted from coal to natural gas would have no traceable impact on regional economic activity although it is assigned a dollar value.

Although the values of recovered chemicals (e.g., elemental sulfur, ammonium nitrate fertilizer) are included as negative purchases from the chemical, petroleum, and rubber sector, this was not done in the case of sulfuric acid (see row 8) obtained as a result of controlling sulfur dioxide from power generation and lead smelting. It is assumed here that this output represents additional sales of sulfuric acid and has no impact on regional activity other than the inputs required for the operation of the recovery processes.

One of the sectors which should be included as a primary demand sector is local government. It is a limitation of this model that the costs of government regulation and enforcement have been omitted. These could effect the optimal control solution both by altering relative control method costs and through the pollution feedbacks from the expansion of the local government sector.

The opportunity cost of capital (see row 9) represents ten per cent of the total investment in control equipment from the machinery and transportation equipment sectors.[13] For example, the upgraded electrostatic

12. One control method which eliminates local labor is the transfer of a portion of power generating capacity to a mine mouth location outside of the region. So that the results of this paper may be as general as possible, this particular interregional transfer of labor is ignored here.

13. The sensitivity of the model to the opportunity cost of capital is examined in Robert E. Kohn, "Air Quality, the Cost of Capital, and the Multi-Product Production Function," *Southern Economic Journal* Volume 38 (October 1971): 156–160.

TABLE 1

Value of Inputs per Activity Unit for Selected Control Methods

(dollars)

Control Method	*Crankcase and Exhaust Controls for 1970 through 1975 Model Automobiles* (1)	*Maximum Sulfur Content of 1% for Residual Fuel Oil Burned by Industry* (2)	*Conversion of Traveling Grate Stokers with Cyclone Dust Collectors from Coal to Natural Gas by Industrial Firms* (3)	*Maximum Sulfur Content of 2% on Coal Burned by Industry in Traveling Grate Stokers with Electrostatic Precipitators* (4)	*Wet Scrubbing of Stack Gases at Meramec Power Plant with Dolomite* (5)	*Cat-Ox System to Convert Sulfur Dioxide at Portage des Sioux Power Plant to Sulfuric Acid* (6)	*Hauling of Waste to a Sanitary Landfill Instead of Burning It On-Site* (7)	*Upgraded Electrostatic Precipitators for Cement Plant with Current Collection Efficiency of 96.4%* (8)
Control Method Activity Unit →	*One thousand gallons of gasoline burned*	*One thousand gallons of residual fuel oil burned*	*One ton of coal burned*	*One ton of coal burned*	*One ton of coal burned*	*One ton of coal burned*	*One ton of waste disposed*	*One barrel of cement produced*
(1) Value of direct inputs from chemical, petroleum, and rubber sector	0	5.00	0	0	.22	0	.30	0
(2) Value of direct inputs from nonelectric machinery sector	0	0	0	0	.19	.58	0	.04
(3) Value of direct inputs from transportation equipment sector	9.23	0	0	0	0	0	1.50	0

(4) Value of direct inputs from mining sector	0	0	−5.27	2.50	0	0	0	0
(5) Value of direct inputs from transportation, communication, and utilities sector	0	0	11.27	0	.20	.36	0	*n**
(6) Value of direct inputs from household sector	3.70	0	−1.57	0	.15	.24	7.20	.01
(7) Miscellaneous unallocated inputs	.91	0	−.06	0	.05	.27	1.16	.01
(8) Sulfuric acid produced from sulfur dioxide	0	0	0	0	0	−1.34	0	0
(9) Opportunity cost of capital	4.62	0	0	0	.39	1.16	1.00	.05
(10) Scarcity premium for natural gas	0	0	3.80	0	0	0	0	0
Total Control Method Unit Cost	18.46	5.00	8.17	2.50	1.20	1.27	11.16	.11

* *n* less than one half cent.

precipitator for a specific category of cement plant (see column 8) requires an incremental capital investment of $.50 per barrel of cement produced. It is assumed here that capital investments in control equipment are additional expenditures in the region and do not replace other planned investments. Hence, there is no feedback on polluting levels in the airshed.[14]

Another economic cost which is assumed to have no current impact on regional activity is the scarcity premium for natural gas (see row 10). This represents the excess of the social value of natural gas used for pollution control over its regulated market value.[15] The costs in the final row of Table 1 are unit control costs from the vector *c*. Each is the sum of input values in that column.

The G matrix

The *G* matrix is taken directly from Liu's input-output study and contains intersectoral multipliers for the St. Louis region. A portion of this matrix is reproduced in Table 2. Each entry shows, per dollar of direct sales of inputs for pollution control by the sector at the top, the total dollar value of production directly and indirectly required from the sector on the left. These multipliers are based on a model in which the household and local government sectors are endogenous.

It is assumed here that the technology, trading patterns, and relative prices for 1967 are applicable to 1975, although Miernyk notes that an input-output structure with fixed coefficients should be projected no more than two or three years.[16]

The F matrix

The *X* vector in Leontief's model represents sector activity variables and his emission factors, a_{gi}, relate directly to sector levels.[17] However, it is

14. In their regional impact model, Norsworthy and Teller, "Regional Interindustry Effects of Pollution Control," p. 18, use the opportunity cost of capital as a proxy for purchases from the machinery and construction sectors. They treat depreciation as a cash flow to households and imports. In the present study, only the impact of actual production activities are considered.

15. The scarcity premium for natural gas is discussed in Robert E. Kohn, "Application of Linear Programming to a Controversy on Air Pollution Control," *Management Science* Volume 17 (June 1971): 609–621.

16. William H. Miernyk, *The Elements of Input-Output Analysis* (New York: Random House, 1965), p. 33.

17. Leontief, "Environmental Repercussions," p. 271.

convenient in air pollution control models to identify polluting activities, *s*, across conventional sector classifications. Many of the 94 separate polluting activities defined in this model occur in each of the industrial sectors (e.g., the combustion of coal in various types of stokers, evaporation of industrial solvents, combustion of diesel fuel in trucks, disposal of refuse by open burning, etc.). To define each of these pollution sources according to sector as well as type would multiply the control method set enormously. In order then to relate polluting levels, as defined in the present model, to input-output multipliers based on a conventional economic sector classification, a conversion matrix is required. The *F* matrix fills this need.

The coefficients of the *F* matrix indicate changes in polluting levels associated with changes in sector sales. They are illustrated in table 3 for selected pollution sources.[18] In the case of gasoline burned in motor vehicles (see row 1), it was assumed that the ratio of gasoline consumption in 1967 to the value of retail trade services in that same year, which was 643.52 thousand gallons per million dollars of sector activity, is a constant.[19] The production of cement, primary steel, grain handling, and sulfuric acid were assumed to be proportional to the sales of the respective standard industrial code classification sector, with the constant of proportionality based on 1967 levels.[20]

The combustion of coal in industrial furnaces was based on the sales of thirteen industry sectors. It was estimated, for example, that the food, tobacco, and kindred products sector used 96.3 tons of coal per million dollars of sales in 1967.[21] It was projected that in 1975, given the same level of pollution control as existed in 1963, which was the base-year in

18. For convenience, the coefficients in Table 3 are related to *millions* of dollars of sector activity.

19. Gasoline consumption in the St. Louis region in 1967 was an estimated 857,143 gallons (Kohn, "A Linear Programming Model," p. 553) and retail trade services in that year totaled $1,350,862,000 (Liu, *Interindustrial Structure,* Table IV–1). The reader is cautioned that the *F* matrix, like the G matrix, is based on the assumption that the technological relationships and relative prices that prevailed in 1967 are applicable for 1975.

20. Four categories of cement plants are included in the model. It is assumed that the older plants (see, for example, column 8 in table 1) are at capacity, so that any increases in cement production will take place in newer plants equipped with 99.4% efficient electrostatic precipitators. To this limited extent, the present study incorporates a normal advance in abatement technology.

21. Extrapolated from data in *1963 Census of Manufactures, Volume 1, Summary and Statistics* (Washington, D.C.: U.S. Department of Commerce, 1966), pp. 45, 7–92 and in *1967 Census of Manufactures, Missouri* (Washington, D.C.: U.S. Department of Commerce, 1970), pp. 26–14, 26–15, 26–16.

TABLE 2

Selected Elements of the Matrix of Intersectoral Multipliers for the St. Louis Region

	Chemicals, Petroleum, and Rubber Products (5)	*Machinery (except electrical)* (10)	*Transportation Equipment* (12)	*Mining* (15)	*Transportation, Communication, and Utilities* (17)	*Household* (22)
(1) Food, tobacco, and kindred products	.0243	.0513	.0216	.0544	.0525	.0892
(2) Textiles and apparel	.0047	.0101	.0043	.0108	.0102	.0173
(3) Lumber and furniture	.0011	.0026	.0010	.0024	.0024	.0037
(4) Paper and printing	.0135	.0158	.0079	.0183	.0207	.0233
(5) Chemicals, petroleum, and rubber products	1.0613	.0126	.0063	.0137	.0119	.0201
(6) Leather products	.0007	.0017	.0007	.0017	.0016	.0027
(7) Stone, clay, and glass	.0138	.0038	.0017	.0040	.0040	.0063
(8) Primary metals	.0019	.0346	.0166	.0071	.0046	.0022
(9) Fabricated metals	.0066	.0271	.0057	.0163	.0050	.0060
(10) Machinery (except electrical)	.0008	1.0255	.0062	.0067	.0018	.0010
(11) Electrical machinery	.0014	.0026	.0022	.0017	.0036	.0020
(12) Transportation equipment	.0048	.0122	1.0165	.0114	.0121	.0181
(13) Miscellaneous manufacturing	.0018	.0037	.0083	.0025	.0029	.0034
(14) Agriculture	.0013	.0028	.0012	.0030	.0029	.0049

(15) Mining	.0028	.0014	.0008	1.0262	.0029	.0014
(16) Construction	.0057	.0112	.0048	.0128	.0189	.0166
(17) Transportation, communication, and utilities	.0887	.1168	.0561	.2865	1.2178	.1676
(18) Wholesale trade services	.0118	.0306	.0094	.0213	.0219	.0234
(19) Retail trade service	.0571	.1419	.0561	.1332	.1309	.2082
(20) Finance, insurance, and real estate	.0680	.1429	.0613	.1807	.1619	.2245
(21) Business, personal, and other services	.0540	.1153	.0484	.1285	.1309	.1833
(22) Household	.3938	.8674	.3665	.9249	.8756	1.5313
(23) Local government	.0240	.0491	.0199	.0535	.0672	.0745

Source: B. C. Liu, *Interindustrial Structure of the St. Louis Region, 1967* (St. Louis: St. Louis Regional Development Corporation, 1968), Table V–1.

TABLE 3

Detail of the Matrix for Relating Increased Polluting Levels to an Increase in Sector Sales of One Million Dollars

	Food, Tobacco, and Kindred Products (1)	*Textiles and Apparel* (2)	*Lumber and Furniture* (3)	*Paper and Printing* (4)	*Chemicals, Petroleum, and Rubber Products* (5)	*Leather Products* (6)	*Stone, Clay, and Glass* (7)	*Primary Metals* (8)	*Fabricated Metals* (9)	*Machinery (except electrical)* (10)
(1) Thousands of gallons of gasoline burned in automobiles	0	0	0	0	0	0	0	0	0	0
(2) Tons of coal burned by industry in traveling grate stokers with cyclone controls	6.26	1.24	3.75	22.83	20.59	6.59	53.16	19.55	2.26	4.45
(3) Tons of coal burned by industry in pulverized coal units with 90% efficient electrostatic precipitators	37.56	7.42	22.38	136.35	123.02	39.34	317.57	116.79	13.48	26.55
(4) Tons of coal burned in the Meramec Power Plant	7.93	2.55	4.95	10.60	20.01	6.06	55.06	38.58	9.18	4.44
(5) Tons of coal burned in the Portage des Sioux Power Plant	32.59	10.48	20.33	43.71	82.23	24.91	226.31	158.54	37.71	18.26
(6) Tons of waste disposed of by on-site open burning	13.24	15.42	60.11	10.44	39.56	43.57	181.31	23.38	19.38	26.93
(7) Barrels of cement produced in cement plant with 99.4% efficient electrostatic precipitators	0	0	0	0	0	0	43210.	0	0	0
(8) Tons of primary steel produced in basic oxygen furnaces	0	0	0	0	0	0	0	1470.6	0	0
(9) Tons of grain handled and processed in elevators	1869.2	0	0	0	0	0	0	0	0	0
(10) Tons of sulfuric acid produced by the contact process	0	0	0	0	701.55	0	0	0	0	0
(11) Tons of dry cleaning solvents used	0	0	0	0	0	0	0	0	0	0

(*continued*)

the pollution model, approximately 6.5% of the industrial coal burned in the St. Louis region would be burned in traveling grate stokers with cyclone controls. Accordingly, for every million dollars in sales by the food, tobacco, and kindred products sector, (96.3) × (.065), or 6.26 tons of coal would be burned in this furnace category (see row 2, column 1). Be-

TABLE 3 (*concluded*)

Electrical Machinery (11)	*Transportation Equipment* (12)	*Miscellaneous Manufacturing* (13)	*Agriculture* (14)	*Mining* (15)	*Construction* (16)	*Transportation, Communication, and Utilities* (17)	*Wholesale Trade Services* (18)	*Retail Trade Services* (19)	*Finance, Insurance, and Real Estate* (20)	*Business, Personal, and Other Services* (21)	*Household* (22)	*Local Government* (23)
0	0	0	0	0	0	0	0	643.52	0	0	0	0
2.78	3.37	3.35	0	0	0	0	0	0	0	0	0	0
16.59	20.10	19.99	0	0	0	0	0	0	0	0	0	0
10.46	6.23	9.19	3.91	17.46	3.66	2600.[a]	8.03	9.22	13.42	16.06	12.12	7.18
42.98	25.60	37.77	16.08	71.74	15.03	10710.[a]	33.01	37.89	55.16	66.01	49.83	29.50
26.93	11.68	2.40	0	38.05	5.51	8.79	12.19	18.92	5.63	17.32	24.97	31.60
0	0	0	0	0	0	0	0	0	0	0	0	0
0	0	0	0	0	0	0	0	0	0	0	0	0
0	0	0	0	0	0	0	0	0	0	0	0	0
0	0	0	0	0	0	0	0	0	0	0	0	0
0	0	0	0	0	0	0	0	3.29	0	0	0	0

[a] Tons of coal burned per million dollars of electric power sold.

cause the combustion of industrial coal in pulverized coal furnaces equipped with 90% efficient electrostatic precipitators represents 39.0% of the projected coal combustion in 1975, the corresponding coefficient for this polluting activity is larger (row 3, column 1).

It could have been assumed that coal combustion in individual power plants is proportional to sales of the transportation, communications, and utilities sector. However, this would mean that a million dollars in sales of natural gas for air pollution control would have the same impact on power plants as would the sale of a million dollars of electricity. To avoid a serious distortion in the model, it was assumed that the sale of natural gas had no direct effect on power plant activity. This required a separate accounting of natural gas and electricity inputs for air pollution control. The coefficient relating coal combustion at the Meramec Power Plant to a million dollars of electricity sales (row 4, column 17) was obtained as follows. It was assumed that a million dollars in sales to industrial users represents 93,620,000 kilowatt hours of electricity.[22] The Meramec Power Plant, which burns approximately .000427 tons of coal per kilowatt hour produced will supply an estimated 6.5% of the area's power requirements in 1975. Accordingly, the appropriate element of the F matrix is (93,620,-000) × (.000427) × (.065), or approximately 2,600 tons of coal per million dollars of sales of electricity.

The sale of natural gas as well as electricity would affect power plant activity via the impact on other economic sectors. For example, it was estimated that 381,950 kilowatt hours of electricity are required by the paper and printing sector per million dollars of sales.[23] The f_{ik} coefficient of the Meramec Power Plant for this industrial sector is therefore (381,-950) × (.000427) × (.065), or 10.6 tons of coal per million dollars in sales by the paper and printing sector (see row 4, column 4).

It was estimated that for each million dollars of sales by the primary metals sector, there would be 96.2 tons of nonrecycled solid waste generated.[24] Assuming, in the absence of regulatory activity, that 24.3% of solid waste in the St. Louis region is disposed of by on-site open burning,[25] it was estimated that (96.2) × (.243), or 23.38 tons of waste would be burned on industry property per million dollars of metal sales (see row 6, column 8).

22. The average price of electricity sold to industrial users by Union Electric Co. in 1967 was $.010682 per kilowatt hour. See *1967 Annual Report* (St. Louis: Union Electric Co., 1967), p. 26.

23. A breakdown of power shipments to St. Louis industrial sectors was provided by Union Electric Co. and Illinois Power Co. The breakdown of power shipments to non-industrial sectors was estimated from data in Liu, *Interindustrial Structure,* table IV–1.

24. Waste production data are based on a study by the Institute of Industrial Research, University of Louisville, *Louisville, Ky.—Ind. Metropolitan Region Solid Waste Disposal Study, Volume I, Jefferson County, Kentucky* (Cincinnati, Ohio: U.S. Department of Health, Education and Welfare, 1970), pp. 6, 7, 25, 95, 96.

25. Relative polluting activity levels in the St. Louis airshed in 1975 are taken from Kohn, "A Linear Programming Model."

The effect of the F matrix is to relate directly the level of a polluting activity to the production level of either a single economic sector, or in the case of coal and refuse burning, to a linear combination of sector levels.

The FGH matrix product

The ijth element of the FGH matrix product represents the change in source level i per unit of control method j activity. This element has the form,

$$\sum_{k=1}^{L} \sum_{K=1}^{L} f_{ik} g_{kK} h_{Kj}.$$

For example, when one ton of waste, customarily burned on-site, is hauled away for landfill disposal (see table 1, column 7), the increased combustion of gasoline in automobiles is an estimated (643.52) (.0571 $\times$ .30 + .0561 $\times$ 1.50 + .2082 $\times$ 7.20)(10^{-6}), or approximately .001 thousand gallons (see table 2, row 19 and table 3, row 1).[26]

It should be noted that there are other feedbacks of abatement activity on polluting levels which are not included here. If the private costs of control are added to the selling prices of pollution-related intermediate and final goods and services, there may be substitutions which reduce the levels of polluting activities.[27] Furthermore, the abatement of pollution may result in technical efficiencies which reduce the level of polluting activities.[28]

26. The previously noted exception for the transportation, communication, and utilities sector can be formally stated. For elements, $\Sigma_k \Sigma_K f_{ik} g_{kK} h_{Kj}$, in which both subscripts k and K denote the transportation, communication, and utilities sector, the coefficient h_{Kj} is the value of the electricity input only. When subscript k refers to any of the other economic sectors, h_{Kj} represents the combined value of natural gas, water, and electric power inputs.

27. This type of feedback is programmed into the model in Robert E. Kohn, "Price Elasticities of Demand and Air Pollution Control," *Review of Economics and Statistics* Volume 54 (November 1972): 392–400.

28. Raphael and Enscore, "Impact of Regional Air Pollution," have modified an input-output model for Clinton County, Pennsylvania so that the technical coefficients describing the production structure are altered by air pollution levels. In addition, certain sector levels, e.g., the production of cleaning services are a function of pollutant concentrations. Because the technological and human effects of air pollution are not as well known as the costs of abatement, cost-effectiveness models for air pollution control are more satisfactorily implemented with empirical data than are benefit-cost models. However, when the costs of abatement are thoroughly investigated it becomes apparent that the damage effects of air pollution cannot be ignored, even in a cost effectiveness model.

Results: The Abatement Multiplier

The abatement multiplier is a measure of the increase in the cost of pollution control caused by input-output feedbacks of the control technology on the vector of polluting activities. It is defined as the ratio of Z' to Z. If we let x_o represent the optimal control method solution of the feedback model, (4), the minimized cost of pollution abatement, Z', is the product cx_o. This cost was found to be $36.1 million, which compares to a cost of $35.3 million for the original model (1) in which the feedbacks were ignored. The abatement multiplier is therefore equal to 1.023.

The factors which determine the size of the abatement multiplier will be investigated here. It will be useful to examine the following: x_o, the optimal set of control methods; Hx_o, the total value of direct inputs for abatement; GHx_o, the increase in economic sector production levels; $FGHx_o$, the changes in polluting source levels; and E^*FGHx_o, the additional emissions which must be eliminated.[29]

Optimal control method activity levels, x_o

Selected activity levels of the x_o vector are listed in table 4. These are compared to corresponding activity levels from the solution of the original model. The fact that control method activity levels do not increase in the same proportion demonstrates that there is a shift in abatement technology. Because the efficient control method set is altered as the level of control is increased, it is clear that the abatement multiplier as defined in this paper differs from conventional input-output multipliers, which are based on fixed technological relationships.

The major change in the control method solution is a 15 per cent overall increase in the quantity of natural gas required for pollution control. For traveling grate stokers equipped with cyclone collectors, there is a sharp increase in the substitution of natural gas for coal (see table 4). Another control method, representing the conversion of a different category of stoker from coal to natural gas, entered the basis in the second solution.

Direct inputs for pollution control, Hx_o

Of the $36.1 million in annual costs of pollution control, only $24.1 million are assumed to have an input-output feedback. This is the sum of the

29. The matrix E^* will be defined subsequently.

TABLE 4

Optimal Activity Levels of Selected Control Methods in the Original Model and the Feedback Model and Total Cost of Abatement

Control Method	*Control Activity Level in the Original Model*	*Control Activity Level in the Feedback Model*	*Ratio of Control Levels (feedback model to the original model)*
Exhaust control device for 1970 to 1975 model automobiles	545,638,000 gallons of gasoline combustion controlled	548,319,000 gallons of gasoline controlled	1.005
Crankcase evaporation control device for 1970 to 1975 model automobiles	25,885,000 gallons of gasoline combustion controlled	40,249,000 gallons of gasoline controlled	1.555
Upgraded electrostatic precipitators for industrial pulverized coal furnaces	444,000 tons of coal combustion controlled	444,615 tons of coal combustion controlled	1.001
Conversion of traveling grate stokers with cyclone controls from coal to natural gas	106,030 tons of coal combustion controlled	173,185 tons of coal combustion controlled	1.633
Desulfurization process for the Meramec Power Plant	730,000 tons of coal combustion controlled	740,330 tons of coal combustion controlled	1.014
Conversion of burning dumps to sanitary landfill	455,000 tons of refuse burning controlled	455,516 tons of refuse burning controlled	1.001
High energy wet scrubber for blast furnaces	1,417,500 tons of iron production controlled	1,417,846 tons of iron production controlled	1.000
Annual cost in 1975 for all air pollution control method activity levels combined	$35.3 million	$36.1 million	1.023

nonzero elements of the matrix product Hx_o, which are listed in the upper part of table 5. The larger the portion of Z' which reflects direct pur-

TABLE 5

Value of Direct Inputs for an Efficient Set of Air Pollution Control Methods in the St. Louis Airshed in 1975

(millions of dollars)

Inputs		*Value of Inputs*
Assigned to an Input-Output Sector (Hx_o)		
Chemicals, petroleum, and rubber products sector	.4	
Machinery sector	6.7	
Transportation equipment sector	8.7	
Mining sector	−13.0	
Transportation, communication, and utilities sector		
Electric power only	3.2	
Natural gas, etc.	13.3	
Household sector	4.8	
Subtotal		24.1
Not Assigned to an Input-Output Sector		
Miscellaneous unallocated inputs	1.6	
Credit for by-product sulfuric acid produced from sulfur dioxide	−11.9	
Opportunity cost of capital	19.4	
Scarcity premium for natural gas	2.9	
Subtotal		12.0
Total value of inputs		36.1

chases in the region the greater will be the abatement multiplier. In a more elaborate model all of the "miscellaneous unallocated inputs" would be assigned to appropriate input-output sectors and the abatement multiplier would be larger.

If the recovered by-product sulfuric acid were to diminish current production of sulfuric acid in the chemical sector, this would reduce the sum of elements of the product Hx_o, and the abatement multiplier accordingly. A sensitivity test with the model indicated that if the $11.9 million worth of recovered sulfuric acid were to replace an equal valued quantity of commercial sulfuric acid production in the airshed, the abatement multiplier would decline to 1.011.[30]

Increases in sector production, GHx_0

As a consequence of the direct demand for inputs, Hx_0, there are secondary or derived demands as well. The equilibrium set of increased activity levels, GHx_0, is presented in table 6. Note that it requires $54.1 million in increased production levels to supply the $24.1 million of direct inputs for abatement. The $19.0 million increase in demand for household services suggests that pollution abatement could create employment for 2,500 people.[31]

A sensitivity test was performed to determine the relative significance of the *derived* demand for inputs on the size of the abatement multiplier. When the G matrix was omitted (or, in effect, replaced by a $[23 \times 23]$ identity matrix), the abatement multiplier declined from 1.023 to 1.014. While the feedbacks associated with indirect inputs account for less than half of the abatement multiplier, it can still be observed that the larger the input-output multipliers, the larger will be the abatement multiplier.

Increases in polluting activity levels, $FGHx_0$

The increases in pollution source levels, assumed proportional to increases in corresponding economic sector activity levels, are represented by a

30. Recovered sulfuric acid is valued at one-half to one-third the value of commercially produced sulfuric acid depending on whether it is a by-product of power generation (and relatively pure) or of lead smelting. Assuming a fixed dollar demand for sulfuric acid, it would take two to three tons of recovered acid to replace one ton of commercial acid production. This feedback effect was implemented by treating by-product sulfuric acid as a negative input independently of the chemical, petroleum, and rubber products sector. The savings in indirect inputs associated with a dollar reduction in the projected level of commercial acid production are then included as negative direct inputs. These include a reduced demand by sulfuric acid producers for labor, machinery, water, power, and elemental sulfur.

31. This is based on the annual income per manufacturing employee in the St. Louis SMSA in 1967 (*1967 Census of Manufactures, Missouri*). This does not include any decreases in employment due to higher operating costs and prices. For a study of adverse impacts of abatement on employment, see Robert J. Kohn, "Labor Displacement and Air Pollution Control," *Operations Research* Volume 21 (September-October 1973): 1063–1070.

TABLE 6

Increased Indirect and Direct Economic Activities Associated With an Efficient Set of Air Pollution Control Methods in the St. Louis Airshed in 1975

(millions of dollars)

Economic Sector	*Increased Activity Levels* (GHx_0)
Food, tobacco, and kindred products	1.1
Textiles and apparel	.2
Lumber and furniture	.1
Paper and printing	.4
Chemicals, petroleum, and rubber products[a]	.6
Leather products	[a]
Stone, clay, and glass	.1
Primary metals	.4
Fabricated metals	.1
Nonelectric machinery	6.9
Electrical machinery	.1
Transportation equipment	9.1
Miscellaneous manufacturing	.1
Agriculture	.1
Mining	−13.3
Construction	.3
Transportation, communication, and utilities	18.5
Wholesale trade services	.5
Retail trade services	2.9
Finance, insurance, and real estate	3.0
Business, personal, and other services	2.6
Households	19.0
Local government	1.3
Total of all sectors	54.1

Note: The $11.9 million in sales of recovered sulfuric acid, a by-product of pollution control, are not included in this table.

[a] Less than $50,000.

(94×1) matrix product, $FGHx_0$. Selected elements of this matrix product are contained in table 7.[32] It will be observed here that the largest per-

32. Some of the values in table 7 can be checked by the reader. The increased combustion of gasoline in automobiles and light duty trucks (row 1) is the product of the f_{ik} coefficient, 643,520 gallons, in table 3 (row 1, column 19) and the equilibrium increase in the value of retail trade services in table 6, $2.9 million. (The discrepancy in results is due to rounding.) The increase in the combustion of coal in pulverized coal furnaces equipped with electrostatic precipitators (see row 3, table 7) is verified by multiplying

TABLE 7

Estimated Production Levels for Selected Pollution Sources and Increases in These Levels Associated with an Efficient Set of Air Pollution Control Methods in the St. Louis Airshed in 1975

Pollution Source	Estimated Production or Consumption Level for 1975	Increase Because of Pollution Control ($FGHx_0$)	Percentage Increase
(1) Combustion of gasoline in automobiles and light duty trucks	1,137,000,000 gallons of gasoline	1,841,000 gallons of gasoline	.2
(2) Diesel fuel used by railroads	40,800,000 gallons of fuel	385,000 gallons of fuel	.9
(3) Combustion of coal by industry in pulverized coal furnaces equipped with electrostatic precipitators	583,000 tons of coal	625 tons of coal	.1
(4) Combustion of coal in residential stokers	428,000 tons of coal	1,330 tons of coal	.3
(5) Combustion of coal at the Meramec Power Plant	730,000 tons of coal	10,330 tons of coal	1.4
(6) Combustion of coal at the Labadie Power Plant	5,500,000 tons of coal	77,840 tons of coal	1.4
(7) Refuse burned in municipal incinerators	357,000 tons of refuse	405 tons of refuse	.1
(8) Grain handled and processed in elevators	2,400,000 tons of grain	2,110 tons of grain	.1
(9) Crude oil processed in refineries	137,606,000 barrels of crude oil	50 barrels of crude oil	[n]
(10) Rock and gravel crushed, screened, conveyed, and handled	4,000,000 tons of rock	1,225 tons of rock	[n]

[n] Less than .05%.

---

the f_{ik} coefficients in row 3 of table 3 by the corresponding sector increases in table 6 and summing. To verify the increased coal combustion at the Meramec Power Plant, the corresponding f_{ik} coefficients in row 4 of table 3 and sector increases in table 6 are multiplied and summed. However, the coefficient in row 4, column 17 must be multiplied by the product of $3.17 million (the value of direct electrical inputs for pollution control in the feedback model) and the transportation, communication, and utilities sector self-multiplier, 1.2178 (see table 2). This special case is explained in footnote 26.

centage increases are for the power plants which supply the electricity needed for pollution control.

The percentage increases in pollution source levels are substantially less than the 2.3 per cent increase of Z' over Z. Essentially, this is because a portion of emissions associated with the original pollution source levels is allowable, whereas all emissions associated with the increased levels must be eliminated. However, the comparatively small percentage increases in table 7 help to explain why the abatement multiplier in the present study is as small as it is.

Additional emissions, E^*FGHx_0

The increase in air pollutants associated with the vector of increased pollution source levels, $FGHx_0$, is found by premultiplying the latter by a $(P \times M)$ matrix, E^*. The element, e_{ij}^*, of this matrix is the emission flow of pollutant i per activity unit of control method, j, where j is the existing or base year control state for pollution source j. The E^* matrix is contained in the E matrix and is used here for explanatory purposes only. These incremental emissions, elements of the (5×1) matrix product, E^*FGHx_0, are contained in table 8. It is not surprising that the largest percentage increases in the pollution reduction requirements are for nitrogen oxides and sulfur oxides, which are the major pollutants from the larger power plants in the St. Louis airshed. As noted earlier, the most significant impact of pollution control will be the increase in power generation.

The percentage increases in required emission reductions, which range from .3 to 2.1 per cent, are less than the 2.3 per cent increase in abatement costs (of Z' to Z). This is in contrast to the Leontief example, where the cost of pollution abatement increases by the same per cent as the increase in the quantity of pollution which must be eliminated. Because pollution control is represented by Leontief as a constant cost industry, the marginal cost of eliminating one gram of pollutant does not change. In the present model, the cost of abatement increases more than the pollution reduction requirements because of increasing costs. This would not have been the case if each of the nonzero control method activity levels had increased by the same proportion (see table 4). Because of the rising cost of pollution control, the abatement multiplier is larger than it would otherwise be.

Summary of factors which affect the size of the abatement multiplier

The abatement multiplier has been introduced as a device by which to measure the feedbacks of pollution abatement on the flow of emissions. It

TABLE 8

Projected Emissions in the St. Louis Airshed in 1975 in the Absence of Additional Abatement, Allowable Emissions, and Incremental Emissions from Pollution Control
(emissions in millions of pounds)

Pollutant (*1*)	*Projected Emissions in 1975 in the Absence of Additional Abatement* (*2*)	*Allowable Annual Emission Flows*[a] (*3*)	*Required Reductions in Emission in 1975* (*4*)	*Incremental Emissions Because of Abatement* (E^*FGHx_0) (*5*)	*Percentage Increase in Required Reductions* (*6*)
Carbon monoxide	4202.2	2335.2	1867.0	6.1	.3
Hydrocarbons	1518.8	994.5	524.3	2.3	.4
Nitrogen oxides	415.4	303.5	111.9	2.4	2.1
Sulfur dioxide	1389.6	400.4	989.2	11.3	1.1
Particulates	299.6	135.8	163.8	.9	.5

Note: Emissions from stacks higher than 600 feet are adjusted down to ground level equivalent emissions.

[a] The allowable flows are based on the following air quality goals (annual averages at the St. Louis Continuous Air Monitoring Program Station): carbon monoxide, 5.0 ppm; total hydrocarbons, 3.1 ppm; nitrogen oxides, .069 ppm; sulfur dioxide, .02 ppm; suspended particulates, 75.0 $\mu g/m^3$.

can be concluded from the above analysis that the abatement multiplier is larger:

(1) the greater the portion of pollution control costs which represent current direct purchases of inputs;
(2) the less the replacement of existing production by recycled pollutants;
(3) the larger the input-output multipliers;
(4) the larger the ratios of polluting activities to sector levels (the less a region imports the larger these ratios will be);
(5) the greater the emissions associated with polluting activities;
(6) the more steeply rising are the costs of pollution abatement.

It should be stressed that this study of the abatement multiplier is based on a specific model of a specific airshed. Any conclusions must be viewed as tentative because they may be sensitive to parameters and data unique to the particular model. It is likely, however, that the cost of pollution abatement and the optimal set of control method activity levels are more sensitive to factors other than the abatement multiplier. There are important cost and emission parameters in the model which are only estimates. These include data which characterize the technologies for desulfurizing power plant stack gases and controlling nitrogen oxides from automobiles. Relatively small changes in these would have a more substantial impact on the optimal solution than do the abatement feedbacks. In addition, minor changes in certain air quality goals or in the formulas which describe the relationship of emission flows to pollutant concentrations would have a more important impact on the control solution.

It is not clear whether the size of the abatement multiplier might not also be sensitive to such changes in parameters. One such sensitivity test was performed. The allowable emission flows (see table 8, column 3) were reduced 10 per cent for each pollutant in the model. The new values of Z and Z' were respectively $55.5 million and $56.7 million. While this test confirmed the increasing costs of pollution abatement, the abatement multiplier changed very little, and in fact, declined slightly.[33]

Although the value of 1.023 for the abatement multiplier for the St. Louis model appears to be small, it should be noted that the incremental control costs of $.8 million are 1.5 per cent of the sum of incremental economic activities, which would be $54.1 million. In contrast, the total cost of abatement from this model is only .1 per cent of the projected total value of economic activity in the St. Louis region in 1975. Thus the ratio of control costs to economic activity is far greater at the margin than are the corresponding totals. It is apparent that the assumption of fixed maximum allowable pollution flows implies that increased economic activity will require significantly higher expenditures for environmental control.

Results: Shadow Prices

The pollutant shadow prices presented in table 9 indicate the increase in the total cost of abatement associated with a decrease of one pound in the corresponding allowable annual emission flow. The pollutant shadow

33. The decline in the multiplier should not be too surprising. None of the first five factors which explain the size of the abatement multiplier are necessarily related to the level of abatement. Although the marginal costs of pollution control are likely to increase as abatement levels are increased, they could, in a linear programming context, be fairly constant for any specific small range equal to E^*FGHx.

TABLE 9

Shadow Prices Generated by the Linear Programming Models

Constraints	*Shadow Price in the Original Model*	*Shadow Price in the Feedback Model*
	Pollutants	
Carbon monoxide requirement	$.00428 per pound	$.00432 per pound
Hydrocarbon requirement	.02476 per pound	.02482 per pound
Nitrogen oxides requirement	.32639 per pound	.33333 per pound
Sulfur dioxide requirement	.02193 per pound	.02220 per pound
Particulate requirement	.07748 per pound	.07941 per pound
	Inputs from Input-Output Sectors	
Chemical, petroleum, and rubber products sector	0	.01811 per dollar
Nonelectric machinery sector	0	.01110 per dollar
Transportation equipment sector	0	.00494 per dollar
Mining sector	0	.01236 per dollar
Transportation, communication, and utilities sector		
Electricity only	0	.17132 per dollar
Natural gas, etc.	0	.01422 per dollar
Household sector	0	.01423 per dollar

prices from the feedback model incorporate the incremental pollution control costs associated with abatement. To the extent that control costs in the model correspond to control costs that would be borne by polluters, these shadow prices functioning as emission fees would theoretically achieve the optimal control solution x_o via decentralized decision making.[34]

The merger of linear programming and input-output analysis produces the unique set of shadow prices at the bottom of table 9. These indicate the pollution control costs in the St. Louis airshed associated with an increased production of $1.00 by the corresponding economic sector.[35] If

34. The reader who is interested in calculating the government revenue from these emission taxes can multiply the rates in table 9 times the corresponding allowable flows in table 8. He may be surprised to find that the total annual revenue is more than four times the annual cost of abatement.

35. The shadow prices for inputs were obtained as follows. The constraint, $[U - FGH]x = s$, was incorporated in the model in two equations, $Ux - FGy = s$ and

for example, the chemical, petroleum, and rubber products sector would increase its sales by $1.00, pollution control costs in the airshed would rise by 1.8 cents. The sale of an additional dollar of electricity would increase control costs by 17 cents.[36] A dollar increase in annual sales by the transportation equipment sector, which imports a large per cent of its inputs, would raise total costs of abatement in the airshed by half a cent. These costs reflect the fact that final demand sales by any sector increase the production levels of other sectors.

The shadow prices of the inputs have a second interpretation. If the pollutant shadow prices were used as emission fees, an increased production of $1.00 by an economic sector would involve incremental control costs and emission fees in the airshed equal to the shadow price.

Implications of This Research for Cost-Effectiveness Models for Environmental Planning

Abatement feedbacks

It is appropriate that the feedbacks of pollution abatement on the levels of polluting activities be included in cost-effectiveness models. Not only is the cost of abatement higher because of these feedbacks, but adjustments in the control solution may result. This was illustrated in this paper by the revisions in the optimal control method set (table 4) when feedbacks were incorporated.

While the inclusion of input-output multipliers improves the model, there is some question as to whether the increased accuracy is sufficient compensation for the immense computational effort involved. It was observed in this paper that 60 per cent of the feedback impact could be captured by incorporating only the direct inputs and not the indirect inputs to abatement (i.e., by omitting the G matrix).[37] Moreover, a substantial

$Hx - y = 0$, where the elements of y are values of direct inputs for abatement supplied by the separate economic sectors. The shadow prices of the elements of the null vector represent the incremental cost of abatement associated with a dollar increase in sales for the corresponding economic sector.

36. Alternatively, the additional cost of abatement associated with the sale of one kilowatt hour of electricity to industrial, commercial, or residential customers would be .18 cents.

37. It should be noted that the shadow prices for the inputs (see table 9) may in some cases be largely attributable to multiplier effects. If, for example, derived demands are excluded from the model (this is the case where the G matrix is omitted), a dollar in sales by the nonelectric machinery sector, would increase total cost of abatement by only .1 cents, far less than the 1.1 cents noted in table 9.

portion of the primary feedback could be incorporated through electricity inputs alone, thereby further simplifying the model.

If it were anticipated that large quantities of recovered sulfuric acid were to replace existing commercial acid production, it would be advisable to incorporate this abatement feedback into a cost of control method. The present research suggests that certain inputs have a more significant feedback effect than do others, and that the latter might, for simplicity, be ignored.

Measurement units for pollution source levels

Emission factors are generally based either on inputs (i.e., tons of coal burned, gallons of diesel fuel consumed, etc.) or on outputs (i.e., tons of steel manufactured, number of airplane landing and take-off cycles, barrels of cement produced, etc.).[38] As a result, it is typical to measure polluting levels in terms of both inputs and outputs. This asymmetry, apparent in tables 1, 3, 4, and in the example used in the appendix, is in contrast to the uniformity found in input-output analysis.

Some thought should be given to expressing the levels of polluting activities in future cost-effectiveness models in terms of either inputs or outputs, but not both. If output units are used, the cost-effectiveness model could more readily be related to input-output tables as well as other data arranged according to a standard industrial classification. Although the possibility of basing pollution coefficients on output units would eliminate the need for the F matrix used in the present model, it would also increase the dimensions of the control method vector.[39]

Implications of This Research for Input-Output Models for Environmental Planning

Aggregation of economic sectors

One of the problems encountered in this research relates to the aggregation of industries in the input-output model. The aggregation of all utilities in a single sector required special handling to separate the very

38. See *Compilation of Air Pollutant Emission Factors* (Revised), (Research Triangle Park: Environmental Protection Agency, 1972).

39. This has been done in Wassily Leontief and Daniel Ford, "Air Pollution and the Economic Structure: Empirical Results of Input-Output Computations," *Fifth International Conference on Input-Output Analysis,* Geneva, Switzerland, January 1971.

different impacts of natural gas and electricity purchases. The fact that the chemical and petroleum industries, both major sources of air pollution in the St. Louis airshed, were included in the same sector of Liu's input-output model was a distinct limitation. The input-output models being developed for environmental studies should avoid aggregating industries with significantly different pollution characteristics.

The pollution abatement sector

Leontief has expanded the input-output structure with an additional row for pollution output and an additional column for pollution abatement. The feasibility of treating air pollution control as a constant cost industry is challenged in the present paper. Whereas there are no capacity constraints on interindustry sales in an open input-output model, there are significant capacity constraints on pollution control processes *when abatement occurs at the source*.[40] Thus there are only so many underfeed stokers which can be converted from coal to natural gas, so many new automobiles which can be factory equipped with the latest pollution control equipment, etc. As these upper limits become binding, successive levels of abatement are attained at rising marginal costs. If, for example, the pollutant shadow prices for the original model (see table 9) were *average* costs, the cost of pollution control in the original model would be the vector product of these costs and the corresponding required reductions in pollutant emissions (see column 4 of table 8), or more than $90 million a year. This demonstrates the extent of increasing costs, for clearly, a substantial amount of pollution abatement would have to occur at much smaller costs than these shadow prices for the annual cost of abatement to be $35.3 million. If, because of increasing costs, it is not feasible to incorporate pollution control sectors in input-output models, it may be that future research relating economic activity and pollution control costs will depend on interfaced input-output and cost-effectiveness models such as the one presented in this paper.

Appendix: Numerical Illustration of the Model

To clarify the model, consider the following example with two pollution sources, three pollutants, four economic sectors, and five control meth-

40. This may be more applicable to air pollution than water pollution control.

ods. This hypothetical airshed contains two sources of air pollution; a steel mill producing 1,000,000 tons of steel a year and a power plant whose annual consumption of coal is 2,000,000 tons. The vector of polluting production levels is,

$$s = \begin{bmatrix} 1{,}000{,}000 \\ 2{,}000{,}000 \end{bmatrix}.$$

Desirable air quality can be achieved in this airshed if total annual emissions do not exceed 8,000,000 pounds of particulates, 40,000,000 pounds of sulfur dioxide, and 35,000,000 pounds of nitrogen oxides. The vector of allowable emission flows is

$$a = \begin{bmatrix} 8{,}000{,}000 \\ 40{,}000{,}000 \\ 35{,}000{,}000 \end{bmatrix}.$$

The steel mill currently emits 7 pounds of particulates, 13 pounds of sulfur dioxide, and 2 pounds of nitrogen oxides per ton of steel produced. These emissions occur in the operation of basic oxygen furnaces, blast furnaces, sintering machines, coke ovens, and during the combustion of fuel oil, natural gas, and coke oven gas. The power plant currently emits 3 pounds of particulates, 118 pounds of sulfur dioxide, and 20 pounds of nitrogen oxides per ton of coal burned. Thus annual emissions in the airshed are well in excess of allowable flows for all three pollutants.

The present state of control (x_1) at the steel mill includes electrostatic precipitators for the basic oxygen furnaces, primary cleaners for the blast furnaces, and dry cyclone collectors for the sintering operations. The present pollution control method (x_4) at the power plant is an electrostatic precipitator.

The alternative control methods for the steel mill are (x_2) high energy wet scrubbers for the blast furnace, which would cost an additional $.10 per ton of steel output, and (x_3) the high energy wet scrubbers for the blast furnace plus electrostatic precipitators for the sintering operations, which would add incremental costs of $.25 per ton of steel output. The alternative control method (x_5) for the power plant is a desulfurization process costing an additional $1.20 per ton of coal burned. The row vector of control method costs is, $c = [\$.00\ \$.10\ \$.25\ \$.00\ \$1.20]$. Each of the alternative control methods would be used in combination with the existing control method. However, because it is the *incremental* cost of pollution control which is being minimized, the existing control methods are, for convenience, assigned zero costs.

The alternative control methods for the steel mill reduce particulate emissions from 7 to 4 pounds per ton of steel for the first alternative (x_2) and from 7 to 3 pounds for the second alternative (x_3). The desulfurization process (x_5) would reduce emissions from the power plant to 2 pounds of particulates, 12 pounds of sulfur dioxide, and 16 pounds of nitrogen oxides per ton of coal burned. The matrix of emission factors is therefore,

$$E = \begin{bmatrix} 7 & 4 & 3 & 3 & 2 \\ 13 & 13 & 13 & 118 & 12 \\ 2 & 2 & 2 & 20 & 16 \end{bmatrix}.$$

The distributive matrix which equates the sum of control method activities for each pollution source to the production level of that source is,

$$U = \begin{bmatrix} 1 & 1 & 1 & 0 & 0 \\ 0 & 0 & 0 & 1 & 1 \end{bmatrix}.$$

The linear programming model in standard form is,

$$\begin{array}{llllllll} \text{minimize} & Z = \$.00x_1 + & \$.10x_2 + & \$.25x_3 + & \$.00x_4 + & \$1.20x_5 & & \\ \text{subject to} & x_1 + & x_2 + & x_3 & & & = & 1{,}000{,}000 \\ & & & & x_4 + & x_5 & = & 2{,}000{,}000 \\ & 7x_1 + & 4x_2 + & 3x_3 + & 3x_4 + & 2x_5 & = & 8{,}000{,}000 \\ & 13x_1 + & 13x_2 + & 13x_3 + & 118x_4 + & 12x_5 & = & 40{,}000{,}000 \\ & 2x_1 + & 2x_2 + & 2x_3 + & 20x_4 + & 16x_5 & = & 35{,}000{,}000 \\ & x_1, & x_2, & x_3, & x_4, & x_5 & = & 0 \end{array}$$

The optimal solution is $x_1 = 0$ tons of steel, $x_2 = 971{,}698$ tons of steel, $x_3 = 28{,}302$ tons of steel, $x_4 = 28{,}302$ tons of coal, $x_5 = 1{,}971{,}698$ tons of coal and $Z = \$2{,}470{,}283$.[41]

41. The solution of this example problem is awkward. It would be difficult to install control devices for an arbitrary fraction of a plant's production. Although an integer programming solution would be more realistic, it was found that in the standard linear programming model, divisibility occurs in no more rows than there are binding pollutant requirements (in this example, the nitrogen oxides requirement is not binding). The larger the number of pollution sources, M, in comparison to the number of pollutants, P, the smaller will be the relative importance of the problem of divisibility. In the actual model, the operation of basic oxygen furnaces, blast furnaces, sintering machines, coke ovens, the combustion of coke oven gas, the combustion of fuel oil, and the combustion of natural gas by industry are all treated as individual pollution sources, each with separate production levels and with control method coefficients based on the units in which the corresponding production is measured (i.e., tons of pig iron, tons of sinter, millions of cubic feet of coke oven gas, gallons of fuel oil, etc.). These various activities were combined so as to limit the size of the x vector in the example.

The input-output feedbacks associated with pollution control are now included. In this simple example, there are only four economic sectors: (1) a primary metals sector, (2) a machinery sector, (3) an electric power sector, and (4) a household sector. Assume that the annual purchases of local inputs for pollution control are as follows. Each activity unit of control method x_2 requires \$.03 worth of inputs from the machinery sector, \$.01 from the electric power sector, and \$.02 from the household sector. Each activity unit of control method x_3 requires \$.10 worth of inputs from the machinery sector, \$.02 from the electric power sector, and \$.04 from the household sector. Each activity unit of control method x_5 requires \$.20 worth of inputs from the machinery sector, \$.20 from the electric power sector, and \$.15 from the household sector. The matrix of input requirements is accordingly,

$$H = \begin{bmatrix} 0 & 0 & 0 & 0 & 0 \\ 0 & .03 & .10 & 0 & .20 \\ 0 & .01 & .02 & 0 & .20 \\ 0 & .02 & .04 & 0 & .15 \end{bmatrix}.$$

Because no direct inputs are purchased from the primary metals sector, the first row of the H matrix contains only zeros. Because no *incremental* inputs are required for the two existing control methods, columns 1 and 4 contain only zeros. The input requirements for the existing control methods are already incorporated in sector production levels.

The matrix of intersectoral multipliers is determined from a regional interindustry flow model. For this example, it is assumed that,

$$G = \begin{bmatrix} 1.010 & .040 & .005 & .002 \\ .002 & 1.030 & .002 & .001 \\ .100 & .120 & 1.220 & .170 \\ .600 & .870 & .880 & 1.530 \end{bmatrix}.$$

Polluting activities are related to sector levels as follows. For every dollar of sales by the primary metals sector, .0015 tons of steel are produced and .0010 tons of coal are burned at the power plant to provide electricity to the primary metals sector. For every dollar's worth of sales by the machinery sector, by the electric power sector, and by the household sector, .0002, .07, and .0004 tons of coal, respectively, are burned at the power plant. The matrix of coefficients relating pollution source levels to sector sales is,

$$F = \begin{bmatrix} .0015 & 0 & 0 & 0 \\ .0010 & .0002 & .0700 & .0004 \end{bmatrix}.$$

The model with input-output feedbacks is the same as the previous model except that the U matrix is replaced by a $[U - FGH]$ matrix. In the present example, this matrix is,

$$[U - FGH] = \begin{bmatrix} 1 & .999998 & .999994 & 0 & -.000014 \\ 0 & -.001378 & -.003115 & 1 & .979173 \end{bmatrix}.$$

The optimal solution is $x_1 = 0$ tons of steel, $x_2 = 889{,}257$ tons of steel, $x_3 = 110{,}774$ tons of steel, $x_4 = 23{,}357$ tons of coal, $x_5 = 2{,}020{,}290$ tons of coal, and $Z' = \$2{,}540{,}967$. As a consequence of the feedback effect annual steel production rises 31 tons and coal combustion at the power plant increases 43,647 tons a year. The abatement multiplier in this example is $2,540,967/$2,470,283, or 1.03.

COMMENT

Frederick M. Peterson, University of Maryland

Using input-output analysis, Leontief showed that pollution abatement activities generate some pollutants themselves by requiring inputs.[1] For instance, the fans and pumps needed to clean the air use electricity, and the production of this electricity causes additional air pollution. Leontief's illustration raised two empirical questions. Is a significant amount of pollution caused by abatement activities? Do planners have to consider the Leontief effect?

For Kohn's air pollution study of St. Louis, the answer is no. If Kohn's results are supported by other findings, the Leontief effect will be reduced to a theoretically interesting, but empirically unimportant phenomenon. Planners will be able to ignore the effect or dispose of it with a few back-of-the-envelope computations.

Kohn's computations were exhaustive. He included the Leontief effect in a linear programming model of the St. Louis airshed. The model picked the control techniques that achieved a set of emission standards at least cost.[2]

1. Wassily Leontief, "Environmental Repercussions and the Economic Structure: An Input-Output Approach," *Review of Economics and Statistics* Vol. LII (August 1970): 262–71.

2. The model is hard to master. There is much unorthodox terminology, such as an "air pollution control method activity level," which is an amount of some input consumed or output produced that causes pollution. To understand the model, it is sug-

The Leontief effect added only 2.3 per cent to the cost of achieving the standards, a small percentage compared to the other errors and uncertainties that an environmental planner faces. Half of this percentage was achieved without an input-output model, by considering only direct inputs to abatement activities. Kohn showed this by replacing his *G* matrix with an identity matrix. Kohn assumed that sulfuric acid recovered from power plants and lead smelters was additional production rather than a substitute for existing production. When he tried the alternate assumption that sales were constant and that virgin production was reduced, the Leontief effect was cut from 2.3 per cent to 1.1 per cent.

Kohn's estimates of the Leontief effect may be low, but the bias is probably small. The effects of abatement activities were fed back through only six of the twenty-three sectors in the model, as is reflected by the seventeen zero rows in the *H* matrix. This means that inputs from the seventeen sectors had no direct or indirect effect on pollution. To the extent that abatement activities used inputs from these sectors and caused additional pollution, the Leontief effect was understated. It is probably true, as Kohn argued, that these sectors are not important, but it would be nice to have enough details in the paper to check his argument. Generally, the paper lacks sufficient detail for the reader to find out what is happening.

Another area where more information is needed is Kohn's treatment of interregional imports. Kohn ignored the effect of pollution control in the St. Louis airshed on pollution levels and pollution control costs in other airsheds, an omission that probably decreased the observed Leontief effect. If St. Louis imports abatement machinery from Cleveland and increases Cleveland's pollution control bill, the additional cost to Cleveland must somehow get back to St. Louis in the form of higher machinery prices. Even if the costs are not passed back, the effect in other regions should be estimated. It seems that Kohn could do this with knowledge of the import sector in Liu's interindustry model.

If Kohn wanted to estimate the size of the Leontief effect, one would think that a reasonable estimate could have been obtained with back-of-the-envelope calculations. By making a crude guess at the direct inputs needed for abatement, estimating the pollution generated, and doubling the figure to account for indirect flows, he would probably have gotten an estimate between 1 per cent and 3 per cent, low enough to forget

---

gested that the reader study the numerical illustration in the appendix, or see Robert E. Kohn, "Optimal Air Quality Standards," *Econometrica* Vol. 39 (November 1971): 983–87.

about elaborate modeling and computation. Back-of-the-envelope calculations are very useful for environmental problems. Claims are constantly being made about the importance of this or that environmental effect, and many of these claims can be disposed of by a few calculations with approximate engineering data that are readily available.[3]

The fact that the Leontief effect was small and might have been estimated with simpler computations does not totally erase the importance of Kohn's paper. He did not build the model just to estimate the size of the Leontief effect. He also wanted to advance the art of environmental modeling, which he did. He included pollution abatement activities in an input-output model, demonstrated how linear programming can be used to find the least cost way of achieving ambient standards, and calculated some interesting shadow prices that could be used to achieve the least-cost solution with a set of taxes.

3. For an example, it has been claimed that insulating homes does not save energy because energy is required to make the insulation, but simple calculations show otherwise. With typical temperature differentials between the inside and the outside of the home, the insulation can be shown to save more energy in a single year than was required to make it.

Studies of Residuals Management in Industry

Blair T. Bower, Resources for the Future

Introduction and History

For more than ten years Resources for the Future (RFF) has been studying residuals generation and management in industry. The roots of these efforts lie in the original concern of RFF with problems relating to water quantity, and with the corresponding problem of estimating industrial water demand. Hence, the first efforts focused on water intake and water utilization within the individual industrial plant, rather than on the residuals stemming from production activities.

There were two basic reasons why RFF became involved in these detailed industry water studies. First, in 1960 industrial water withdrawals comprised the largest type of water use in the United States, excluding water power. Since that time industrial water withdrawals have become even more predominant quantitatively, as industrial output has continued to increase in magnitude and complexity. Both water withdrawals for, and liquid residuals from, industrial operations have major impacts on the economics, technology, and institutional arrangements for water resources planning and management, including water quality management. Second, past estimates of future industrial water use, traditionally termed "needs" or "requirements," had been done on a very rudimentary and naive basis. This became particularly apparent in the studies under the aegis of the Senate Select Committee on Water Resources.

Traditionally estimates of future industrial water use were based on

Note: I am much indebted to my colleague, Clifford S. Russell, for many helpful comments on the original version of this paper.

historical data on gallons per employee, gallons per unit of raw product processed, gallons per unit of final product output, and even gallons per acre of "type" of industrial activity, i.e., light manufacturing. The basic data used were aggregate data, i.e., across a given industry, based on nationwide mail questionnaire surveys, such as those of the Bureau of the Census. Thus, regional differences in industrial water utilization patterns were very often obliterated. In general, all estimates by planning agencies—both public and private—failed to consider changes in technology relating not only to production processes but also to raw material inputs and product mix. The efforts also failed to consider the price of water at both intake and outlet, as price affected industrial water utilization through the many substitution possibilities available in industrial water utilization systems and between such systems and other factor inputs. For these reasons RFF undertook to develop more rational bases for estimating future industrial water demands, where demand refers to economic demand. The Cootner-Löf study of the steam electric generation industry is the major example of published work from this first period of research.[1]

The next industry studies continued the concern with water, but the focus was broadened to include not only questions of water intake and in-plant water utilization, but also liquid residuals generation, modification, disposal, and discharge. The liquid residuals and water quality parameters of concern were biochemical oxygen demand (BOD) and suspended solids (SS). The Löf-Kneese study of the beet sugar industry is the primary product of this period of research.[2] Even though the beet sugar industry is simple, in terms of process technology, and hence of the residuals generated, much was learned in the study about methodology, and the study generated information which has been widely used.

The study of residuals management in the New York region for the Regional Plan Association, although not directly a research project of RFF, provided the stimulus for a further expansion of the framework of the RFF industry studies.[3] This study made clear the basic technologic, physical, and economic interrelationships among the two basic types of

1. P. H. Cootner and G. O. G. Löf, *Water Demand for Steam Electric Generation: An Economic Projection Model* (Washington, D.C.: Resources for the Future, 1966).

2. G. O. G. Löf and A. V. Kneese, *The Economics of Water Utilization in the Beet Sugar Industry* (Washington, D.C.: Resources for the Future, 1968).

3. B. T. Bower, et al., *Waste Management: Generation and Disposal of Solid, Liquid and Gaseous Wastes in the New York Region,* A Report of the Second Regional Plan (New York: Regional Plan Association, 1968).

residuals—materials and energy, and the three states of the former—liquid, gaseous, and solid. The result was that subsequent RFF industry studies focused simultaneously on the management of all residuals generated in the industrial plant. Water was still of concern as a factor input, and in terms of various types of possible trade-offs, both among components of the total water utilization and residuals management subsystems and between those subsystems and the production process. Current studies of petroleum refining,[4] pulp and paper manufacture,[5] and the coal electric energy industry[6] are examples of this expanded focus.

The presently conceived objectives of the industry studies have evolved over the years. They are:

1. to determine the factors which influence residuals generation in an industry and to determine the quantitative responses, to the extent possible, to variations in those factors;
2. to determine the range of options available in an industry to respond to increasingly stringent constraints placed upon the discharge of residuals to the environment, i.e., constraints on the use of common property resources as inputs to production processes;
3. to determine the proportion of total production costs represented by net residuals management costs taking all impacts on costs into consideration, under increasingly stringent constraints on residuals discharges and in relation to different sets of factor input costs, such as fuel and raw materials, production variables, technology of production, and product output specifications;
4. to develop models of production-residuals generation for different industries for use in analyses of regional residuals management; and
5. to determine the extent to which the physical, technological, and economic interrelationships among the types and states of residuals require that all residuals be considered simultaneously in order to determine the optimal residuals management strategy for an industrial plant.

All except the last are discussed to some degree herein.

4. C. S. Russell, *Residuals Management in Industry: A Case Study of Petroleum Refining* (Baltimore: Johns Hopkins Press, 1973).
5. B. T. Bower, G. O. G. Löf, and W. M. Hearon, "Residuals Management in the Pulp and Paper Industry," *Natural Resources Journal* 11: 4 (1972): 605–623.
6. J. K. Delson, R. J. Frankel, and B. T. Bower, "Residuals Management in the Coal Electric Energy Industry," Resources for the Future, unpublished.

Conceptual Frameworks and Analytical Methods

Various conceptual-analytical frameworks could be utilized in studying residuals management in industry. The following describes both the evolution and the essence of the approach adopted.

A production process—manufacturing, mining, logging, agriculture—operates on one or more raw materials via physical, chemical, and biological transformations by use of capital equipment and inputs of human and nonhuman energy to produce one or more desired outputs. However, no production process can be designed for 100 per cent conversion of inputs into desired outputs.[7] Thus there are material and energy outflows in addition to the desired outputs of products and/or energy.[8] The former are termed "nonproduct outputs" of the production process. They consist of: (1) nonproduct materials *formed* in the production process; (2) raw materials not transformed in the production process, such as catalysts; and (3) nonused or nondesired energy outputs from the production process.

It is assumed that the objective of the firm undertaking the production process is, at least loosely, to maximize the present value of net profits in relation to prices of inputs and outputs and subject to whatever constraints are relevant. Even if there are no constraints on the use of common property resources, i.e., atmosphere, biosphere, water bodies, it is economically necessary in many cases to recover and reuse substantial portions of the nonproduct outputs, both material and energy. Although the discussion immediately below is couched in terms of materials, it is equally relevant to energy flows.

The extent to which materials recovery is practiced at any point in time at a particular industrial plant is a function of the relative costs of recovered materials versus new (makeup) materials, the latter usually being purchased in the market or from another section of the plant at a price which may or may not be close to the open market price. The costs of recovery are a function of the technology of the production process

7. Almost always, if not always, several processes and operations are involved in transforming the inputs into the outputs. For convenience the *set* of such activities is referred to as a production process. For a useful classification of production processes see R. U. Ayres, "A Materials-Process-Product Model," *Environmental Quality Analysis: Theory and Methodology in the Social Sciences,* A. V. Kneese and B. T. Bower, editors (Baltimore: Johns Hopkins Press, 1968), pp. 35–67.

8. In the production processes of energy generation and heating, the desired outputs are electric energy and heat energy, respectively.

and the technology of materials recovery. Trade-offs are possible between the design of the production process to reduce the formation of non-product materials and the extent of utilization of materials recovery technology. In effect, the plant optimizes the *combination* of the production process *plus* the materials recovery system, in the absence of constraints on residuals discharges. (When constraints of one type or another are imposed on residuals discharges, the "total system" is optimized—production process, materials and energy recovery, residuals management measures, as is discussed below.)

Although essential in terms of describing the "ground rules" for studies of residuals management in industry, the above provides no operational framework. A first attempt to become operational was based on a formulation adapted from studies of industrial water utilization.[9] Thus, residuals generation in the absence of constraints on residuals discharges is expressed in terms of the primary variables as follows:

$R_{git} = f\,(RM, PP, PO)$, where
R_{git} = quantity of residual i generated per unit time, t;
RM = type of, and hence characteristics of, raw materials;
PP = technology of production process, including technology of materials and energy recovery and technology of by-product production; and
PO = product output specifications.

There are other variables which affect residuals generation, and may be of major importance in specific cases, particularly in the short run.[10] Examples are operating rate, i.e., output per unit time, the cost of in-plant water recirculation, and the physical layout of the plant—which in turn affects other variables such as the cost of water recirculation.

Residuals *discharge,* i.e., into the various environmental media, is then a function of the same factors plus the effluent controls imposed on the plant and the technology of residuals modification. Thus:

$R_{dit} = f\,(RM, PP, PO, EG, TR)$, where
R_{dit} = quantity of residual i discharged per unit time, t;
RM, PP, PO are the same as above;

9. B. T. Bower, "The Economics of Industrial Water Utilization," *Water Research,* A. V. Kneese and S. C. Smith, editors (Baltimore: Johns Hopkins Press, 1966), pp. 143–173.

10. Bower, "Industrial Water Utilization," p. 153.

EC = controls imposed on discharge of liquid, gaseous, solid, and energy residuals (heat and noise), i.e., standards, charges; and
TR = technology of residuals modification.

However, these formulations are inadequate, particularly in failing to make explicit the role of: (1) prices of factor inputs, i.e., chemicals, electric energy, and heat; and (2) exogenous variables such as tax policies, import quotas, postal rates, technological changes in other production processes which utilize the outputs of the production process under consideration, and the factors which influence final demand in terms of the characteristics of final products, i.e., Madison Avenue, internal R & D for product development, sales departments. Russell has proposed an excellent conceptual model which includes such factors; it is shown in figure 1.[11] As Russell states, even though this is only a qualitative framework, it serves two useful purposes:

> First, it focuses attention separately on the influences outside the firm which indirectly and those which directly affect residuals generation patterns. And second, it emphasizes that within the production process itself other inputs can frequently be substituted for primary residuals generation. That is, it illustrates the fundamental sense in which it is misleading to assume that analyses based on fixed coefficients (such as pounds of BOD generated per unit of output) are conceptually valid.[12]

Basically this is the conceptual framework for the ongoing RFF studies of residuals management in industry.

Given a conceptual framework, several procedures have been used to make it quantitative, in terms of (1) method of analysis, (2) focus of study, and (3) source of information. With respect to the first, simulation and linear programming have been used, and simulation-linear programming combinations could have been used. The individual plant and the total system, i.e., set of spatially separate component activities, to produce a specified output, are the two foci which have been used. The following list shows the combinations of analytical method-focus used in the industry studies thus far:

Beet sugar: Plant—Simulation
Pulp and paper: Plant—Simulation

11. C. S. Russell, "Models for Investigation of Industrial Response to Residuals Management Actions," *Swedish Journal of Economics* 73: 1 (1972): 136–138.
12. Russell, "Models for Investigation," p. 138.

A Proposed Model of Industrial Residuals Generation and Discharge

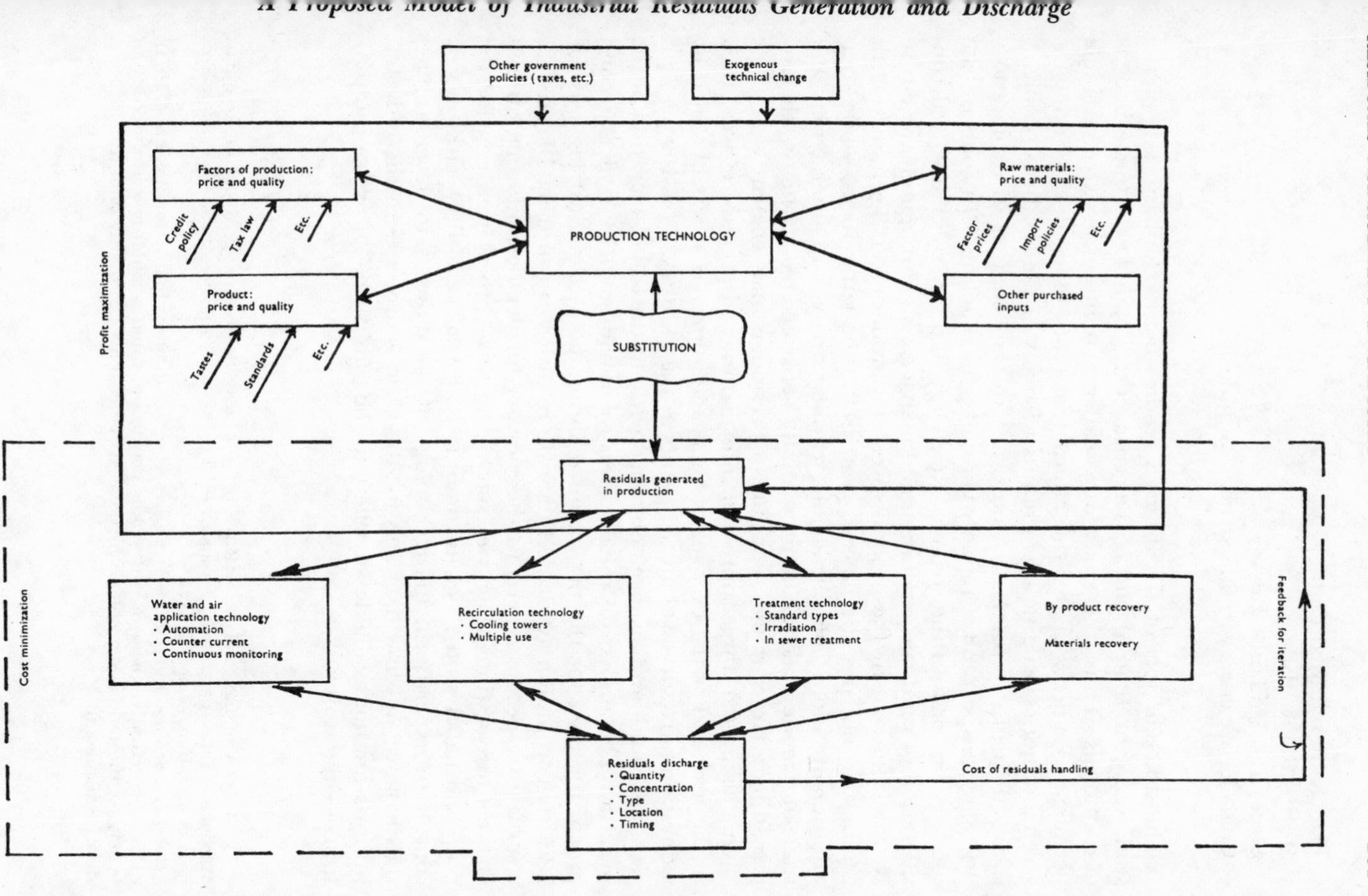

Source: C. S. Russell, "Models for Investigation of Industrial Response to Residuals Management Actions," *Swedish Journal of Economics* Vol. 73, No. 1 (1972): 137.

Petroleum refining: Plant—LP
Steel:[13] Plant—LP
Steel scrap:[14] Plant—LP
Coal-electric energy: Total system—Simulation

The different foci of the industry studies merit emphasis. All of the studies except that of the coal-electric energy can be characterized as micromodels. Thus, the study of petroleum refining analyzed a single petroleum refinery, taking the alternative types of crude oil inputs as given and not associating those inputs with any of the residuals problems involved in providing those inputs, i.e., in the activities of exploration, drilling, transport to the refinery via tanker or pipeline.[15] In contrast, the total system focus of the coal-electric energy industry study involved consideration of residuals generation and management throughout the entire system from coal in the ground to energy produced at the high side of the busbar, not just the power plant itself. The latter focus, while computationally more difficult, enables explicit analysis of interactions and trade-offs among the different spatially separated components, thereby providing a larger range of options for residuals management.

One or more of three sources of information and procedures were used in the studies. For the beet sugar study, the information was obtained by a questionnaire survey of all the plants in the industry. This 100 per cent response could not have been obtained without the full cooperation of the industry. The analysis of the data was then made by RFF staff. A second procedure, used for the petroleum refining and steel industries, was for RFF staff to plumb available literature in order to construct LP models, resorting to specific experts in the industry for detailed information and answers to questions when necessary. The third procedure, used for the pulp and paper study, was to combine RFF staff capability with outside consultant expertise in the technology of the industry to develop materials balance and flow diagrams for the various processes in the industry, information not available anywhere in the published—or even in unpublished—literature.

13. W. J. Vaughan and C. S. Russell, "A Linear Programming Model of Residuals Management for Integrated Iron and Steel Production," *Journal of Environmental Economics and Management* (July 1974).
14. J. W. Sawyer, *Automotive Scrap Recycling* (Baltimore: Johns Hopkins Press, 1974).
15. Theoretically, the price of crude petroleum input should reflect the costs of managing the residuals problems in these activities, but it is highly doubtful if this is the case at present.

Some Difficulties Facing Industry Studies

Although relatively few benefits are likely to be derived from presenting a complete listing and discussion of the difficulties facing the researcher undertaking a study of residuals management in industry, some discussion of at least a few difficulties is warranted. The difficulties relate to the availability of data and to the interpretation of whatever data are available.

First, data typically are available on only some of the various residuals of interest, because few plants measure regularly even all of the major residuals. Further, most of the data published consist of residuals discharges, which discharges are the result of some combination of residuals generation and residuals modification after generation. Such data preclude any analysis—statistical or otherwise—of the effects of the variables identified previously.

Even where data on residuals generation are available, they rarely are published in relation to the variables which have determined that generation. Rather, the data are published in terms of, for example, pounds of BOD per ton of paper for the total integrated paper mill, or perhaps an integrated kraft paper mill, or pounds of BOD per barrel of crude petroleum charge to a "petroleum refinery," not by individual operations. In fact, there are many combinations of type(s) of raw material used, production operations, and detailed product output specifications for a single 4-digit SIC category, such as paper mills. Typically in a single integrated paper mill there are: multiple cellulose-containing raw materials used—logs, chips, wood products residues, sometimes waste paper; multiple pulping processes—sulfate, mechanical, perhaps sulfite, with both batch and continuous digesters; multiple product outputs—dozens of types of paper and/or paper products. Even for a mill producing only linerboard, typically a dozen or more weights (grades) of board are produced.

Second, and closely related to the first, is that an adequate industry study requires the calculation of almost complete materials, electric energy, and heat balances for the production processes involved. Such materials and energy balances are rarely available, even in unpublished form. The "almost complete" stems from two caveats. One, some energy and materials quantities representing less than 1 or 2 per cent of the total energy or materials involved can usually be neglected, except where even the small amount can have substantial adverse effects when discharged, as in the case of toxic or malodorous materials. Two, heat, water vapor, and carbon dioxide discharged to the atmosphere can be ignored, at least in the short run. Thus, keeping track of these residuals is not necessary, other than for insuring consistency in the total materials and energy balances. It should be emphasized that the determination of all of the residuals associated with producing a given product requires determining the amount of purchased energy or fuel required, over and beyond the energy generated within the plant. This is simply to say that, in producing a ton of paper, SO_2 is SO_2, whether it is generated at the paper mill or in the fossil-fuel power plant of the utility serving the area and the paper mill. Similarly, the materials and energy used in residuals modification in response to effluent controls must be included in the analysis.

A third data problem is a lack of data on costs of factor inputs, process units, and residuals modification measures—both capital and operation/maintenance, where relevant. For example, depending on the bookkeeping "policy" of the company's financial officer, wood products residues obtained from another company mill and shipped to the company's paper mill in the same region for use as input in paper production, may be priced at the going market price for such raw material in the region, or at zero (excluding transport costs). The "cost" of chemical inputs to pulping depend on whether the source is a captive chemical plant at the paper mill or the open market, and the bookkeeping policy. This becomes particularly important in evaluating residuals management costs where nonproduct material outputs are the same as the chemical inputs, as is discussed in some detail below.

Many costs are site specific, or at least significantly affected by site factors. Site topography, access to water, energy or fuel costs, raw material availability, et al., affect plant design and costs, and hence total production costs. For a single company producing the same product by the same process in two different paper mills, the ratio of raw wood costs between the mills can be as much as two to one. In absolute terms, the difference is several times greater per ton of paper output than total residuals management costs per ton. Published data on capital costs of some residuals management facilities show ranges of ten to one for the identical type of facility for the same volume throughput and residuals loading. Published data on basic production costs are often misleading because they rarely relate to a totally new plant, but rather to some component thereof. Usually there is a mix of technologies in a given plant, because of the multiplicity of processes involved in producing the output and the changes which have been made in different process units at different points in time.

A fourth difficulty is the short-run variability in residuals generation. To what extent such variation affects, or is reflected in, published data on residuals is not known. Only rarely are ranges of residuals published; even more rare is a frequency distribution of residuals discharge or generation per ton.

A given production process—such as the manufacture of paper—is normally designed in the engineering sense to produce a range of types of outputs, i.e., grades of paper, with one particular grade likely to be dominant in terms of proportion of total output. Maximum production efficiency in the physical sense is achieved when producing this grade. But once in operation several variables affect residuals generation from day to day, and seasonally, such as the quality of incoming logs and/or chips, the sharpness of the saws in the wood preparation operation, and demand for different product outputs. Pressure for increased daily output can result in "pushing" certain components of the production process, such as digesters, beyond design capacity, thereby resulting in larger than "normal" residuals generation per ton. Variation in BOD residuals generation per ton of almost two to one has been recorded within a single month for a given linerboard mill.

Similar short-run variations occur in other industries.[16] One illustration, from the canning industry, is shown in figure 2. The daily variation in pounds of BOD generated per ton of tomatoes processed results primarily from variation in the quantity and quality of incoming raw material, but also because of daily variations in product output mix.

A fifth problem is that of determining what the particular production process to produce a specified product with a given basic technology would be in the absence of pollution controls. Such information is *essential* for determining both residuals generation and costs attributable solely to residuals management. Most existing manufacturing installations have varying amounts of residuals modification facilities, which have been added over the years in response to various pressures for effluent control. Neither the production process itself nor the residuals modification measures would likely be identical for the given plant, if a new plant were going to be constructed at the same site with today's constraints and prices. The available data typically do not differentiate between basic production costs and residuals management costs.

The problem is illustrated in figure 3, which shows an hypothetical

16. It should be noted that short-run variations in residuals generation also occur as a result of accidental spills and breakdowns.

Figure 2

Daily Variation in Residuals Generation During Processing of Tomatoes

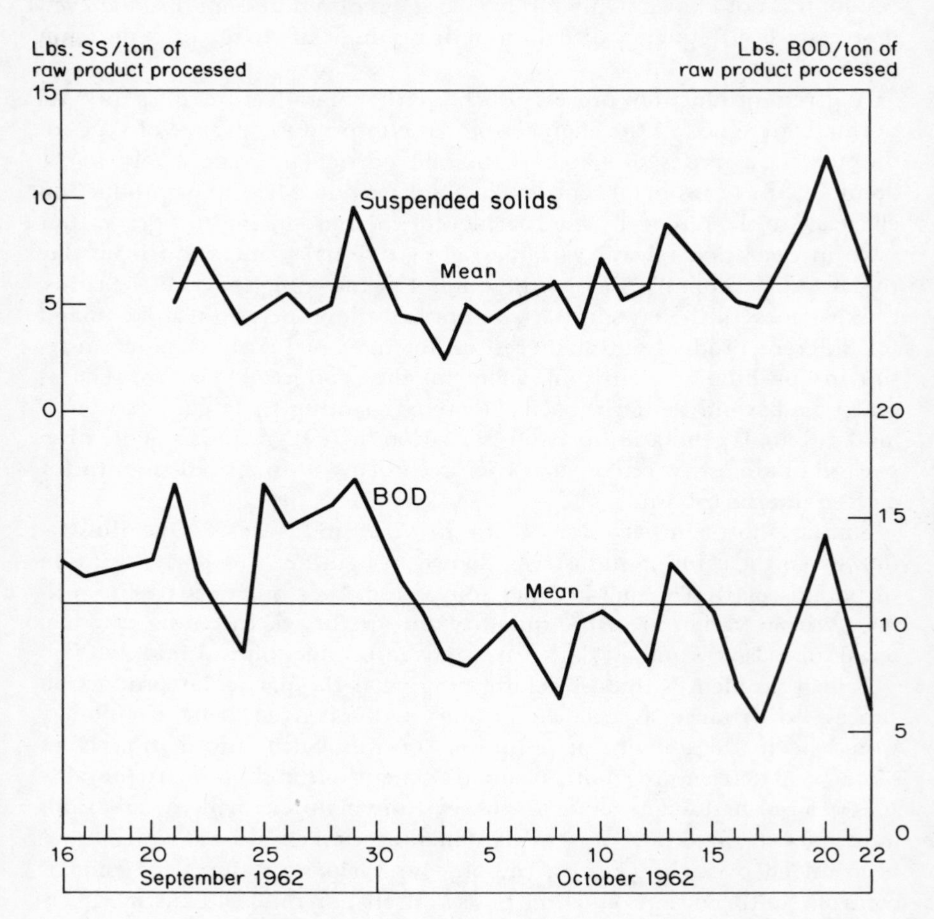

relationship between degree of modification of nonproduct material outputs and net annual cost, for cases—relatively numerous it should be noted—in which these outputs have a positive value. In terms of profit maximization for the individual plant, materials recovery is undertaken to the level where the marginal benefits from materials recovery equal the marginal costs of recovery.[17] This level defines "residuals generation,"

17. The same is true for by-product production. The degree to which such production would be undertaken in the absence of environmental quality controls is a func-

Figure 3

Hypothetical Cost Curve for Modification of Nonproduct Material Which can be Used In-plant

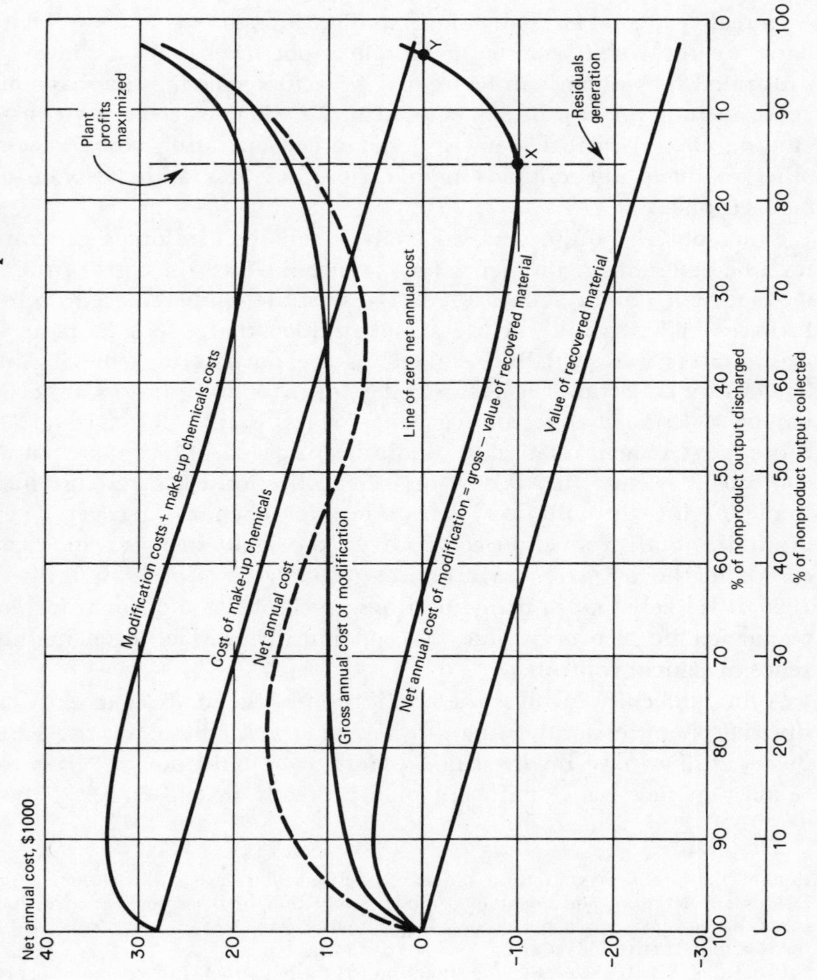

Source: Löf, Hearon, and Bower, "Residuals Management."

and is designated by *X* in figure 3. The range in materials recovery from zero to this level is termed economic materials recovery. Thus, the quantity of this particular material residual generated equals the quantity of *nonproduct output* material minus the quantity of economically recovered materials.[18]

In reality the net annual cost of modification curve is likely to be much flatter in the vicinity of the maximum profit level than as shown, as is illustrated subsequently in figure 6. The factors affecting the shape of the curve include the cost of makeup chemicals, chemical recovery costs—including capital facilities, energy, and other operating costs, effects on other residuals and residuals modification measures, and effects on other factor inputs.

Thus, only by quite detailed analysis can the "residuals generation" level be determined, and hence the base from which the costs of residuals management *should* be measured. To illustrate, with respect to recovery furnaces in kraft mills, Kittle stated in 1966 that, ". . . in past years, precipitators were usually installed for the purpose of achieving about 90 per cent collection efficiency. Today few new precipitators are ordered having a designed collection efficiency of less than 97.5 per cent." [19] Because the statement referred to conditions prior to significant air pollution controls, it is clear that the 90 per cent collection efficiency installations were *not* for air pollution control but for chemical recovery, i.e., to minimize total production costs. Given the small increases in chemical costs and the *reduction* in chemical recovery costs at kraft mills since 1966, it is likely that in many locations much of the additional increment of particulate recovery would be economically justified even in the absence of effluent controls.

A final difficulty involves the application of the analyses to existing industrial operations rather than to new plants. In the main, the RFF industry studies have involved new plants to be built, i.e., a "grass roots" refinery or new paper mill, not plants already in existence.[20] However,

---

tion of the profit maximization objective. Beyond that level, which corresponds to residuals generation, additional by-product production involves costs greater than returns, hence represents a net economic cost to the firm. However, it may still be the least-cost alternative for reducing residuals discharge.

18. Minus, where relevant, the nonproduct output used in *economic* byproduct production. Economic by-product production is analogous to economic materials recovery.

19. R. W. Kittle, "Current Status and Future Prospects—Pulp Mill Air Pollution Control," *Proceedings: The Third National Conference on Air Pollution,* PHS Publication 1649 (Washington, D.C.: Government Printing Office, 1967), pp. 232–235.

20. The exception is the steel industry study.

much of what has been learned is valid for existing plants. This is demonstrated by the analysis of actual behavior in various cases, where many of the alternatives available to new plants to reduce residuals discharges have been used at existing plants, albeit at sometimes higher costs than would be the case for the same alternative at a new plant. Residuals management costs at existing plants are of course affected by existing conditions, particularly the physical layout of the plant and the existing technology of production. Focusing on new plants has the advantage of being able to make explicit the base conditions for the analysis, and has led to many insights which would not have otherwise been gained.

Results of Industry Studies

Having discussed the conceptual framework for, and some of the problems of, the industry studies, the next task is to illustrate the types of outputs and what has been learned from these studies. Basically the studies have illuminated in detail: (1) the major factors influencing residuals generation, particularly the interrelationships among type of raw material, nature of production process, and product output; (2) the major significance of final demand, in terms of the relatively large changes in residuals generation stemming from changes in product characteristics; and (3) the range of options available for responding to effluent controls, and the costs associated therewith.

Individual plant: residuals generation

Paper manufacture. The types and quantities of residuals generated in the production of a specific type of paper are a function primarily of the raw material(s) employed, the pulping process used, the extent of bleaching, and the characteristics desired in the paper product. All of these factors are of course interrelated. The desired brightness of the final product determines the amount of bleaching required, for a given raw material and pulping process. Similarly, the desired strength, or any other product characteristic, limits the combinations of type of raw material and pulping process which can be used.

More specificity is given to these statements by considering an integrated mill producing jumbo rolls of tissue paper, i.e., *excluding* the converting operation. Table 1 shows the residuals generated in producing one ton of tissue paper for different combinations of raw material, pulping, bleaching, and product brightness. Not shown are by-products, and normally innocuous nonproduct outputs of water, carbon dioxide, and nitrogen. The residuals generated in the combustion of purchased fuel necessarily used in generating process steam and electric energy are included. Because the output under consideration is tissue *paper,* rather than tissues, the converting operation is not considered, although residuals are generated in converting. To the extent that there are losses of paper in the converting operation, *more* than one ton of tissue paper is required to produce one ton of *tissues.* This of course implies that residuals generation is higher when expressed per ton of final-user product.

The effects of the major variables are readily apparent: for example, pulping process—Ti 3 (magnefite) vs. Ti 4 (kraft); brightness—Ti 31 (GEB 25) vs. Ti 4 (GEB 80); type of raw material—Ti 34 (waste paper) vs. Ti 31 (soft wood). For these comparisons, respectively, the magnefite process generates no reduced sulfur compounds and less than half the particulates, but almost 2.5 times the SO_2, compared with the sulfate process. Reducing the product brightness, and hence the extent of bleaching—all other specifications remaining the same—from 80 to 25, the brightness of unbleached kraft, cuts SO_2 in half, dissolved solids by over 85 per cent, and BOD by almost 80 per cent. Using No. 1 mixed waste paper as the raw material instead of softwood logs and kraft pulping, results in an increase of almost 50 per cent in SO_2, no reduced sulfur compounds, essentially no particulates, some increase in dissolved solids, and many times more suspended inorganic solids. For other types of waste paper the quantities would be different.

The data in table 1 are based on materials, energy, and heat balances for each of the steps in the production processes involved, and represent the sums of all of the individual residuals streams generated, after taking into consideration economic amounts of recirculation of water and heat. As noted previously, there is considerable variation among plants using the same process and producing the same product, and day-to-day in the same plant.

Steel production. Similar to the manufacture of paper, the types and quantities of residuals generated in the production of steel are a function of the raw materials used, the production process, and the product output specifications.[21] There are three basic types of furnaces for the production of steel: open hearth, basic oxygen (BOF), and electric arc. Each of these is physically capable of producing, but at different costs, three ge-

21. This section is based on the Vaughan-Russell study, cited in footnote 13. Vaughan made the specific computer runs and prepared the background material for this section.

TABLE 1

Residuals Generation in the Production of One Ton of Tissue (or Napkin) Paper

(pounds per ton)

	P.C. No. and Pulp Mix (Brightness 80–82 GEB)							
	Ti 1	*Ti 2*	*Ti 3*	*Ti 4*	*Ti 5*	*Ti 6*	*Ti 7*	*Ti 8*
Residual	*(100% Ca/ CEH)*	*(100% NH_4/ CEH)*	*(100% Mg/ CEH)*	*(100% K/ CEHD)*	*(50% K/ CEHDED; 50% SG/Zn)*	*(50% Mg/ CEHH; 50% SG/Zn)*	*(50% K/ CEHD; 50% Mg/ CEH)*	*(75% K/ CEHDED; 25% WPN/ FIB)*
Cl_2	1.1	1.1	1.1	1.2	0.6	0.6	1.1	0.9
ClO_2	0	0	0	0.6	0.6	0	0.3	0.9
SO_2[1]	125/34.0	114/29.0	48.7/15.0	5.6/20.0	2.8/25.0	24.6/23.0	27.1/17.0	4.2/19.0
RS	0	0	0	25.5	12.8	0	12.7	19.1
Part.[1]	27.5/1.7	27.1/1.4	27.7/0.8	57.5/1.0	34.6/1.3	19.8/1.2	42.6/0.9	43.2/1.0
DIS[2]	127	130	108	263	159	96	185	244
DOS[2]	2970	2900	190	244	193	178	217	261
SS–O	113	112	109	113	97	100	111	139
SS–I	4.4	4.4	4.1	4.5	3.2	3.0	4.3	3.3
BOD_5	820	804	92	147	105	84	120	151
So–I	55.4	46.0	77.9	82.0	50.3	49.7	79.9	65.1
So–O	83.0	63.1	63.1	0	0	31.9	31.5	3.2

(*continued*)

TABLE 1 (*concluded*)

	P.C. No. and Pulp Mix (Brightness 70–72 GEB)					P.C. No. and Pulp Mix (Brightness 25 GEB)			
Residual	*Ti 21 (100% Mg/H)*	*Ti 22 (100% K/CEH)*	*Ti 23 (50% Mg/H; 50% Sg/Zn)*	*Ti 24 (60% K/CEHD; 40% WPM/FIB)*	*Ti 25 (35% K/CEHD; 65% WPN/FIB)*	*Ti 31 (100% K/O)*	*Ti 32 (50% Mg/O; 50% SG/O)*	*Ti 33 (35% K/O; 65% WPN/F)*	*Ti 34 (100% WPM)*
Cl_2	0	1.2	0	0.7	0.4	0	0	0	0
ClO_2	0	0	0	0.3	0.2	0	0	0	0
SO_2[1]	48.1/25.0	5.6/20.0	24.0/23.0	3.4/20.0	2.0/19.0	5.1/7.0	23.4/19.0	1.8/17.0	0/17.0
RS	0	25.4	0	15.3	8.9	23.2	0	8.1	0
Part.[1]	27.4/0.7	57.4/1.0	19.5/1.1	34.5/1.0	20.1/0.9	52.4/0.3	19.1/1.0	18.4/0.6	0/0.8
DIS[2]	103	263	74	235	192	22	17	15	21
DOS[2]	144	227	140	311	278	41	108	29	63
SS–O	108	113	94	145	178	107	93	105	92
SS–I	4.1	4.4	3.0	101	1.6	4.1	2.9	1.4	202
BOD_5	79	143	69	163	148	31	46	27	36
So–I	79.0	81.8	48.7	55.6	37.7	73.7	44.1	32.8	13.8
So–O	62.2	0	31.1	6.0	8.3	0	30.3	6.9	12.8

Note: Raw material is softwood logs, except where waste paper is used. Specifications other than brightness are 14 basis weight (single ply), low wet strength (25%). Output is as air-dry paper, equivalent to 1,880 pounds on a bone-dry basis. Abbreviations follow the table.

[1] 1 per cent sulfur fuel oil is assumed for the purchased fuel used to generate heating steam and electric energy for plant use. Right-hand figure in each column of these rows is the quantity generated associated with fuel combustion.

[2] The division of total dissolved solids into organic and inorganic portions is to an extent arbitrary. For example, a dissolved compound comprised of a metal and a wood ingredient or derivative might be considered organic, inorganic, or partially in each category. In the analysis represented by this table, most of the dissolved excess and wasted chemical agents are classified DIS; most of the dissolved organic fractions of wood or waste paper as DOS.

Abbreviations for Table 1

Ti = tissue paper
P.C. = production combination
Ca = calcium base sulfite pulping
NH_4 = ammonium base sulfite pulping
Mg = magnefite (sulfite) pulping
K = kraft (sulfate) pulping
SG = stone groundwood pulping
WPN = waste paper, No. 1 News (raw material)
WPM = waste paper, No. 1 Mixed (raw material)
CEH, CEHD, etc. = kraft or magnefite bleaching sequences, where C = chlorination; E = caustic extraction; H = hypochlorite bleaching; and D = chlorine dioxide bleaching
Zn = groundwood bleaching, zinc hydrosulfite
O = no bleaching
F = waste paper processing-defibering
FIB = waste paper processing-defibering, deinking and bleaching

Residuals—Gaseous

Cl_2 = chlorine
ClO_2 = chlorine dioxide
SO_2 = sulfur dioxide
RS = hydrogen sulfide and organic sulfides
Part. = particulates

Residuals—Liquid

DIS = dissolved inorganic solids
DOS = dissolved organic solids
SS–I = suspended inorganic solids
SS–O = suspended organic solids
BOD_5 = 5-day biochemical oxygen demand

Residuals—Solid

So–I = inorganic solids
So–O = organic solids

neric types of steel: drawing quality, commercial quality, and alloy, which are defined principally on the basis of the contents of alloy elements—copper, chromium, nickel, molybdenum, and tin. Drawing quality steel has a total alloy content ≤ 0.13 per cent; commercial quality ≤ 0.21 per cent. Alloy steel must have exactly 1.75 per cent of these elements, subject to a specified distribution among them.

The three principal steel furnace types are most fundamentally distinguished on the basis of their heat sources. In the BOF the heat required for melting any cold metal charged and for carrying forward the refining reactions is contained in the molten iron charged. For the open hearth, combustion of an outside fuel (oil, coke oven or natural gas, etc.) is the heat source. In the electric arc, as the name implies, electrical energy, transformed into heat by an arc, is used. This heat source distinction implies, in turn, technological upper limits on the percentage of cold metal (scrap) in the furnace charge and has, under historically prevailing relative costs for ore, scrap, electricity, coal and other fuels, produced a range of normal charging practices. For the BOF, the upper limit on cold metal input is 30 per cent of total metallic input in the absence of such refinements as natural gas lancing for scrap premelting. For the open hearth and electric arc, since the heat source is external, 100 per cent scrap may be charged, but historically, integrated mills (i.e., those with blast furnace capacity), have seldom gone below a 50 per cent hot metal charge in the open hearth.[22] The electric arc, on the other hand, is normally operated on a 100 per cent cold metal charge, though hot metal may be used.

These differences have important implications for residuals generation and management. The electric arc process is free of the problems associated with by-product recovery, treatment and disposal of residuals generated in coke oven operations, specifically BOD, oil, phenols, cyanides, and sulfur. On the other hand, the cold metallic charge to the electric arc can result in a very high level of particulate generation per ton of molten steel when the charge contains a significant portion of No. 2 steel scrap bundles—the proportion depending on the type of steel being produced. In addition, the electric arc furnace requires much more electric energy per ton of steel produced, with the consequent generation of larger quantities of gaseous residuals in the associated energy generation than for the open hearth and basic oxygen furnaces.

22. Open hearths are sometimes used in "cold-melt" shops with a 100 per cent cold metal charge, but these operations are not part of integrated mills. This practice is relatively rare.

Table 2 shows the pounds of residuals generated per ton of semifinished steel shapes—blooms, billets, and slabs, for a daily output of 2,000 tons, for the three types of furnaces and the three types of steels. The quantities would vary slightly with different mixes of shapes. (Note that any loss in fabrication, i.e., converting, if done at the steel mill, is not included, similar to the analysis of paper manufacture.) In addition to the assumptions indicated in the table, it is assumed that only 66 per cent of the ammonia produced per ton of coal charged to the coke ovens is contained in the coke oven gas, the remainder being contained in a raw ammonia liquor. Ammonia and phenol recovery from coke plant liquid residuals streams is possible, but will be undertaken only if the market prices of the ammonium sulphate and sodium phenolate by-products are sufficient to cover the costs associated with their production. The prices assumed for the analysis reflected in table 2 result in no recovery being undertaken. Finally, some of the slag generated in the OH and BOF steel furnaces could be recycled, depending on the relative costs of processing and disposal and on steel content. No recycling is assumed economically justified in this analysis.

The most pronounced effect of process and product variables on residuals generation is with respect to particulates. For a given type of steel, the BOF results in more than twice as many particulates per ton as the OH and EA furnaces. It should be emphasized however, that the scrap prices assumed resulted in a 50 per cent hot metal/50 per cent scrap charge to the OH furnace. Some higher level of absolute scarp prices would induce a 70 per cent hot metal/30 per cent scrap charge, thereby resulting in substantially higher generation of particulates, and of all other residuals as well, stemming from the corresponding increase in ancillary operations, i.e., coke ovens, blast furnaces.

Comparing the different types of steel for the same steel making process shows that again particulates are most affected. For all three furnace types, generation of particulates increases for the sequence of drawing quality, commercial quality, and alloy steel production. But the increase is significantly different among the furnace types, the relative quantities for the three types being, respectively: 1.0/1.55/1.87 for the open hearth; 1.0/1.04/1.11 for the basic oxygen furnace; and 1.0/1.20/4.26 for the electric arc. The increase in particulate generation reflects the fact that more No. 2 bundles of steel scrap can be used per ton of output, moving from drawing quality steel to alloy steel, because of increasingly higher total alloy content permitted. No. 2 bundles have higher dirt and organic matter, as well as alloy, content compared to other scraps; hence the higher particulate generation per ton as their use increases. The relatively smaller

TABLE 2

Residuals Generation in the Production of Three Types of Semifinished Steel Shapes, 2,000 Tons Per Day Output

(pounds per ton except heat)

Residual	*Drawing Quality Steel*			*Commercial Quality Steel*			*Alloy Steel*		
	OH[a]	*BOF*[b]	*EA*[c]	*OH*	*BOF*	*EA*	*OH*	*BOF*	*EA*
Gaseous									
Particulates[d]	20.9/1.4	55.3/2.1	9.5/13.3	33.1/1.5	57.6/2.1	14.6/12.7	40.0/3.3	61.4/4.4	84.4/12.7
SO_2[d]	7.9/2.3	11.7/3.4	0/22.0	8.1/2.4	11.8/3.4	0/21.0	8.1/2.6	11.7/3.4	0/21.0
Liquid									
BOD_5	1.76	2.62	0	1.75	2.63	0	2.62	1.76	0
Oil	0.21	0.31	0	0.21	0.31	0	0.31	0.21	0
Phenols	0.51	0.77	0	0.52	0.77	0	0.77	0.52	0
Ammonia	1.17	1.74	0	1.20	1.75	0	1.74	1.20	0
SS[e]	0.10	0.14	0	0.10	0.14	0	0.14	0.10	0
Sulfur	0.19	0.28	0	0.19	0.28	0	0.28	0.20	0
Heat, 10^6 BTU	1.01	1.54	0.08	0.79	1.57	0.08	1.47	1.00	0.08
Solids									
Slag	457.2	683.2	235.9	460.3	686.3	53.2	691.5	448.3	48.9
Solids OTS[f]	0.8	1.2	19.7	0.8	1.2	53.4	1.2	0.8	53.4

Assumptions: Raw material for ironmaking: iron ore—high iron (65 per cent Fe), low sulfur (0.6 per cent S), fines; coal—predominantly low sulfur (0.6 per cent S); charge to OH is 50 per cent hot metal/50 per cent scrap; to BOF is 70 per cent hot metal/30 per cent scrap. The OH charge reflects the low absolute level of scrap price assumed, i.e., relative price of scrap/hot metal = 0.59.

Particulate removal facilities assumed in place only at the sinter strand and the blast furnace, operating at removal efficiencies of 90 per cent and 99.2 per cent, respectively. This level of removal is justified in the absence of environmental controls, because of the value of the recovered gas as fuel.

3 per cent sulfur coal is assumed for use in energy generation.

[a] Open hearth furnace.

[b] Basic oxygen furnace.

[c] Electric arc furnace.

[d] Right hand figure in each column is the quantity associated with energy generation.

[e] Suspended solids.

[f] Solids other than slag, i.e., unrecycled mill scale, bottom ash, unburned forerunnings.

increase in particulate generation for the BOF reflects the technological constraint on the quantity of scrap which can be used in that type of furnace.

Individual plant: response to effluent controls

Petroleum refining. The response to effluent controls is first illustrated by the analysis of a 150,000 barrels per day (crude charge) "grass roots" refinery with the sizes of process units and magnitudes of product outputs shown in table 3.[23] For this refinery, the residuals generated and discharged under conditions of no effluent controls—other than those reflected by existing standard American Petroleum Institute oil-water separators and sour-water scrubbers—are shown in table 4. Given this refinery and the specified conditions, what will be the response to an effluent charge, such as that imposed on oxygen demanding organics, expressed as BOD_5?

The response is shown in figure 4. Discharges of BOD, sulfide, phenols, and ammonia decrease rapidly over the range of charges from 1 cent to 7 cents per pound. At the latter level of effluent charge on BOD, almost 70 per cent of the BOD generation, i.e., the load after the oil-water separators and sour-water scrubbers, has been reduced. Note that, simultaneously, the following reductions have occurred in other liquid residuals: phenols, about 82 per cent; ammonia, about 48 per cent; sulfide, about 74 per cent. Reduction of oil is very slight; of heat not at all. No further reduction occurs until the effluent charge is between 14 and 15 cents per pound, when BOD reduction reaches about 80 per cent. As the charge increases between 15 and 25 cents per pound, the BOD reduction increases to about 95 per cent at the latter figure.

It is important to emphasize that reduction in discharge of BOD results in the generation of secondary residuals, particularly "solids" formed in the standard activated sludge process for modifying BOD, i.e., at an assumed rate of 0.75 pounds of dry sludge solids per pound of BOD removed.[24]

23. The material in this section is based on Russell, *Residuals Management in Industry,* especially Chap. VI.

24. Sulfur released as hydrogen sulfide or SO_2 and vaporized hydrocarbons are quantitatively insignificant. Some BOD removal at each level is accounted for by recirculation, e.g., effluent from secondary treatment to desalter water makeup. In these cases sludge generation depends on the steady state concentrations attained in the streams involved. This complication is ignored, and it is assumed that no increment to sludge generation results from such recirculation alternatives.

TABLE 3

Process Units and Product Outputs 150,000 Barrels Per Day Petroleum Refinery

	Process Units Per Barrel of Crude Charged (barrels)
Desalting	1.00
Atmospheric distillation	1.00
Coking	.133
Hydrotreating	.139
Reforming	.139
Catalytic cracking	.466
Alkylation	.076
Sweetening	.393
	Product Outputs (quantity per day)
Products sold	
Refinery gas	2.944×10^6 lbs.
Kerosene/diesel oil	15,760 barrels
Distillate fuel oil	17,400 barrels
Low sulfur	8,880 B
Medium sulfur	8,230 B
High sulfur	290 B
Polymer	660 barrels
Premium gasoline[a]	35,100 barrels
Regular gasoline[b]	51,150 barrels
Residual fuel oil	3,000 barrels
Straight run gasoline sold as petrochemical feed	16,360 barrels
Recovered sulfur	40.0 long tons
Products used internally	
Hydrogen (burned)	100,250 lbs.
Sweet coke (burned)	1,180,000 lbs.
Sour coke (burned)	260,000 lbs.
Coke burned in catalyst regeneration	1,540,000 lbs.

Note: Crude charged: 111,000 barrels East Texas (low sulfur) plus 39,000 barrels Arabian Mix (high sulfur) totals 150,000 barrels per day.

[a] Octave $\geq$ 100; tetraethyl lead content $\leq$ 2.5 cc/gal.

[b] Octave $\geq$ 94; tetraethyl lead content $\leq$ 2.5 cc/gal.

TABLE 4

Residuals Generation Per Barrel in a 150,000 Barrels Per Day Petroleum Refinery

Residual	*Generation (lbs. per barrel)*
Gaseous	
Particulates	0.423
SO_2	1.429
Liquid	
BOD_5	0.060
Oil	0.047
Phenols	0.032
Ammonia	0.021
Sulfide	0.003
Heat (10^6 BTU)	0.300

Assumptions—

Cost of water withdrawals: cooling, \$.015/1000 gallons; desalter, \$.025/1000 gallons; process steam, \$.15/1000 gallons.

Cost of purchased fresh heat: 2.0% sulfur, \$.477/10^6 BTU; 1.0% sulfur, \$.593/10^6 BTU; 0.5% sulfur, \$.661/10^6 BTU.

Price of recovered sulfur: \$20/long ton.

At the 7 cents per pound of BOD effluent charge, the dry weight of sludge generated is 4,020 pounds per day, in the form of a dilute sludge, i.e., 5 per cent solids. Thus the total weight of raw sludge generated represents about 80,400 pounds, or 40.2 tons per day. At the 15 cents and 25 cents per pound charge levels, the weights of raw sludge generated are 45.4 and 52.9 tons per day, respectively. Disposal of this sludge requires thickening and incineration. Assuming the cost of thickening and incineration is about \$2.00 per ton of raw sludge, sludge handling increases the costs of BOD reduction by \$80 per day, \$91 per day, and \$106 per day, at charge levels of 7 cents, 15 cents, and 25 cents, respectively. In addition, the generation of particulates is increased by 1.2 per cent, 1.3 per cent, and 1.6 per cent at the corresponding three charge levels. The increase in solids, i.e., incinerator residues, is shown on figure 4.

Translating these BOD reductions into cost impacts on the refinery is illustrated in figure 5, in terms of the increase in daily cost (capital plus

Figure 4

Response to BOD Effluent Charge, 150,000 Barrels Per Day Petroleum Refinery

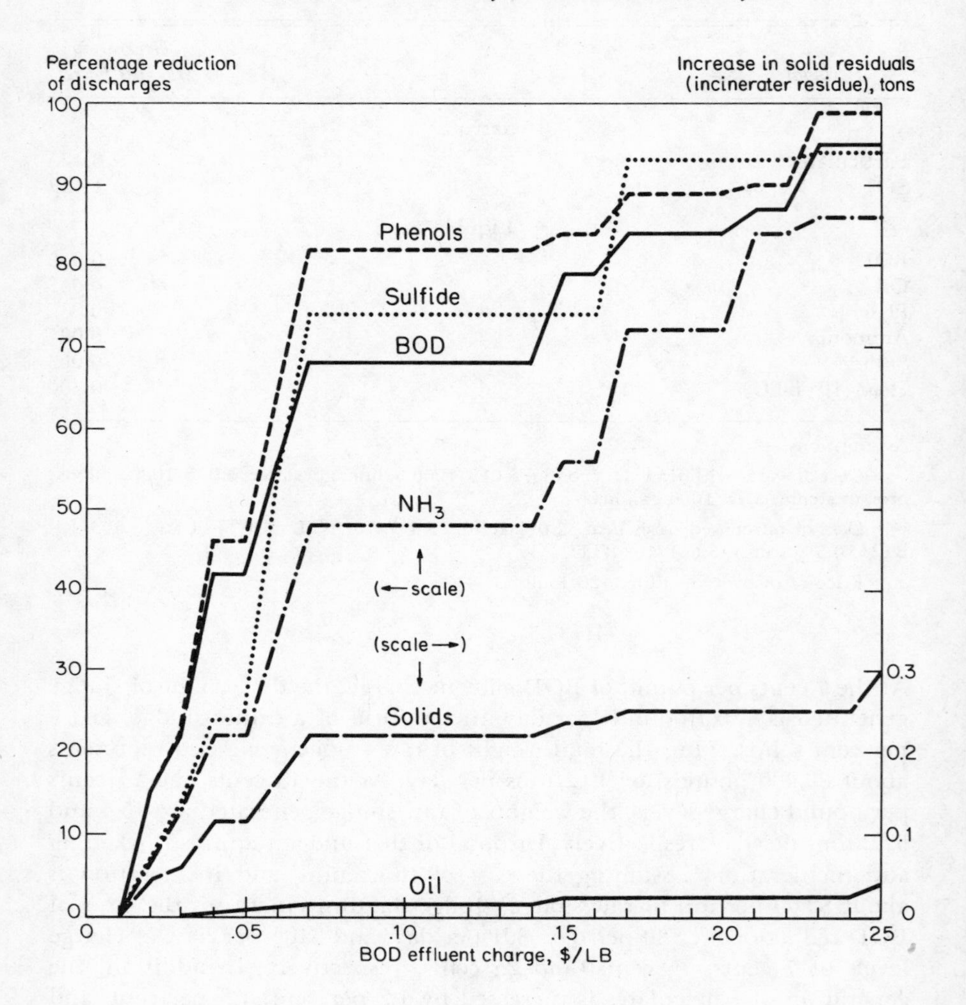

operating) per barrel as the degree of BOD reduction increases. It is clear that the total cost of reducing BOD discharge is small relative to daily refinery costs, for example, 75 per cent BOD reduction costs only about $0.002 per barrel of crude processed or about 0.045 per cent of daily costs. For 100 per cent reduction the cost would be about 3.5 times as much.

Figure 5

Total Cost of BOD Discharge Reduction, 150,000 Barrels Per Day

(including disposal of sludge)

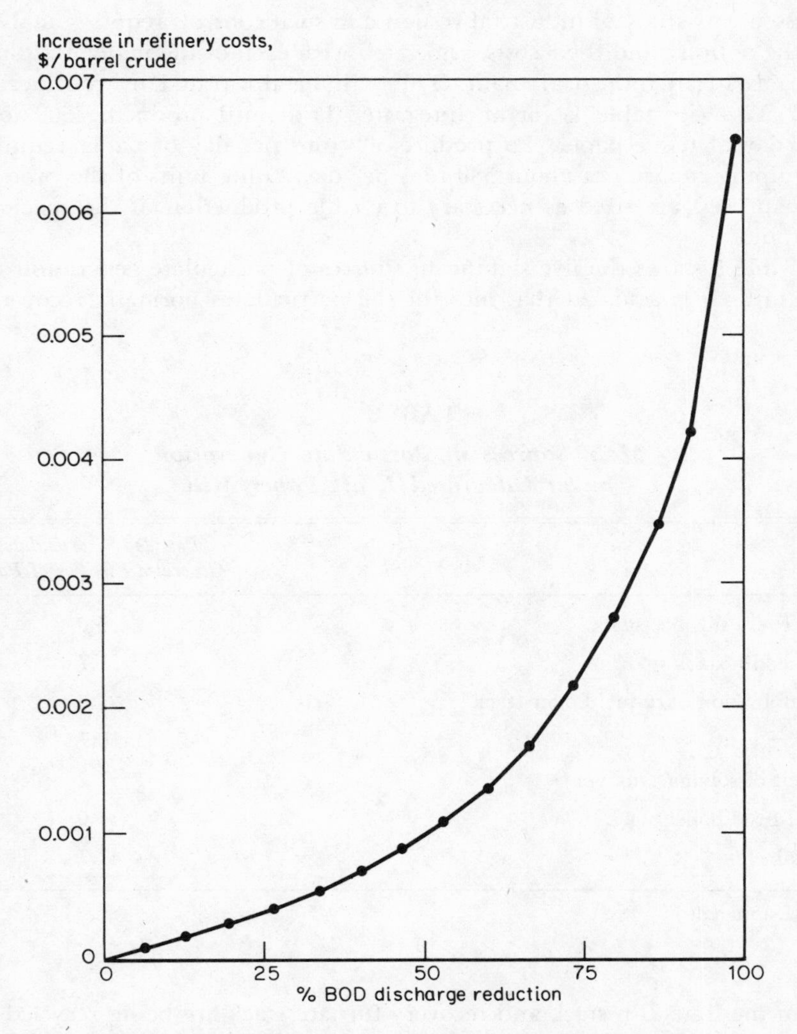

The costs are of the same order of magnitude for phenols, sulfide, and ammonia.

Although the preceding discussion of the impacts of effluent charges on a grassroots petroleum refinery is only partial, it serves to illustrate clearly the method of analysis and the useful results which can be obtained. A complete discussion is of course available in the previously cited book by Russell.

Paper manufacture. Regardless of how the effluent controls are expressed, any study of industrial response to such controls requires analysis of the options, and their costs, associated with each of the major residuals streams in an industrial plant. This will be illustrated by considering P.C. Ti 4 (see table 1) for an integrated kraft mill producing 500 tons per day of tissue paper. To produce 500 tons per day of paper requires a pulping capacity of about 580 tons per day. Other units of the production process are sized as necessary to enable production of the specified output.

Table 5 shows the five significant sources of particulate generation for the mill. It is assumed that most of the particulates normally recovered

TABLE 5

Main Sources of Particulate Generation in an Integrated Kraft Paper Mill

Source	*Pounds of Particulates Generated Per Ton of Paper*
Recovery furnace stack	5.0
Lime kiln stack	11.7
Combination bark-fuel boiler stack	27.6
Slaker vent	0.7*
Smelt dissolving tank vent	1.2
Fuel-fired boiler stack	1.0
Total	47.2

* Estimated.

from the lime kiln stack and recovery furnace stack are being recycled to the chemical system and that the quantities shown in the table are those in excess of the economically recoverable amounts, represented by point X in figure 3.

Figure 6 shows the net annual cost for different degrees of modification

Figure 6

Net Annual Cost of Recovery Furnace Particulate Modification: Integrated Kraft Mill Producing 500 Tons Per Day of Tissue Paper

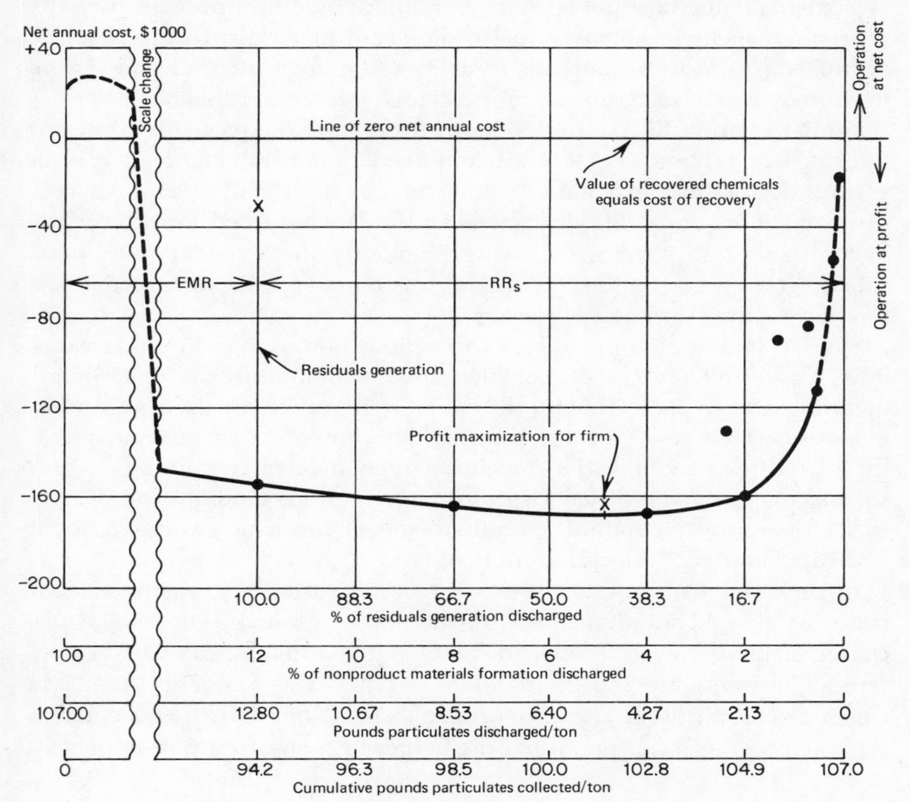

Source: Bower, Löf, Hearon, "Residuals Management."

of the nonproduct particulate materials formed in the flue gases from the recovery furnace, the first entry in the preceding list.[25] The points shown above the curve represent alternative measures for the same degree of particulate modification, but with higher costs. Up to the level of modification designated by *X*, particulate modification represents economic

25. G. O. G. Löf, W. M. Hearon, and B. T. Bower, "Residuals Management in Pulp and Paper Manufacture," *Forest Products and the Environment,* AIChE Symposium Series Vol. 69, No. 133 (Dec. 1973): 141–149.

materials recovery (EMR), i.e., measures which would be undertaken in the absence of effluent controls because of the value of the recovered chemicals.[26] From level X to 100 per cent removal, if that were possible —particulate modification represents residuals modification with partially offsetting benefits from the value of recovered materials. In this case the residuals modification-materials recovery technology involves end-of-pipe measures, such as electrostatic precipitators and wet scrubbers.

Similar relationships were developed for each of the particulate streams indicated in table 5. These were combined to develop the least cost of various degrees of particulate modification, as shown in table 6. Analysis of wet methods for modifying gaseous residuals indicated that the incremental costs associated with having to modify the secondary liquid residuals from wet scrubbers were higher than "dry" gaseous residuals modification measures for all streams.

Similar analyses were made for the various nonproduct liquid streams. Two major differences between the gaseous and liquid residuals merit mention. One is that few options exist for liquid residuals modification measures which result in any returns. The other is that minimum cost liquid residuals modification measures often involve combining various streams of the same residual for simultaneous (joint) modification. Rarely is it economical to combine several streams of the same gaseous residual generated at different locations in the plant.

In order to demonstrate the impact of increasingly stringent effluent controls imposed simultaneously on the major residuals, four sets of discharge standards were developed, involving the major residuals of concern. These sets are shown in table 7. Level III represents standards which are currently in effect in various states. Note that these standards are framed in terms of permitted discharges per ton of output. Standards

26. Note the difference in the totals of particulates generated between table 1 and table 5. The reason is that the data available at the time the flow diagrams and materials balances were made, for table 1, indicated 12 per cent discharge as the profit maximizing point. Additional information obtained subsequently provided a better basis for the analysis and showed that at present the profit maximizing level of discharge for a new plant would be about 5 per cent. As can be seen, the curve is relatively flat between about 12 per cent and about 2 per cent. It does not take much of a change in factor input costs to shift the profit maximizing level in this range by several per cent. This demonstrates the difficulty in determining, from available data, the quantity of residuals which would be generated in a "grass roots" plant in the absence of environmental controls.

It should be emphasized that the costs of particulate removal should be interpreted as illustrative. For example, variation in gas flow rates, particle sizes, equipment costs, amortization schedules, and other factors may result in substantial differences from plant to plant for the same process and product.

TABLE 6

Net Annual Cost of Particulate Modification, Integrated Kraft Mill Producing 500 Tons Per Day of Bleached Tissue Paper

	No Modification		*"Low" Modification*		*"Medium" Modification*		*"High" Modification*	
	Lbs. Part. Per Ton	*Annual Net Cost in Dollars*	*Lbs. Part. Per Ton*	*Annual Net Cost in Dollars*	*Lbs. Part. Per Ton*	*Annual Net Cost in Dollars*	*Lbs. Part. Per Ton*	*Annual Net Cost in Dollars*
Particulate discharge	47.2		30		8		4	
Particulate removal:								
Bark boiler stack	0	0	17.2	8,940	25.9	14,300	26.0	19,500
Lime kiln stack	0	0	0	0	10.3	17,000	11.6	32,900
Recovery furnace stack	0	0	0	0	2.5	5,000	3.9	32,400
Lime slaker vent	0	0	0	0	0.5	750	0.6	2,390
Smelt tank vent	0	0	0	0	0	0	1.0	10,560
Fuel boiler stack	0	0	0	0	0	0	0	0
Total removal	0	0	17.2	8,940	39.2	37,050	43.2	97,750
Cost of solids disposal		0		3,150		4,540		4,560
Cost of liquid residual disposal		0		0		0		0
Total net annual cost of particulate control		0		$12,090		$41,590		$102,310

Notes:

1. Costs are in 1970 dollars and are based on estimates of operating labor, maintenance labor and supplies, power and material requirements, 12.5 per cent annual charge on estimated capital investment, and are credited with chemical recoveries at typical market prices. Operation 350 days per year was assumed.

2. The gas streams to be treated and the extent of particulate removal from a stream were selected so as to obtain the lowest cost of removing the required quantity of particulates, regardless of particulate composition.

3. Cost of solids disposal includes only the solids resulting from removal of particulates not recycled to the process, and excludes disposal costs for other solid residuals. It is based on a nominal $2 per ton hauling and landfill cost.

4. None of the particulate removal processes used results in generation of a liquid residual.

TABLE 7

Specification of Increasingly Stringent Residuals Discharge Standards

(pounds per ton)

Residual	*Level 0*	*I*	*II*	*III*	*IV*
SO_2	No Control[1]	50	35	20	10
Particulates	No Control	30	8	4	2
Reduced sulfur compounds	No Control	10	2	0.5	0.2
Suspended solids	No Control	50	20	10	5
BOD_5	No Control	60	35	20	10

Note: Standards apply to total mill operation, i.e., from all sources.
[1] No restrictions on discharges; reflects basic production costs.

so expressed ignore the size of a plant, and hence implicitly the assimilative capacity of the environment of the plant, but they represent the approach which the federal government has adopted, i.e., "standard effluent levels."

The residuals management costs to meet levels I, II, and III are shown in table 8, in terms of dollars per ton, for an integrated kraft mill producing 500 tons per day of tissue paper (P.C. Ti 4).[27] The much higher costs for liquid residuals modification than for gaseous residuals modification reflect the substantial value of materials recovered from gaseous residuals streams, comparable alternatives not existing for the liquid residuals streams. Also shown in table 8, for comparison, are the residuals management costs per ton for a paper mill of the same output capacity but producing *unbleached* tissue paper (P.C. Ti 31). The impact of changing just one product specification is very significant, residuals management costs for Ti 31 being about 15 per cent, 24 per cent, and 21 per cent of those of Ti 4 for levels I, II, and III, respectively.

Implications for future behavior and technological change can be drawn from this type of detailed analysis of residuals management in an industry, implications which could not be obtained by any other approach, i.e., statistical analysis (even if data were available, which they

27. These levels correspond to "low," "medium," and "high" residuals modification in table 6.

TABLE 8

Net Residuals Management Costs Per Ton of Output, Integrated Kraft Mill Producing 500 Tons Per Day of Tissue Paper

Costs	*Level of Discharge Standards* I	II	III
P.C. Ti 4: Unbleached Tissue Paper			
Gaseous residuals modification, $/ton	0.16	0.59	1.66
Liquid residuals modification, $/ton	3.07	4.09	6.76
Solid residuals disposal, $/ton	0.38	0.38	0.38
Total, $/ton	3.61	5.06	8.80
P.C. Ti 31: Unbleached Tissue Paper			
Gaseous residuals modification, $/ton	0.12	0.46	0.83
Liquid residulas modification, $/ton	0.10	0.40	0.72
Solid residuals disposal, $/ton	0.33	0.33	0.33
Total, $/ton	0.55	1.19	1.88

Note: Costs are in 1970 dollars and are based on estimates of operating labor, maintenance labor and supplies, power and material requirements, 12.5 per cent annual charge on estimated capital investment, and are credited with chemical recoveries at typical market prices. Operation 350 days per year was assumed.

The costs of any secondary solid residuals generated in liquid and gaseous residuals modification, i.e., sludge, are included in the liquid and gaseous residuals modification costs.

are not). For example, the fact that liquid residuals management costs are so large relative to gaseous residuals management costs suggests the logical direction of plant responses and research and development efforts. In fact this is what has occurred in the industry. To reduce the former costs, paper mill water systems have been tightened and in-plant water recirculation has been increased, thereby reducing the hydraulic load on liquid residuals modification systems, and hence costs. Additional effort is being expended to develop "dry" paper-making processes. Because the effluents from bleaching by traditional methods of bleaching represent the major source of liquid residuals and of liquid residuals management costs, as the comparison of Ti 31 with Ti 4 shows clearly, research continues on new methods of bleaching, such as oxygen bleaching, to eliminate or drastically reduce the bleaching residuals that are expensive to modify. In addition, research is underway on new methods of (wet) pulping, which would result in fewer and/or more easily modified liquid residuals from pulping. The other logical response, indicated clearly by the above analysis, namely, changing product brightness to eliminate the need for bleaching—and minimize total resource use—has been given little consideration.

Total system: residuals generation and response to effluent controls

Coal-electric energy industry. Up to this point the focus has been on the individual industrial plant. In this section the analysis shifts to a focus on the total system, i.e., several spatially separate activities necessary to produce a given output. The study of the coal-electric energy industry illustrates this approach. Instead of analyzing a power plant as the "residuals-environmental quality management system," as has traditionally been done, the activities involved from mining the coal in the ground through energy delivered at the load-center substation are included. The four elements of this system, shown in figure 7, are: coal mining; coal preparation; coal or energy transport, the latter being relevant when energy is generated at the mine; and energy generation via coal combustion. Coal preparation generally takes place at or very near the mine. Also shown in figure 7 are the major residuals streams generated throughout the system.

The problem is formulated as follows: assuming a given energy demand at the load-center substation, determine the minimum cost system of coal production, processing, transport, coal use in the power plant, energy transport, and residuals handling activities which will meet specified and comparable quality levels with respect to all environmental media, air, water, land, *throughout* the system.[28] Trade-offs are possible among raw coal quality, degree of coal preparation, transport of coal or energy, combustion technology, and residuals handling technology at the mine-preparation plant and the power plant. For example, if an objective is to reduce the ambient concentration of sulfur dioxide in a specified region, there are several combinations of alternatives among the many elements of the system which can be utilized, such as high or low sulfur

28. The ambient environmental quality standards, and hence discharge standards, should rationally vary among the different spatially separated elements of the system, if there are differences in assimilative capacity and damages from discharges. Thus, the early development of mine-mouth plants assumed that location in relatively isolated, i.e., sparsely settled, areas would permit lower discharge standards. The current controversy with respect to power plants in the southwest suggests that this was not an optimal policy. Hence, essentially the same standards have been assumed throughout the system.

Figure 7

Alternative Residuals Generation and Modification in the Use of Coal to Produce Electric Energy

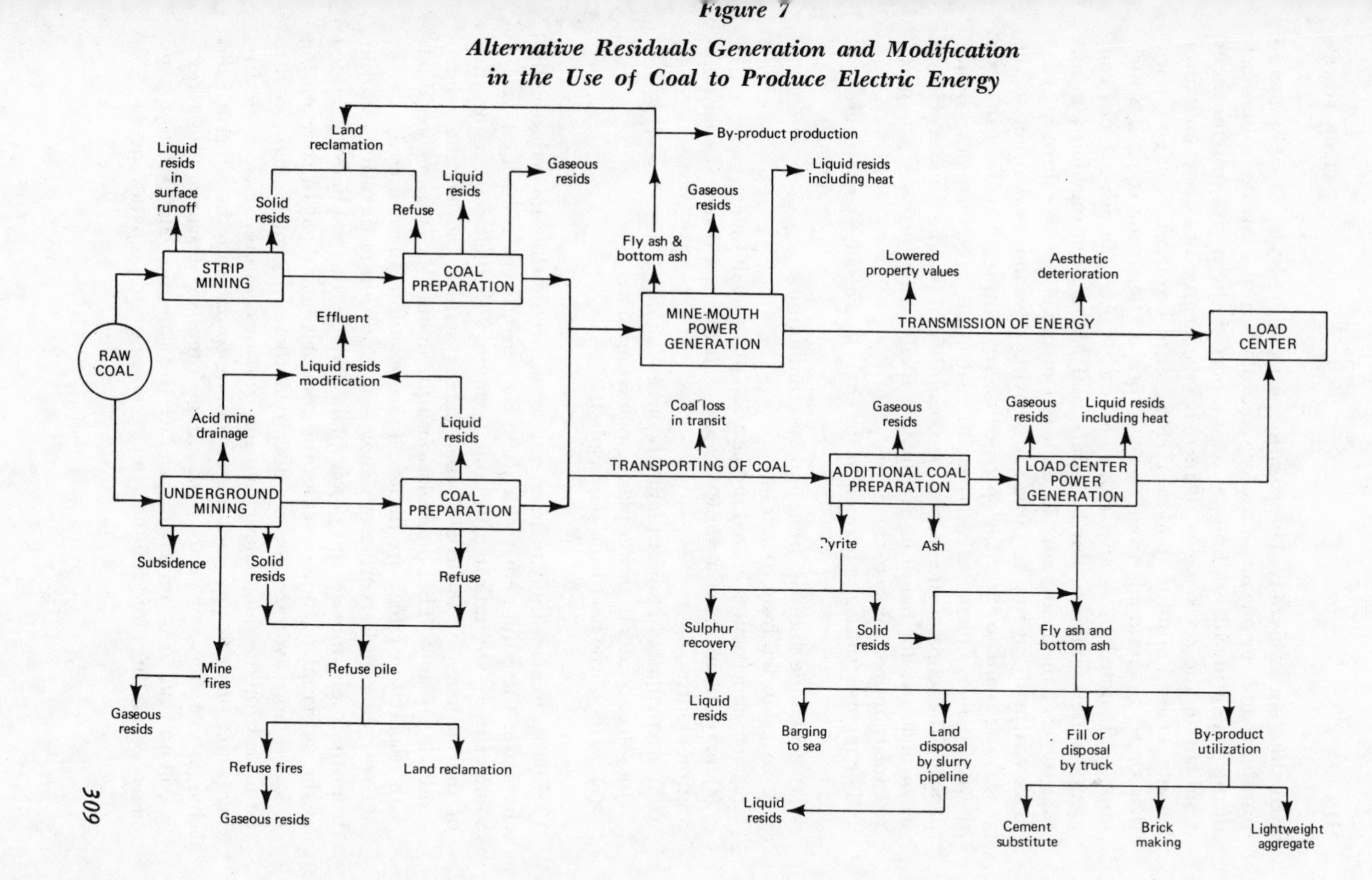

coal, different levels of coal preparation, various methods of coal or energy transport, and various measures for removal of sulfur dioxide in the stack of the generating plant. Thus, at least a partial alternative to removal of residuals in the stack gas is to increase the degree of coal preparation for the removal of sulfur (and ash). However, it is important to emphasize that there are residuals from the coal preparation process itself which must be handled in a satisfactory fashion to preclude adverse environmental quality impacts. Further, since coal preparation inevitably means that some of the heat value in each ton of mined coal is thrown away, more coal will have to be mined to provide the same heat input to the generating plant to meet the specified energy output. This may increase the residuals problem with respect to coal mining. Nevertheless, by expanding the scope of the system the possibility of finding a more efficient set of measures for handling residuals is likely to be increased compared to a focus limited solely to the power plant.

The major residuals generated in the coal-electric energy system are:

(1) acid mine drainage from underground mining;
(2) overburden from strip mining;
(3) suspended solids in coal preparation plant wash water;
(4) particulates from air-flow cleaners and thermal driers at coal preparation plants;
(5) particulates in power plant gaseous emission;
(6) sulfur oxides in power plant gaseous emission;
(7) water-borne heat from power plant.

Various strategies, singly or in combination, are possible for reducing discharges of these different residuals to the various environmental media. Some of these are indicated in table 9, along with their impacts on each of the elements of the coal-electric energy system and on the related residuals streams. The interrelationships among the various residuals with respect to strategy are indicated. For example, increasing the degree of coal preparation in order to remove some portion of the ash and sulfur from the fuel input to the power plant reduces the residuals generated at the power plant, but increases the residuals generated in coal mining (because more hydrocarbons are discarded, thereby requiring more coal to be mined for the same energy output) and in coal preparation. Modifying power plant flue gas and use of cooling towers to reduce thermal discharges to water courses are both energy-intensive measures, thus requiring more coal to be mined to produce the same net energy output and hence generating more residuals in coal mining and preparation.

TABLE 9

Impacts of Strategies for Improving Ambient Environmental Quality in the Utilization of Coal to Produce Electric Energy

Strategy	*1*	*2*	*3*	*4*	*5*	*6*	*7*	*8*	*9*	*10*
Impacts on system										
Quantity of coal mined	0[1]	0	0	0	−[1]	+[1]	−	+	0	+
Quantity of raw mine drainage	−	0	0	0	−	+	−	+	0	+
Quantity of refuse at mine	0	+	+	+	−	+	−	+	0	+
Quantity of coal transported	0	0	0	0	−	−	−	+	0	+
Quantity of solid residuals at power plant	0	0	0	0	−	−	−	+	0	+
Impacts on residuals										
Useful land	+	0	0	0	+	−	+	−	0	−
Acid and iron effluent	0	−	0	0	−	+	−	+	0	+
Suspended solids discharged from preparation plant	0	0	−	0	−	+	−	+	0	+
Particulate discharged from preparation plant	0	0	0	−	−	+	−	+	0	+
Particulate and sulfur oxides discharged from power plant	0	0	0	0	−	−	−	−	0	+
Heat discharged from power plant to water courses	0	0	0	0	−	0	0	+	0	−
Suspended solids discharged from power plant to water courses	0	0	0	0	−	−	−	+	−	+
Solid residuals from power plant	0	0	0	0	−	−	−	+	+	+

Strategies:

1. Grade and replant land
2. Treat acid mine drainage
3. Treat waste water from preparation plant
4. Collect particulates from preparation plant
5. Increase generator efficiency
6. Use more coal preparation
7. Use higher quality *raw* coal
8. Treat power plant flue gas
9. Treat suspended solids from power plant
10. Use cooling towers

[1] 0 = no change; − = less; + = more; all + changes represent negative impacts except for land, i.e., increasing quanity of coal mined *increases* amount of acid and iron effluent, a negative impact.

A strategy to reduce the discharge of one residual may add to the costs of handling another residual or generate additional residuals which result in other environmental quality problems. Removing sulfur in power plant flue gas increases the resistivity of the fly ash, thereby making electrostatic precipitation less efficient.[29] The use of cooling towers to reduce thermal discharges into water courses involves a transfer of the residual heat from a liquid discharge to a gaseous discharge. This emission from a cooling tower may result in such undesirable environmental effects as local fogging and icing, cloud formation, and increased precipitation.[30] In turn, various alternatives are available for modifying these secondary effects, such as by superheating the plume or by using finned heat exchangers.[31] Any such further modification adds to the residuals modification costs.

The impact on energy cost of effluent controls imposed on various residuals throughout the coal-electric energy system was assessed by positing three levels of control on effluents. The characteristics of these three levels for the two cases investigated, mine-mouth power plant and load center power plant, are listed in table 10.

Level I basically reflects no modification of nonproduct outputs except what would be done in the absence of effluent controls, i.e., economic materials recovery, plus what additional control was being done in the early 1960's before the major push for improving air quality began. Level II approximates current (1972–73) standards; Level III reflects a still larger reduction in discharges.

The least-cost combination of activities throughout the system to meet these control levels was determined. The results for the three levels are shown in table 11, in terms of the cost of energy and the quantities of residuals discharged into the environment.

The relative costs, compared with Level I, are shown in table 12. These cost increases are distributed approximately as follows for the mine-mouth plant: from Level I to Level II—about one-third to coal mining and preparation; about two-thirds to power plant for higher

29. A shift from 2 per cent sulfur coal to 1 per cent sulfur coal reduces collection efficiency of electrostatic precipitators from 99 per cent to 98 per cent. See W. J. Cahill, Jr., and R. G. Ransdell, Jr., "Low Sulfur Coal Cuts Precipitator Efficiency," *Electrical World* 168, 20 (1967): 111–112.

30. See E. Aynsley, "Cooling Tower Effects: Studies Abound," *Electrical World* 173, 19 (1970): 42–43.

31. H. Veldhuizen and J. Ledbetter, "Cooling Tower Fog: Control and Abatement," *Journal Air Pollution Control Association* 21, 1 (1971): 21–24.

Characterization of Environmental Quality Control Levels: Coal-Electric Energy System With Mine-Mouth Power Plant and Coal-Electric Energy System With Load-Center Power Plant

Phase of Production	Environmental Control Level: *Minimal–I*	*Moderately High–II*	*High*[1]*–III*
Coal-Electric Energy System With Mine-Mouth Power Plant			
Coal mining[2] (strip mining)	No grading; minimal erosion control	Grading; high degree of erosion control	Grading; high degree of erosion control; reclamation
Coal preparation	No control	Closed circuit water system; scrubbers for particulate control	Closed circuit water system; scrubbers and air recirculation for particulate control
Power plant	Relatively low efficiency particulate control; no SO_2, NO_x, thermal controls	High efficiency particulate control; limestone scrubbing for SO_2 control, high stack + flue gas reheat for NO_x; natural draft wet type cooling towers	Very high efficiency removal of particulates, SO_2, NO_x, i.e., by two scrubbers in tandem or scrubber plus wet-type electrostatic precipitator; dry-type cooling towers
Energy transport[3]	Standard lattice-type transmission towers throughout; no underground line	About 10 per cent longer route; about 45 per cent tubular pole-type transmission towers; no underground line	100 per cent tubular pole-type transmission towers; 10 per cent of line underground
Coal-Electric Energy System With Load-Center Power Plant			
Coal mining[4] (underground)	No control	Prevention of contact between ground water and oxidized pyritic material	Prevention plus neutralization of acidic mine drainage, if formed
Coal preparation	Same as for mine-mouth plant	Same as for mine-mouth plant	Same as for mine-mouth plant
Coal transport	Negligible residuals generated	Negligible residuals generated	Negligible residuals generated
Power plant	Same as for mine-mouth plant except for moderate efficiency particulate control plus somewhat higher stack than at mine-mouth plant	Same as for mine-mouth plant	Same as for mine-mouth plant
Energy transport[3]	Standard lattice-type transmission towers throughout	About 10 per cent longer route; 100 per cent tubular pole-type transmission towers	100 per cent of line underground

Note: See page 315 for notes to table 10.

TABLE 11

Residuals Management Costs and Residuals Discharges for Three Levels of Environmental Quality

(coal-electric energy system)

	Case I: Mine-Mouth Plant,[1] High-Impurity Seam, Area-Strip Mine			*Case II: Load-Center Plant,[2] Low-Impurity Seam, Deep Mine*		
Environmental Quality Design Level	I	II	III	I	II	III
Price of power (mills/kwh):						
At busbar	6.00	7.05	8.57	6.94	7.92	9.37
At load-center substation	7.64	9.00	11.78	7.17	8.20	9.98
Residuals Flow to Environment (annual basis, except for transmission)						
Acres of disturbed land/Acres of reclaimed land	$\left[\frac{980}{0}\right]$	$\left[\frac{1040}{1040}\right]$	$\left[\frac{1250}{1250}\right]$	rs	rs	rs
Gross increase in mine drainage:						
Million gallons	rs	rs	rs	110	80	90
Tons of sulfuric acid	rs	rs	rs	360	180	0
Preparation plant water (1 to 8% solids), million gallons	0	0	0	0	0	0
Preparation plant refuse, tons	0	1,600,000	2,600,000	0	0	300,000
Preparation plant air-borne dust, tons	0	4,400	900	0	0	700
Power-plant stack emissions, tons:						
Particulates	140,000	900	100	25,000	700	100
Sulfur oxides (sulfur content)	200,000	29,000	15,000	28,000	7,500	3,500
Nitrogen oxides (nitrogen content)	10,000	8,000	2,000	10,000	8,000	2,000
Power-plant solid waste, tons	1,000,000	950,000	800,000	300,000	450,000	300,000
Thermal discharge to watercourse, billion BTU	6,000	0	0	6,000	0	0
Water consumption (extra evaporation), million gallons	7,500	5,000	0	7,500	5,000	0
Transmission-line towers:						
Lattice type	4,000	2,500	0	120	0	0
Tubular poles	0	2,000	4,000	0	150	0
Underground circuit-miles/Total circuit-miles	$\left[\frac{0}{800}\right]$	$\left[\frac{0}{880}\right]$	$\left[\frac{80}{800}\right]$	$\left[\frac{0}{25}\right]$	$\left[\frac{0}{28}\right]$	$\left[\frac{25}{25}\right]$

Notes for Table 10

[1] Reflects the limit of available or nearly available technology. TVA implies that relatively high removals of NO_x could be achieved by some combination of facilities involving only incremental development beyond present technology. See Tennessee Valley Authority, *Sulfur Oxide Removal From Power Plant Stack-Gas-Use of Limestone Wet-Scrubbing Process* (Springfield, Virginia: National Technical Information Service, 1969) p. 60.

[2] High sulfur and high ash contents coal seam.

[3] To load center substation.

[4] Low sulfur and low ash contents coal, deep seam.

Notes for Table 11

Note: Residuals management costs are based on 1968 prices; 8 per cent rate of return on power plant investment, 10 per cent on coal mining/preparation investment.

Power plant consists of two 800-MW units; other elements of system are sized to provide the requisite input to the power plant.

rs indicates not relevant.

[1] Four-year construction period, with annual outlays of 10 per cent, 40 per cent, 40 per cent, and 10 per cent; half of draft and flue gas equipment replaced after 15 years; insurance and local taxes at 1 per cent of plant investment; 30-day supply of coal; heat rates 9,010, 9,300, 9,690 BTU/net kw, for levels I, II, and III, respectively, reflecting the increased energy required for residuals modification.

[2] Five-year construction period, with annual outlays of 5 per cent, 20 per cent, 50 per cent, 20 per cent, and 5 per cent; half of draft and flue-gas equipment replaced after 15 years; insurance and local taxes at 3 per cent of plant investment; 60-day supply of coal; heat rates 8,850, 9,175, 9,565 BTU/net kw, for Levels I, II, and III, respectively.

TABLE 12

Relative Costs of Energy

	Environmental Quality Control Level		
Type of Cost	*I*	*II*	*III*
Mine-Mouth Plant[1]			
Without transmission cost	1.00	1.18	1.43
With transmission cost	1.00	1.18	1.54
Load-Center Plant[2]			
Without transmission cost	1.00	1.14	1.35
With transmission cost	1.00	1.14	1.39

[1] Area-strip mine, high impurity coal seam.

[2] Deep underground mine, low impurity coal seam.

level of gaseous residuals modification and the shift from once-through cooling to a wet cooling tower; from Level II to Level III—primarily to power plant, where over 90 per cent is attributable to the shift to a dry cooling tower, and secondarily to the more expensive design of the transmission lines. Less than 5 per cent of the increased cost is attributable to the higher level of modification of the gaseous residuals. For the load center plant: from Level I to Level II—virtually all of the increase to the power plant, attributable to the higher level of gaseous residuals modification and to the shift to a wet cooling tower; from Level II to Level III—as with the mine-mouth plant, primarily to the power plant and transmission.

Concluding Observations

The foregoing discussion has illustrated several basic findings from a program of research on residuals management in industry. These findings can be summarized as follows.

First, there are many factors, exogenous to the plant, the company, and even to the particular industry, which affect residuals generation in industry. For some of these factors the linkage is long, but the effects are substantial. Although this may appear obvious to some, prior to such industry studies the linkages have not been traced nor spelled out quantitatively.

Second, there are many factors, endogenous to the plant, which affect residuals generation in industry. There is, of course, overlapping between the exogenous and endogenous factors, for example, with respect to the stimuli of technological change in production processes. In order to predict behavior, this overlapping must be unraveled, which can only be done by detailed microanalysis.

Third, there is a multiplicity of possible responses by plant management to constraints imposed on the discharge of residuals to the various environmental media. whatever the nature of those constraints. These options include changing raw materials, production process, even product output specifications, plus materials recovery, by-product production, and conventional residuals modification.

Fourth, management policies for a single type of residual, i.e., gaseous, liquid, solid, thermal, have often overlooked the effects on the generation of secondary and tertiary residuals and on the additional inputs required for reduction in the discharge of primary residuals, especially energy. The industry studies described herein enable explicit consideration of the physical magnitude and economic costs of these effects.

What remains is to indicate the various uses of these studies (models). At least three major uses should be mentioned. One use is that of determining the possible responses of an industry to constraints, of whatever type, placed on the discharge of residuals to the environment. This includes both the increases in costs as the contraints become more stringent and the *net* effect on total production costs under constraints, for a given product output mix. The results show that up to relatively high levels of reduction in residuals discharge, per unit of product or per unit of raw product processed, the proportion of total production costs represented by residuals management costs is only a few per cent. However, as "zero discharge" of liquid and gaseous residuals is approached, residuals management costs become a substantial proportion of total production costs.[32] Both economies of scale and multi-product outputs reduce the per unit costs of residuals management.

A second use is the corollary one of estimating the effects of variables, such as technological change and public policies, on future residuals discharges from industry.[33] Because of the many variables affecting residuals generation, this use requires analyses of possible changes over time in critical variables such as production process, product mix, tax policies, depletion allowances, secondary materials recovery technology, and costs of inputs (heat, energy, chemicals).

Both of these uses are directly relevant to current discussions of "pollution control" policies and their effects, for example, by the Council on Environmental Quality (CEQ), Environmental Protection Agency (EPA), and the U.S. Congress.[34] Each of the CEQ annual reports has included estimates of the impacts of pollution control on industry production costs. The 1972 Water Pollution Control Act Amendments included a

32. "Zero discharge" of all residuals is of course impossible, even neglecting the residuals normally considered innocuous, i.e., CO_2, water vapor, heat to the atmosphere, and dissolved solids. The inevitable consequence of a zero discharge policy for liquid and gaseous residuals is an increase in the quantity of solid residuals requiring disposal.

33. For such an application see C. W. Howe, et al., *Future Water Demands—The Impacts of Technological Change, Public Policies, and Changing Market Conditions on the Water Use Patterns of Selected Sectors of the U.S. Economy: 1970–1990* (Washington, D.C.: Resources for the Future, 1971), pp. 44–69.

34. For example, see *The Economic Impact of Pollution Control, A Summary of Recent Studies, Prepared for the CEQ, Department of Commerce, and EPA* (Washington, D.C.: U.S. Government Printing Office, 1972).

Figure 8
Components of Systems for Producing a Specified Paper Product

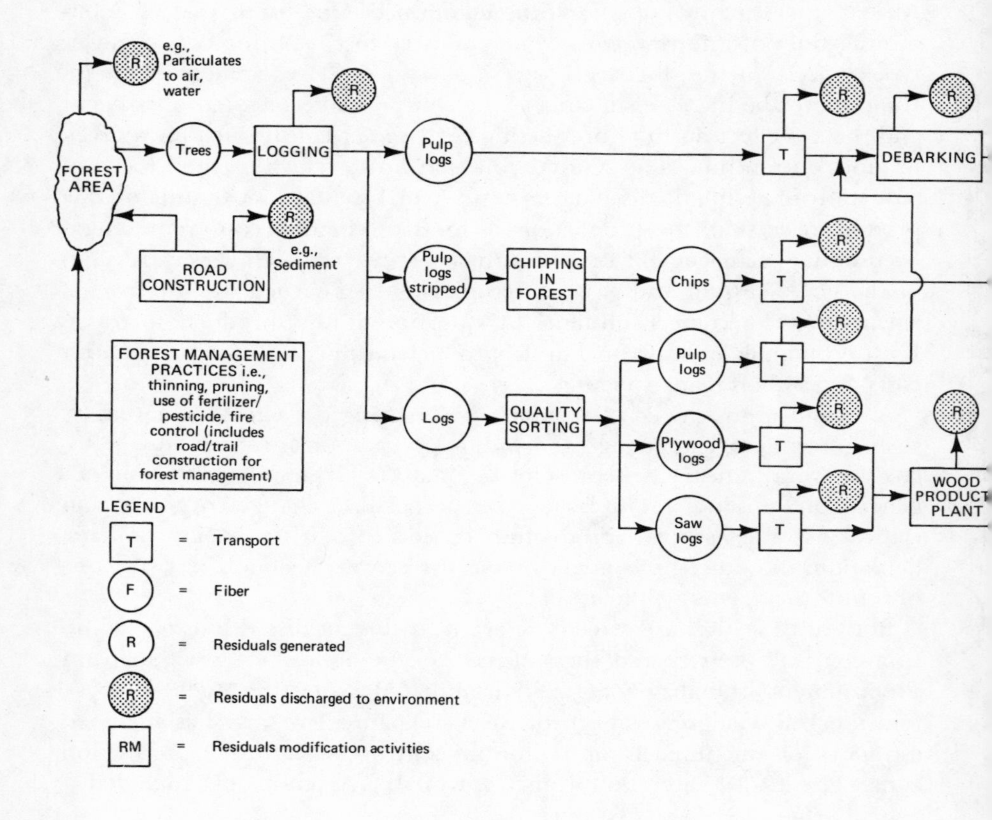

Notes:

1. Not all activities nor all sources of residuals generation are shown.
2. Only major residuals modification activities are shown.
3. Inputs into activities, i.e., electric energy, fuel, chemicals are not shown.
4. Only road construction component of transport on forest lands, which is a function primarily of logging, is shown.
5. The diagram is not meant to imply that there is only one forest area, wood products plant, paper mill, et al, involved in producing the final product.

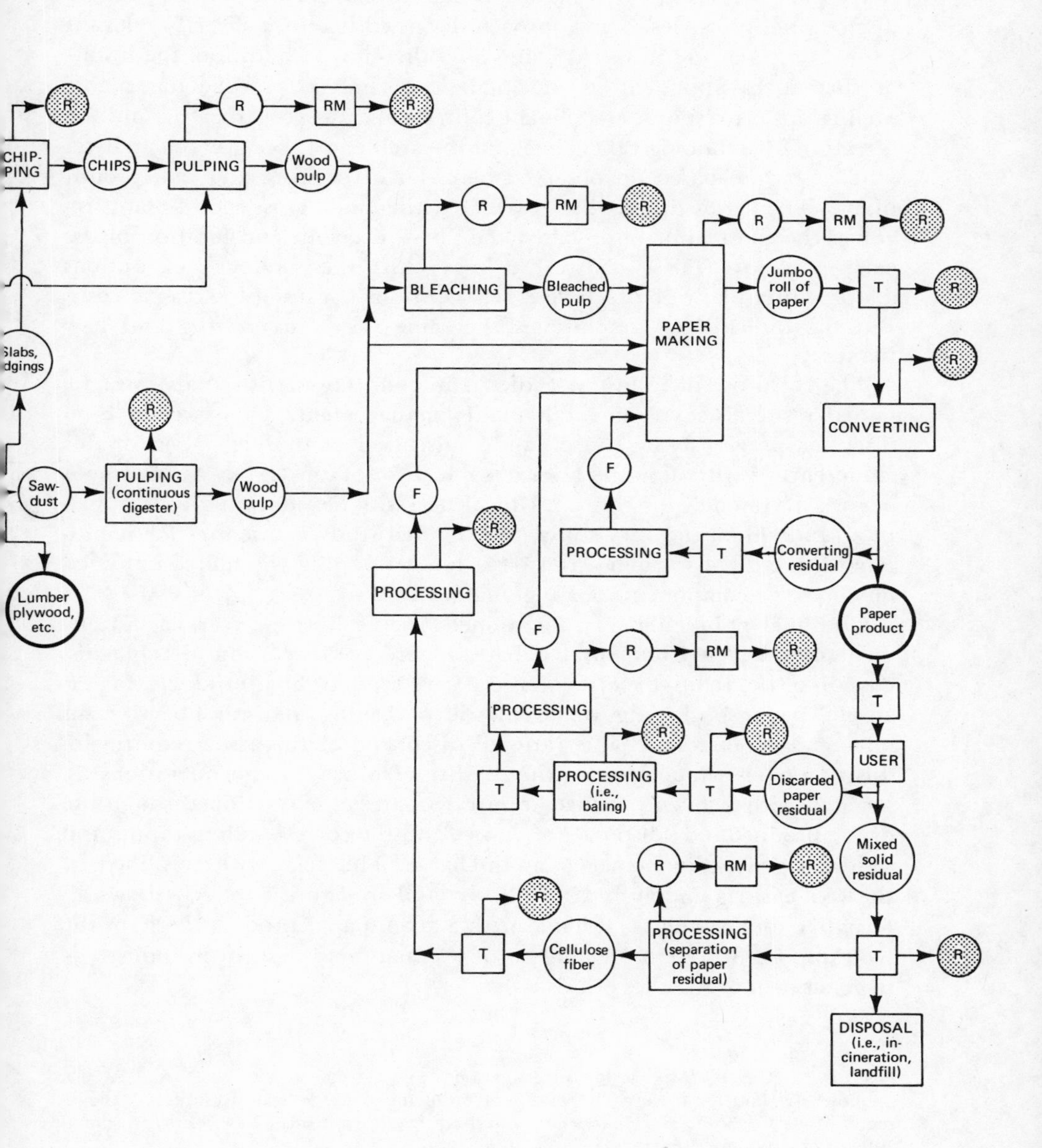

R
R
RM
R
CHIP-PING
CHIPS
PULPING
Wood pulp
R
RM
R
R
RM
R
BLEACHING
Bleached pulp
PAPER MAKING
Jumbo roll of paper
T
R
Slabs, edgings
R
R
CONVERTING
Saw-dust
PULPING (continuous digester)
Wood pulp
F
F
R
PROCESSING
T
Converting residual
Lumber plywood, etc.
PROCESSING
F
Paper product
R
RM
R
PROCESSING
T
R
R
USER
T
PROCESSING (i.e., baling)
T
Discarded paper residual
Mixed solid residual
R
RM
R
R
T
Cellulose fiber
PROCESSING (separation of paper residual)
T
R
DISPOSAL (i.e., in-cineration, landfill)

provision that studies of the cost impacts of the discharge controls specified in the Act be undertaken. In the development of discharge regulations, EPA is developing "standard effluent levels" for the major residuals-discharging industries. The approach discussed herein is directly relevant to this purpose, and in fact is being used therefor. In addition, the results of this micro approach to the analysis of industrial behavior enable predictions of: responses to effluent controls and the costs thereof, and the direction of technological change and research effort, because the analyses (and costs) are based on observed behavior at existing and newly built plants. The major difference between existing and "grass roots" plants involves the constraints imposed by the physical layout and location of the existing plant. These constraints may shift the sequence of options adopted at a plant or preclude certain options because of excessive costs. But the direction of response is the same for both existing and new plants.

The third use of industry studies is in connection with analyses of regional residuals-environmental quality management. This use has been discussed previously by Russell and is discussed in another paper for this conference.[35] Although the level of sophistication or degree of refinement of industry models for this use will of necessity be substantially less than in an individual industry study, the detailed study is essential for identifying the critical variables as a basis for developing the more simplified models to be components of a regional study.

One final point might be mentioned. In the last few years, as an element of the "environmental debate," there has been considerable discussion of the relative merits of alternative types of final products, for example, paper packaging versus plastic packaging, natural fibers versus man-made fibers. To enable rational discussion of this issue requires industry models which include the totality of processes and operations associated with a given product, from raw material extraction through use of the product and "disposal" after use, and the corresponding inputs and residuals generation-management-discharge. This focus is exemplified by the coal-electric energy industry study, and by figure 8, which shows alternative total systems associated with a given paper product. Perhaps the next important step in the analysis of residuals management in industry is to generate more studies of this type.

35. C. S. Russell, "Models for Investigation," pp. 154–156 and C. S. Russell, W. O. Spofford, Jr., and R. A. Kelly, "Operational Problems in Large Scale Residuals Management Models," *Economic Analysis of Environmental Problems* (New York: National Bureau of Economic Research, 1975).

COMMENT

Paul W. MacAvoy, Massachusetts Institute of Technology

Even the most radical environmentalist wants to know the economic impact of more severe antipollution policies. The number of dollars of cost imposed on companies—and perhaps ultimately on consumers of products from these companies—would be a useful figure to have available, if only to anticipate which policies are likely to bring forth a strong political reaction from companies and consumers. The conservative environmentalist wants the same information: base line estimates of costs of various policy alternatives, to put against the "benefits" of carrying through on these alternatives.

The RFF studies of residuals management described by Bower promise estimates of the costs of various policy options in the sugar beet, paper, petroleum refining, electricity and steel industries. The question is whether these studies can deliver useful estimates. The question might be asked in terms of (1) whether the findings are good "predictors" of policy-derived costs likely to be incurred under present circumstances in these industries and (2) in the absence of present findings on the accuracy of predictions, whether the RFF methodology can likely be used to produce predictions in the future of changes in important aspects of corporate behavior from regulatory policy changes. There is reason to doubt the ability of the residuals model to deliver, because of both empirical and methodological problems.

Empirical validation of residuals management models has not begun, at least not in terms of showing accuracy in predicting the steel or petroleum industry-wide costs of new environmental standards. The RFF approach is to build a model of a new or "grass roots" firm along lines of an "input requirements" function where the effluent is the dependent variable or "required input." There is no mention in the Bower paper of comparisons between industry-wide effluent behavior with that of his "representative firm." Moreover, most of the data required for direct comparison are not available. This is because of the extreme specificity of the RFF descriptions of the "firm." These show the marginal costs of waste treatment per physical unit of waste removed in a typical 2,700 ton plant [as in G. O. G. Löf and A. V. Kneese, *The Economics of Water Utilization in the Beet Sugar Industry* (Washington, D.C.: Resources for the Future, 1968), figure 8, page 105]. Companies do not keep statistics on marginal costs, so it is not possible to validate Bower's sensitivity analyses of *costs at the margin* with respect to policy changes. In fact, many companies produce statistics to make such studies impossible. Most firms regularly provide detailed historical cost statistics which—being subject to joint product allocation, revaluation of earlier investments, and arbitrary attribution of "investor's costs"—lead to biased indicators of economic cost changes, as shown in T. R. Stauffer, "The Measure of Corporate Rates of Return: A Generalized Formulation," *The Bell Journal of Economics and Management Science* (Autumn 1971).

The Bower findings on electric power costs, as related to environmental quality standards, are a case in point. The RFF approach of "costing out" a new plant with accompanying mine, rail and storage capacity produces estimates of cost increases from going through three environmental quality levels—with an overall percentage increase in costs per kilowatt hour of only 14 per cent for meeting 1973 standards (Bower, table 11). These are changes in incremental costs in the newest 800 megawatt electric coal fired plant with all of the policy options freely available on location, transportation, and fuel source required to make the least-cost decision in the long run. The RFF "grass roots" firm costs have to be compared with the changes in industry-wide average costs of installed and new plants required to meet new environmental quality levels. These industry-wide changes on average may be many times greater than the RFF grass roots firm changes. In fact, a significant number of East Coast generating companies showed an inability to meet the standards at any cost below regulated prices (which are clearly twice incremental costs or more). They did so by electing to declare shortages at the winter peak *with standards.* Also, others have forecast that the increases in average prices per kilowatt of capacity consequent upon cost increases will be many times larger than those increases in marginal costs shown in Bower (see M. Roberts, "Who Will Pay for Cleaner Power," *Sierra Club Conference on Electric Power Policy,* Johnston, Vermont, January 14–15, 1972). Roberts finds "increase in rates" to be 20 per cent or greater for the cases most similar to the materials management model. See also, Phillip Sporn's article in *The Environment and Economic Growth,* Sam Schurr, editor (Washington, D.C.: Resources for the Future, 1972). The problem is not that the RFF description is inaccurate; rather, the RFF changes in costs with respect to effluent discharge cannot be assessed by measurement on the company level, and industry-wide average cost changes, or price changes from effluent regulation, do not measure the same thing.

The RFF methodology is more troublesome than the predictions. First, there is a problem involved in the motivation for the model. As Bower states at the outset of his paper, the purpose is to delineate important factors which determine the disposition of materials, and to find the cost effects of manipulating these factors to improve environmental quality. This would seem to imply that an initial step in delineating these factors would be to construct models that predict accurately *at the industry level* the behavior of *industry-wide costs and outputs* with respect to environment related factors. But the models which are actually built—or so it seems from the description—are of *brand new firms, not old industries.* The RFF firms do not represent the average state of technology in the industry at the present time, nor even that in firms likely to affect industry output as a result of cost changes brought on by new environmental rules (the largest half-dozen operating firms). The RFF representative firm is free to exercise many more options than would be available to firms already installed in the industry. As a result, the RFF model firm must represent the real firm in the very long run if there were no further technological progress.[1] Thus the difficulty with the method is whether or not the "grass roots" firm is really meant to be predictive of industry behavior.

There are problems with the working procedure as well. Some readers might express displeasure with the totally orthodox nature of the framework—that, underneath all of the new technical words such as "residuals management" there is the classical framework in economic theory of the profit-maximizing multi-product firm. In this RFF formulation, the only new element is that some products (effluents) have negative prices. Going beyond qualitative description, the modeling procedure is either a "simulation approach" or a "linear programming approach." There is a third approach which should have been offered—or at least should not have been ignored, given its wide use at the present time. This is the approach in "industrial organization studies" which calls on the usual models of the multi-product firm in economics, but with cost and demand conditions described at the market level within the framework of prevailing industry institutions. The studies are usually "comparative static," in that predictions are made as to changes of prices or outputs following from changes in government policies. The forecast is tested most often by regression equations with prices and quantities in "reduced forms" as

1. This statement is more or less descriptive of the particular RFF studies—perhaps least descriptive of the sugar beet study, and most of the paper study. Each RFF report is different from each other, because of variation presumably in opportunities to use information as well as because of variations in skills of authorship. But I have taken the view that there is good reason for Bower's report beyond the mere collection of papers from one institutional source—or that there is an RFF modeling "approach," and that Bower's description of that approach is accurate.

functions of the exogenous policy variables. The tests are usually carried out on market-wide industry pricing and production, where these are expressed in ordinary accounting data, and produce statistics of "average" industry-wide effects at the present time.

The procedures proposed in the residuals management approach are in contrast to industrial economics practice. "Simulation" approaches in effect come from the construction of equations expressing engineering or physical relations in the firm where the critical coefficients are posited by those doing an "engineering design" study. The "industrial organization" approach is to posit price, cost and production relations in the market and then fit regression equations to those relations. Both approaches then simulate policy by inserting values of the policy variables in the resulting model. The standard for the simulation model is that it obeys scientific laws, and in some cases that it "optimizes." The model cannot be "tested" against prevailing conditions in the same way that regression analysis replicates present behavior because there are usually significant departures from maximum technical limits and from economic optimization in each company. And there are objections to the simulation approach: the firm being simulated is irrelevant to assessment of industry-wide effects, and the procedure is highly subjective among analysts so that there is no way to tell whether or not it is being conducted "in the correct manner."

Thus the problem is that of a choice between an industrial economic model—roughly a statistical regression equation model of industry pricing practice—and a simulation model of the newest plant. The choice to me would be the former. The industrial economic approach works with variables of the greatest concern for public policy—prices, outputs, changes of product quality where quality is measured by demand. This approach has been widely used, and now with increasing accuracy, in assessments of public policy.[2] In fact, it does not strike me that the residuals management model is likely to survive the present RFF project, because it does not meet "demands" for research findings relevant to setting market or industry-wide standards for environmental quality.

2. The most recent example of the industrial economic approach in electricity, one of the RFF industries, is that of J. M. Griffin, "A Long-Term Forecasting Model of Electricity Demand and Fuel Requirements," Ph.D. dissertation, University of Houston, 1972. This regression equation model has a "demand block," "conversion block" and "fuel share block." The division of the fuel requirements among coal, oil and natural gas is made by equations in which price differentials and local sulfur emission controls are independent variables. In fact, the fitted equations show an 11 per cent contraction in 1970 coal demand in the United States as a result of the imposition of various regional controls. Simulations of future KWH production, prices, and fuel use are made on the basis of various assumed values of GNP, fuel prices, and pollution regulations.

Acute Relationships Among Daily Mortality, Air Pollution, and Climate

Lester B. Lave, Carnegie-Mellon University
and Eugene P. Seskin, Carnegie-Mellon University

Introduction

"Killer fogs" and other acute air pollution episodes have occurred in the past fifty years.[1] The highly industrialized Meuse Valley of Belgium experienced climatic conditions permitting the buildup of abnormally high levels of air pollutants (particularly sulfur dioxide) during December 1930. Over a five-day period, approximately 6,000 people became ill and approximately 60 died (most of whom were elderly persons or those with previous heart and lung conditions). This was more than 10 times the number of deaths which would normally be expected. See [9].

In October 1948, a similar situation took place in Donora, Pennsylvania. Within three days, almost 6,000 people (over 40 per cent of the population) became ill and about 20 deaths were reported. This, again, was approximately 10 times the expected number of deaths and again the aged were most susceptible (the average age of the dead was 65). See [32].

London was enveloped by a dense fog in December 1952 and, in a two-week period, 4,000 excess deaths were attributed to the abnormally high concentrations of sulfur dioxide and smoke. Unlike the previous episodes, all age groups were affected. See [29].

Severe air pollution episodes, such as the three mentioned, do adversely

NOTE: We thank Martin S. Geisel for helpful comments. This research was supported by a grant from Resources for the Future, Inc. and by Fellowship AP48992-03, Air Pollution Control Office, Environmental Protection Agency.

1. For a detailed review of the following three episodes as well as several other air pollution incidents, see [1] and [26].

affect health. Little is known of the consequences of acute episodes of lesser severity. Long-term exposure to low levels of pollution have been investigated statistically by Lave and Seskin [22–24] across 117 U. S. Standard Metropolitan Statistical Areas (SMSAs) in 1960. We have also examined annual observations on a number of cities to explore further the association between air pollution and mortality. A consistent and significant association between air pollution and mortality was exhibited in each of these studies. We shall extend this investigation by looking at the former question of the association between daily air pollution levels and daily mortality. Before presenting the analysis, we review briefly studies which were similar in scope.

A Brief Review and Critique

Greenburg, et al. [16] investigated a November 1953 New York City episode where sulfur dioxide and smokeshade reached unusually high levels. Using analysis of variance to compare the period with six control years and assuming a three-day lag in the effect of air pollution on mortality, they concluded that there was a statistically significant increase in the number of deaths and that the increase was generally distributed over all age groups.[2] In a subsequent study, Greenburg, et al. [14] looked at pediatric and adult clinic visits during this period. Using the same control periods and again assuming a three-day lag, they found an increase in upper respiratory illnesses and cardiac visits at the four hospitals under observation. Other air pollution episodes in New York City were also scrutinized by Greenburg and his associates [15] with less consistent results.

McCarroll and Bradley [27] examined daily mortality, air pollution (sulfur dioxide and smokeshade), and weather data for New York City during a three-year period (1962–1964). The interval of time included five air pollution episodes. Using only graphical techniques, McCarroll and Bradley concluded that the periodic peaks in mortality were associated with periods of high air pollution.[3] They further noted that such peaks were not followed by drops of a sufficient degree to compensate for the excess deaths, hence they inferred that what they were detecting was

2. This was based on investigations of the Donora episode discussed above.

3. In an attempt to control for variations in climatic and other conditions, they included a 15-day moving average as a baseline from which to judge mortality peaks. In addition, graphs of temperature and wind speed were included for comparison.

more than a hastening effect. Finally, they remarked that the influence on death rates occurred in the 45–64 year group as well as the elderly over 65.

During Thanksgiving of 1966 still another episode occurred in New York City. Glasser, et al. [12] investigated the relationship between mortality and morbidity and indices of SO_2 and smokeshade. For a seven-day period, they found a total of 168 excess deaths, using a number of different control periods, some of which took into account possible lag effects. They also found the number of clinic visits for bronchitis and asthma at seven New York hospitals rose. It was concluded that the rise in temperature at the time of the episode could not account for the observed increase in mortality, although it might have been a contributing factor [12; p. 694]. In another study covering this episode, Becker, et al. [2] utilized questionnaires in a study of 2,052 executive and clerical personnel and found that ". . . as the air pollution levels increased, a greater response to symptoms of dyspnea, cough, sputum, wheeze, eye irritation, and general discomfort was elicited from the study subjects" [2; p. 419]. The direct effects of weather conditions were not accounted for, although they did consider the smoking histories of individuals.

These studies are difficult to evaluate for a number of reasons. Informal analyses such as the use of graphical techniques, are inadequate. Ad hoc procedures for analyzing small data sets require extreme care in interpretation; there is not justification for using different ad hoc assumptions (such as different lag structures) in analyzing other data sets.

In addition there are a number of more general problems that should be noted. The first surrounds the use of questionnaires and other voluntary responses to measure the effect of air pollution. Awareness of high pollution levels is increased by news coverage and other publicity. Thus, a major part of the measured response is likely to be due to the type and extent of news coverage.

Another problem surrounds the estimation of the number of "excess" deaths during a pollution episode. Expected mortality is estimated by computing averages for adjoining periods or for a previous period. No adjustments are made for population changes or variations in other factors known to affect the mortality rate. As an example of the magnitude of this problem, extrapolating the experience in Donora to London would have led to a prediction that over one million people would have been made ill by the fog, whereas the actual number was much smaller.

Finally, there are many factors that affect the daily mortality rate (or morbidity rate) which have not been controlled. McCarroll and Bradley found that during one episode there was a sharp increase in the number of deaths although the pollution measures they considered were not unusually high (an atmospheric inversion did take place). They pointed out the need for studying other relevant factors. Regarding the Thanksgiving study by Glasser, et al., Eckhardt remarked in a letter to the editor, "Lung cancer deaths can hardly be related to air pollution unless the following factors have been investigated: Is the case postoperative? How many days postoperative? How long did the patient have lung cancer" [7; p. 837]? In addition to these factors, the researchers were aware of the difficulty in assessing the effects of the holiday itself. Eckhardt again commented, "Increased food intake and relaxation of salt restrictions for cardiacs on festive days like Thanksgiving, I am sure exact their toll. . . . Psychiatrists tell us that suicides increase over holidays, and the National Safety Council tells us that highway deaths go up with four-day weekends" [7; p. 837]. Ideally, all these factors should be accounted for; however, data enabling such detailed analysis is seldom available.

Multivariate statistical analysis offers one approach in coping with the possible effects of a number of "relevant" variables on mortality. It is known that climatological characteristics affect mortality [3, 4, 31], air pollution affects climate [11], and air pollutants and climate may have synergistic effects [25]. A multivariate statistical model can simultaneously consider these interrelationships.

Such a study on short-term effects was undertaken by Hodgson [21]. Using multiple regression, he analyzed mortality, air pollution, and certain meteorological factors between 1962 and 1965 in New York City. Hodgson addressed the question: "How is air pollution during [time] *t* influencing short-run future mortality" [21; p. 590]? He found that mortality from respiratory and heart diseases for all ages was significantly related to the level of air pollution. Mortality from other causes was not so related. Furthermore, the effects of the environmental factors on mortality occurred on the day of increased pollution and extreme temperature with lesser effects on the day following. Finally, he concluded from his results that the increase in mortality observed was not merely the bunching together of deaths of persons already ill.

Some questions arise with regard to Hodgson's analysis. Although the author professes concern with day-to-day variations in the health effects of pollution, most of his analysis is based on monthly averages. Detailed analysis of the acute reaction is lost in this aggregation. In addition, aggregating the daily data into months enhances the multicollinearity problems among the explanatory variables (especially when lagged periods are included) and so Hodgson utilized moving averages of the monthly means. As pointed out by Hodgson, this sacrifices much of the short-term effects. In defense, he states that the index will then reflect the cumulative effect of air pollution. It is true that the moving averages will represent the mean pollution levels for the number of months included, but they will neither reflect longer-term effects nor day-to-day effects.

Hodgson does analyze daily observations at the end of his paper, but again he uses moving averages and the interesting lag effects are masked. The daily results are not found to be significantly different from the monthly regressions. A natural hypothesis with such a study is that air pollution merely hastens death of those who would have died shortly. Hodgson remarks, ". . . if the sole effect of air pollution was to redistribute the deaths within a short time interval, for example, a month or less, then the average daily mortality for a given month would be independent of the concentration of air pollution during the month and no statistically significant relation would be observed between monthly mortality and air pollution" [21; p. 593]. The assertion is incorrect since monthly averages would only mask such an effect, not eliminate it. At both the beginning and end of the month there would be days where one or two-day shifts in deaths would still be present. To examine the redistribution question a more complicated lag structure is needed as well as a comparison with cross-section data.

Another recent study employing regression analysis was conducted by Hexter and Goldsmith [18]. Daily mortality in Los Angeles County was related to carbon monoxide and oxidant concentrations and maximum temperature for a four-year period (1962–1965) via cross-spectral analysis. A significant association between mortality and carbon monoxide pollution was found, while no association was demonstrated between mortality and oxidant.[4]

Although Hexter and Goldsmith included maximum temperature as a factor likely to be important in explaining mortality, they did not adequately take account of other possibly relevant factors. This was emphasized in a subsequent letter to the editor. Mosher, et al. [30] found that for the period corresponding to the Hexter and Goldsmith study the

4. They made no attempt to resolve the conflict with the Hechter and Goldsmith study [17] which found that daily Los Angeles deaths from 1956 to 1958 were not correlated with daily oxidant and carbon monoxide concentrations. Many investigators have detected an association with daily data. Mills [28] found a significant association between Los Angeles smog and day-by-day respiratory and cardiac deaths, as did Brant and Hill [6] and Brant [5] when they examined the effects of oxidants on weekly hospital admissions in Los Angeles for respiratory and cardiovascular dysfunction.

correlations between carbon monoxide and nitric oxide or nitrogen dioxide were significant at a level of more than 99.9 per cent. Thus, one cannot isolate which of the three pollutants is the "true cause" of illness. Ellsaesser [8] questioned the ad hoc procedure employed by Hexter and Goldsmith. He argued that their method combined with their failure to take account of other important factors made the results suspect. For instance, the population (and therefore the population at risk) increased over time. As another example, Ellsaesser discussed the interrelationships among carbon monoxide, traffic fatalities and mortality.

Hexter and Goldsmith [19] replied that further analysis showed that changes in population structure were not affecting their previous results. They also found a negative association between CO and traffic fatalities when a death-specific regression was analyzed.

Glasser and Greenburg [13] examined daily deaths in New York City between the years 1960–1964 (excluding April to September[5]). They attempted to explain deviations in daily deaths from a five-year "normal" by air pollution measures (SO_2 and smokeshade), and climatological variables (temperature deviation from normal, wind speed, sky cover, and rainfall).[6] Using descriptive statistics, cross tabulations, and regression analyses, they found a relationship between daily mortality and air pollution (primarily as measured by SO_2).[7]

One must be curious as to the omission of six months from each year. It is true that seasonal factors and air pollution might be confounded during the summer months, however, regression analyses, which include both the relevant weather variables and pollution variables should sort out the individual effects. If the two semi-annual periods differ in some respects, two models can be developed and these can be tested in order to determine whether the underlying structures are similar. Analysis of this time of year is of particular interest in view of the number of inversions and high pollution episodes occurring during the summertime.

Glasser and Greenburg measured mortality in terms of deviations from a 15-day moving average. The measure is suggestive of cycles in the data

5. These months were omitted ". . . because of the generally low levels of air pollution and generally high temperatures during these months. It was believed that including such data would complicate the analysis since seasonal factors would be confounded with air pollution factors" [13; p. 336].

6. The "normal" measure was defined as the mean for corresponding days in each year of a 15-day moving average centered on the 8th day.

7. After an examination of the individual effects of SO_2 and smokeshade, it was concluded that the exclusion of smokeshade as a measure of air pollution would not substantially alter the results.

and of lags in the effects of pollution, however, no explanations are presented. Lagged variables (particularly pollution) are not utilized, hence the investigation never questions the timing of the effect.

Data

As with previous work in the field, the present study is limited by the availability of pollution data and to a lesser extent mortality data. Daily observations on pollution levels (24-hour averages) were obtained from the Continuous Air Monitoring Program (CAMP). There were seven cities and eleven pollutants for which data was collected; however, the series are far from complete. In fact, a complete series for a given city during a single calendar year for one pollutant was nonexistent. Thus, we restricted ourselves to the most complete set of series available.[8] The cities remaining under consideration were Chicago, Denver, Philadelphia, St. Louis, and Washington, D.C. The pollutants remaining were carbon monoxide (CO), nitric oxide (NO), nitrogen dioxide (NO_2), sulfur dioxide (SO_2), and hydrocarbons (HC).

Corresponding to the daily pollution readings, we obtained daily death counts for the five cities from 1962 to 1966 from a special study by the Environmental Protection Agency. These deaths were classified into 47 different causes ranging from tuberculosis to motor vehicle accidents; however, in this paper we limit ourselves to analyzing total deaths.

Finally, we secured climatological data on daily weather factors for the cities from the Department of Commerce. Air pollution, mortality, and climate information constituted our data base. We selected periods when the air pollution data were complete for analysis. (See table 1 for a list of the individual data sets.)

The Model

According to Hodgson, "Nobody knows how air pollution causes death and, unfortunately, controlled scientific enquiry into the nature and mechanism of the deaths has still to be designed" [20; p. 15]. Our own literature review concurs with Hodgson that there is little theory available to specify the functional form of the relationship. There is no alter-

8. Those daily observations which were missing were estimated by simple interpolation.

TABLE 1

Data Sets With Means and Standard Deviations of Pollutants and Deaths

Dates		Pollutants[a] Mean	Pollutants[a] Standard Deviation	Deaths Mean	Deaths Standard Deviation
			Chicago		
8/63–8/64	NO	0.090	0.038	114.62	13.99
8/63–8/64	NO_2	0.045	0.016	114.62	13.99
9/62–6/63	SO_2	0.167	0.131	121.05	16.94
9/63–6/64	SO_2	0.185	0.146	116.37	13.10
8/63–5/64	HC	3.005	0.819	114.80	13.59
9/63–5/64[b]	NO	0.099	0.040	116.39	12.96
9/63–5/64[b]	NO_2	0.041	0.012	116.39	12.96
9/63–5/64[b]	SO_2	0.198	0.147	116.39	12.96
9/63–5/64[b]	HC	2.992	0.779	116.39	12.96
			Denver		
3/65–7/66	NO_2	0.034	0.010	25.17	5.58
3/65–7/66	SO_2	0.017	0.009	25.17	5.58
4/65–7/66	HC	2.377	0.704	25.18	5.67
			Philadelphia		
9/65–10/66	CO	7.256	2.397	25.21	5.38
9/64–4/65	NO	0.056	0.043	24.85	5.06
9/65–9/66	NO_2	0.037	0.014	25.27	5.41
			St. Louis		
4/64–2/66	CO	6.501	2.526	14.18	4.04
9/64–4/65	NO	0.039	0.024	14.68	3.99
9/64–6/66	NO_2	0.028	0.011	14.32	4.07
1/65–6/66	SO_2	0.048	0.028	14.38	4.13
			Washington, D.C.		
4/65–10/66	CO	3.397	1.532	27.93	9.88
11/64–7/65	HC	2.397	0.758	27.15	5.78

[a] Figures for pollution in parts per million (ppm).
[b] Nine month Chicago data set (see text).

native to investigating a number of possible specifications and to performing a sensitivity analysis.

A function explaining mortality can be written as in equation 1:

$$MR = MR(G, H, SE, P, C, e), \tag{1}$$

where G represents genetic characteristics, H represents personal habits (e.g., exercise, eating), *SE* represents socioeconomic characteristics (e.g., income, age, race, occupation, medical care, etc.), *P* represents environmental pollution, *C* represents climatological characteristics, and *e* is all other factors. Some of these factors are difficult to measure (e.g., genetic factors and nutritional history), while data on other factors have not been collected (e.g., smoking habits).

From day to day, within a single city, we further hypothesize that *G*, *H*, and *SE* remain essentially constant. Thus, for examining day-to-day changes in mortality, the relevant function is:

$$MR = MR'(P, C, e), \tag{2}$$

where these variables are defined above.

We expect mortality at time t to be associated with current pollution (at time t) and pollution levels on immediately preceding days as well as with a number of meteorological variables. The coefficients of the lagged pollution variables should shed light on the short-term effects of pollution.

More precisely, we will be considering models with finite lags which are linear, multiplicative, or quadratic (including interaction terms). One such model is shown as equation 3:

$$\begin{aligned} MR_t = {} & 30.056 + \underset{(1.16)}{0.046\, SO_{2_t}} - \underset{(-1.19)}{0.051\, SO_{2_{t-1}}} - \underset{(-.88)}{0.038\, SO_{2_{t-2}}} + \underset{(.80)}{0.035\, SO_{2_{t-3}}} \\ & - \underset{(-.48)}{0.021\, SO_{2_{t-4}}} - \underset{(-.14)}{0.006\, SO_{2_{t-5}}} - \underset{(-.14)}{0.001\, \text{Wind}} - \underset{(-.68)}{0.009\, \text{Rain}} \\ & - \underset{(-3.81)}{0.057\, \text{Mean } T} - \underset{(-1.72)}{1.585\, \text{Sun.}} - \underset{(-1.18)}{1.088\, \text{Mon.}} - \underset{(-1.81)}{1.676\, \text{Tues.}} \\ & - \underset{(-1.61)}{1.504\, \text{Thurs.}} - \underset{(-1.60)}{1.481\, \text{Fri.}} - \underset{(-.78)}{0.725\, \text{Sat.}} \end{aligned} \tag{3}$$

This model and an explanation of the variables are discussed below.

Method and Results

Having little a priori knowledge of the underlying relationships among mortality, air pollution, and weather factors, we initially fit models such as equation 3 to Chicago, Denver, Philadelphia, St. Louis, and Washington, D. C. using ordinary least squares. In each case we regressed the daily deaths on the level of pollution at time t, at time t − 1, and so on up to time t minus five days. We also included a number of weather factors which were similar to those found in the recent study by Glasser and Greenburg [13] cited above. These consisted of average wind speed (Wind), rainfall (Rain), and mean temperature (Mean T). In addition, we included a set of 1–0 dummy variables for the days of the week, since it was hypothesized that both pollution and mortality are cyclic over the week.[9]

Results are shown in tables 2 through 4 and the first regression in table 2 is reproduced in equation 3 above. The daily number of people dying in Denver from March 1965 through July 1966, is regressed on air pollution (SO_2), climate variables, and day-of-the-week variables and 5.6 per cent of the variance is explained. The t statistics are shown in parentheses below the estimated coefficients (see table 2, footnote 1). The coefficients of the SO_2 variables suggest that there is a contemporaneous effect of high levels on the number of deaths, but the association is not statistically significant. When high levels of SO_2 occur, one, two, four, and five days prior to the day of observed deaths, there is a negative correlation with deaths, although none of these effects is significant. Of the climate variables, only mean temperature is significant (indicating that as the mean temperature increases, the number of deaths decreases). The greatest numbers of deaths seem to occur on Wednesdays and the smallest numbers on Tuesdays.

Results are similar for four of the five cities and the five pollutants in that no consistent pattern of statistically significant coefficients can be discerned, and hence there is no evidence of an effect of air pollution on daily mortality. The only exception is Chicago, especially for the pollutant SO_2.* In St. Louis one coefficient is positive and significant, although the sum of pollution coefficients indicates that the effect was negligible (see the second regression of table 2). For Chicago, three coefficients are significant (and all positive), indicating a close association between daily deaths and SO_2, and the sum of pollution coefficients indicates a strong effect (see the third regression of table 2).[10] Some climate variables are significant in each city. Average wind speed is significant only for St. Louis, indicating a negative relationship with daily mortality. This is plausible since wind cleanses the air.[11] Rainfall never attains

9. It was hypothesized that pollution would be lower on weekends due to decreased industrial activity. Silverman [33] has detected a definite day-of-the-week influence on emergency hospital admissions. We felt this effect might also carry over to daily deaths.

10. A useful way in which to view the magnitude of this association for Chicago is to consider a 50 per cent reduction in the mean level of SO_2 pollution. A decrease of .092 ppm in the mean values of the pollution variables is estimated to lead to a decrease of 6.3 deaths per day (approximately 5.4 per cent).

11. The simple correlation between Wind and SO_{2t} is −.30.

* We have already discussed the results of using the sulfur dioxide measure in Denver.

TABLE 2

Regression Analysis Comparing Daily Deaths in Denver, St Louis, and Chicago

	Denver, 3/65–7/66		*St. Louis, 1/65–6/66*		*Chicago, 9/63–6/64*	
Variable	*Coefficients*[a]	*Means*[b]	*Coefficients*[a]	*Means*[b]	*Coefficients*[a]	*Means*[b]
Dependent variable		25.17		14.38		116.37
Independent variables:						
Constant	30.056		16.849		93.268	
(ppm × 1,000)						
SO_{2t}	0.046 (1.16)	17.45	−0.005 (−0.70)	47.54	0.026 (2.90)	184.79
SO_{2t-1}	−0.051 (−1.19)	17.44	0.003 (0.44)	47.56	0.004 (0.39)	184.24
SO_{2t-2}	−0.038 (−0.88)	17.44	0.017 (2.33)	47.51	0.028 (2.87)	184.00
SO_{2t-3}	0.035 (0.80)	17.43	−0.008 (−1.18)	47.49	−0.007 (−0.71)	183.95
SO_{2t-4}	−0.021 (−0.48)	17.43	−0.004 (−0.51)	47.42	0.028 (2.88)	184.02
SO_{2t-5}	−0.006 (−0.14)	17.40	−0.002 (−0.30)	47.41	−0.011 (−1.38)	183.93
Wind (mph × 10)	−0.001 (−0.14)	78.97	−0.011 (−1.90)	98.49	−0.004 (−0.22)	113.03
Rain (in. × 100)	−0.009 (−0.68)	4.96	0.003 (0.38)	8.53	0.054 (1.25)	6.03
Mean T	−0.057 (−3.81)	51.99	−0.036 (−3.92)	53.63	0.182 (3.38)	47.59
Sunday	−1.585 (−1.72)	0.14	1.173 (1.74)	0.14	2.868 (1.11)	0.14
Monday	−1.088 (−1.18)	0.14	0.486 (0.73)	0.14	5.107 (1.98)	0.14
Tuesday	−1.676 (−1.81)	0.14	0.291 (0.44)	0.14	4.329 (1.70)	0.14
Thursday	−1.504 (−1.61)	0.14	0.262 (0.39)	0.14	0.388 (0.15)	0.14
Friday	−1.481 (−1.60)	0.14	0.724 (1.08)	0.14	−1.451 (−0.56)	0.14
Saturday	−0.725 (−0.78)	0.14	0.767 (1.14)	0.14	3.628 (1.39)	0.14
R^2	.056		.050		.266	

[a] Figures in parentheses are t statistics. A value of 1.65 indicates significance at the .05 level using a one-tailed test on a sample of infinite size.

[b] For computational ease, the means of some variables were multiplied by 10, 100, or 1,000 as indicated next to the relevant variables. This applies to all subsequent tables.

statistical significance. Mean temperature is strongly negative for Denver and St. Louis, whereas it is positive for Chicago (a result which needs further investigation). While many of the day-of-the-week variables are statistically significant, it is difficult to explain the different patterns in the three cities.[12] Finally, the coefficient of determination (R^2) is quite small for Denver and St. Louis (.056 and .050, respectively), while for Chicago almost 27 per cent of the variation in daily deaths is explained by the variables.

A natural question arises as to why the Chicago results differ from the results for the other cities. One possibility concerns the unusually high levels of pollution prevalent in Chicago. For each pollutant under consideration, the mean level in Chicago exceeds the mean value for any other city (see table 1). This is especially noteworthy for SO_2 where Chicago's mean levels are almost ten times the levels of Denver and almost four times the levels of St. Louis. Thus, it may be that at levels of pollution substantially below those found in Chicago, acute effects are not important. A closely related issue involves the relative size of the cities. Because of its larger population, the mean number of deaths per day in Chicago is more than four times as large as any other city in our sample (see table 1). Since deaths occur in discrete units (one at a time), effects of pollution in a smaller city may be lost when scattered over a five-day period.[13] We shall examine these conjectures in more detail below.

For whatever reason, only Chicago regressions showed a significant relationship between daily mortality and air pollution. We have attempted to investigate why this occurred and to examine the Chicago relationship more intensively. Much of the remaining estimation is carried out on a nine-month data set (denoted by superscript b in table 1) which contains measures of four pollutants.

Results for Chicago

In addition to the explanations given above that Chicago has much greater levels of pollution and is a large enough city to be able to observe

12. The day-of-the-week variables were not found to be correlated with the pollution levels.

13. Close associations between daily mortality and air pollution have been found for New York City and Los Angeles, both of which are large cities with high levels of air pollution [13, 17].

the effects of pollution more easily, there is the possibility that we have misspecified the relation. We have examined this possibility in a number of ways. First we attempted to see if the structure was the same for months with low pollution as it was for months with high pollution. We divided the two Chicago data sets with SO_2 as the pollutant on the basis of the high and low pollution levels during the year. F ratios were computed to test the hypothesis that the structures dicered between high and low pollution periods. The F ratios were 1.25 and 1.32 for the two data sets. Since $F_{.05}(30\ \infty) = 1.46$, one cannot reject the null hypothesis that the coefficients are identical. While this test is supportive of the notion that the relationship was correctly specified, one can object that it is not relevant to the comparison between Chicago and the other cities, since even the low Chicago months had pollution levels far greater than those of the other cities.

Second, we attempted to fit a relation which would be more sensitive to different effects of pollution according to the level of pollution. We divided the pollution variable into five new variables, each of which represented a different range of pollution levels.[14] These "piecewise" linear variables can approximate much more complicated functional forms. For both data sets, the coefficients for the smallest two variables were insignificant, both statistically and numerically. This is evidence that daily effects of relatively low levels of air pollution are not important.

Third, we fit other functional forms, quadratic and log-linear and included interaction terms (e.g., $HC \times NO$). The coefficients of the quadratic variables were not statistically significant and added little to the explanatory power of the regression. The log-linear specification exhibited results similar to the linear specification; the elasticities at the mean were quite close for the pollution variables. Of the interaction variables, only $HC \times SO_2$ attained statistical significance.

Fourth, we included all four pollutants simultaneously, as shown in the first regression of table 3. Only NO and SO_2 possessed statistically significant coefficients (both were positive). The coefficients for the weather variables were quite similar to those in the previous Chicago regression in table 2.[15] The day-of-the-week dummy variables were almost identical in magnitude and significance to those reported in table 2. It is noteworthy that almost 36 per cent of the variation in daily deaths was explained by the variables.

14. In this formulation, piecewise lags were excluded.
15. The sign of "Wind" changed, but was quite insignificant.

TABLE 3

Three Alternative Specifications Analyzing Daily Deaths in Chicago

	Original		*Deaths as Deviations from a Moving Average*		*Episodic Pollution*	
Variable	*Coefficients*[a]	*Means*	*Coefficients*[a]	*Means*	*Coefficients*[a]	*Means*
Dependent variable		116.39		0.14		116.39
Independent variables:						
Constant	96.591		−19.181		100.038	
HC_t	0.012					
	(0.72)	299.22				
HC_{t-1}	−0.001					
	(−0.05)	298.91				
HC_{t-2}	0.003					
	(0.14)	299.16				
HC_{t-3}	−0.015					
	(−0.74)	299.10				
HC_{t-4}	0.009					
	(0.45)	298.49				
HC_{t-5}	−0.006					
	(−0.36)	297.90				
NO_t	−0.027		−0.032	100.25	−0.025	99.24
	(−0.86)	99.24	(−1.16)		(−0.82)	
NO_{t-1}	0.007		0.007	100.38	−0.013	99.33
	(0.23)	99.33	(0.27)		(−0.47)	
NO_{t-2}	0.029		0.024	100.61	0.027	99.41
	(0.90)	99.41	(0.96)		(0.96)	
NO_{t-3}	0.058	99.45	0.027	100.62	0.036	99.45
	(1.82)		(1.07)		(1.49)	
NO_{t-4}	−0.007		−0.016	100.48		
	(−0.23)	99.24	(−0.63)			
NO_{t-5}	0.046		0.015	100.56		
	(1.68)	99.14	(0.66)			
NO_{2_t}	0.053					
	(0.58)	40.91				
$NO_{2_{t-1}}$	−1.28					
	(−1.22)	40.94				
$NO_{2_{t-2}}$	0.010					
	(0.10)	40.96				
$NO_{2_{t-3}}$	−0.155					
	(−1.49)	40.94				
$NO_{2_{t-4}}$	0.009					
	(0.08)	40.94				

(*continued*)

TABLE 3 *(concluded)*

Variable	*Original*		*Deaths as Deviations from a Moving Average*		*Episodic Pollution*	
	Coefficients[a]	*Means*	*Coefficients*[a]	*Means*	*Coefficients*[a]	*Means*
$NO_{2_{t-5}}$	−0.079					
	(−0.94)	40.92				
SO_{2_t}	0.022		0.019	205.07	0.023	198.26
	(2.31)	198.26	(2.15)		(2.45)	
$SO_{2_{t-1}}$	0.007		−0.002	204.85	0.003	198.12
	(0.66)	198.12	(−0.25)		(0.29)	
$SO_{2_{t-2}}$	0.020		0.019	204.60	0.017	198.04
	(1.92)	198.04	(1.95)		(1.62)	
$SO_{2_{t-3}}$	−0.006		−0.014	203.88	0.001	198.02
	(−0.61)	198.02	(−1.44)		(0.10)	
$SO_{2_{t-4}}$	0.023		0.026	203.81		
	(2.16)	198.02	(2.72)			
$SO_{2_{t-5}}$	−0.014		−0.018	203.84		
	(−1.52)	198.03	(−2.16)			
$\left(\prod_{k=0}^{3} NO_{t-k}\right)^{b}$					0.002	143.54
					(0.31)	
$\left(\prod_{k=0}^{3} SO_{2_{t-k}}\right)^{b}$					0.0001	6904.47
					(1.01)	
Wind	0.002		−0.011	116.71	−0.008	115.03
	(0.07)	115.03	(−0.46)		(−0.34)	
Rain	0.067		0.049	6.25	0.061	5.98
	(1.48)	5.98	(1.16)		(1.36)	
Mean T	0.171		0.233	43.76	0.073	44.74
	(2.42)	44.74	(3.86)		(1.10)	
Sunday	3.125		1.189	0.14	3.551	0.14
	(1.14)	0.14	(0.46)		(1.32)	
Monday	5.112		4.366	0.14	2.358	0.14
	(1.87)	0.14	(1.67)		(0.87)	
Tuesday	3.659		3.577	0.14	4.135	0.14
	(1.38)	0.14	(1.40)		(1.52)	
Thursday	0.743		0.125	0.14	3.598	0.14
	(0.28)	0.14	(0.05)		(1.35)	
Friday	−1.252		−1.215	0.14	−0.284	0.14
	(−0.46)	0.14	(−0.46)		(−0.11)	
Saturday	3.282		2.951	0.14	−1.467	0.14
	(1.22)	0.14	(1.14)		(−0.54)	
R^2	.357		.164		.297	

[a] Figures in parentheses are t statistics.
[b] Multiplied by 10^{-6}.

Fifth, we estimated the specification of Glasser and Greenburg [13]. We altered the dependent variable, so that instead of examining deaths at time t in terms of (current and lagged) pollution and current weather factors, we examined deviations from a 15-day moving average centered on the eighth day.[16] Inasmuch as only NO and SO_2 were significant in explaining daily deaths in Chicago (results just cited), we limited ourselves to current and lagged values of these two pollutants in rerunning the previous regression. We again included the climatological and day-of-the-week variables. (This is presented as the second regression in table 3.)

As can be seen by comparing this regression with the previous one, there was little difference in the results. The NO variables were no longer statistically significant, although the magnitude and significance of the SO_2 variables remained essentially unchanged.[17] The significant weather variables and dummy variables exhibited little change when the "new" dependent variable was included. The explanatory power of the regression R^2 dropped from .357 to .164. Since the formulation had no theoretical justification and since it was less satisfactory empirically, we rejected it.

It is difficult to compare these results with those of Glasser and Greenburg since they included neither lagged pollution nor day-of-the-week variables.[18] Of their meteorological variables, wind speed displayed mixed signs when it was statistically significant (as did ours, although it was not significant) and rainfall exhibited a positive coefficient when significant (as did ours). Because their temperature variable was expressed as a deviation from normal, we cannot easily compare it with our mean temperature measure.

Sixth, we investigated whether a consecutive period of several days of high pollution was more important than isolated days of high pollution. Many of the episodic studies have found significant effects on mortality during such occurrences [12, 16, 27]. We defined a new SO_2 variable as the product of the SO_2 levels for the current day and the three preceding days.[19] We also defined a similar NO variable.

16. In notational terms, the left-hand side of equation 3 becomes

$$MR_t - \sum_{k=-7}^{7} \frac{MR_{t+k}}{15}.$$

17. The coefficient of $SO_{2_{t-5}}$ was negative and quite significant. This will be discussed below.

18. They did disaggregate their data by day of the week and run separate regressions.

19. In notational terms, $P_t = \prod_{k=0}^{3} P_{t-k}$.

The results of adding these new pollution variables are presented as the third regression in table 3. Comparing the first and third regressions, the magnitude and significance of the separate pollution variables exhibited little change. The new pollution variables were statistically unimportant.[20] In examining the weather variables one finds that mean temperature lost significance while precipitation gained significance. "Wind" remained unimportant. The explanatory power of this regression was .297.

Seventh, we tried other specifications, including moving averages and variables measuring the change in pollution levels on consecutive days. None proved as satisfactory as the simple linear form in terms either of explanatory power or the significance of the coefficients.

Eighth, we tried a final test to judge the linearity of the specification. Both cross-section and time-series data should reflect the same underlying structure. Using measures of particulates and sulfates, we had previously determined that a linear specification fit cross-section data as well as other alternatives. We used these cross-section data to estimate a piecewise linear specification similar to that estimated above for the time-series data. To make the comparison with the daily results, we reran the regression including mean sulfates as the only measure of air pollution since it is related to measures of SO_2. The results were quite similar to the piecewise linear form of the daily data.

Ninth, having settled on the linear form with five days of lags, we attempted to refine the estimates by using the Almon technique, as programmed and discussed by Gaver [10]. This procedure imposes structure on the lag coefficients by constraining them to fit a polynomial curve of a specified degree. The process often results in the reduction of large standard errors in the distributed lag coefficients which may arise from multicollinearity in the lagged values of the pollution variables.[21] The technique allows considerable flexibility; we began by fitting second and third-degree polynomials, using lags of five to ten days. The third-degree polynomial added little to the analysis so we confined our interest to the second-degree polynomial. The only other qualification we imposed was that the current pollution coefficient be positive inasmuch as a negative coefficient was deemed unreasonable.[22] In this case the effect of such a

20. We experimented with other variables, but were unable to detect an episodic effect.

21. For example, the simple correlations between NO_t and NO_{t-1} and between SO_{2t} and SO_{2t-1} were .59 and .78, respectively.

22. In practice, this qualification was unnecessary for SO_2; however initially the zero lag of NO had a negative (although statistically insignificant) coefficient. Thus, only the latter variable had such a restriction.

front-end restriction was simply to shift the polynomial so that the t + 1 lag (one future day) had a zero weight.

In table 4 the results of a regression are shown which employed the

TABLE 4

***Regression of Refined Lag Structures of* NO *and* SO_2**

Independent Unlagged Variables	*Coefficients*[a]	*Means*[b]	*Independent Lagged Variables*	*Weight*[a]	*Independent Lagged Variables*	*Weight*[a]
Constant	85.58		NO_t	0.007 (2.53)	SO_{2_t}	0.017 (2.94)
Wind	0.013 (0.65)	114.55	NO_{t-1}	0.013 (2.61)	$SO_{2_{t-1}}$	0.015 (5.39)
Rain	0.053 (1.22)	6.34	NO_{t-2}	0.016 (2.71)	$SO_{2_{t-2}}$	0.012 (4.33)
Mean T	.154 (2.66)	45.17	NO_{t-3}	0.018 (2.85)	$SO_{2_{t-3}}$	0.009 (2.72)
Sunday	2.409 (.91)	.14	NO_{t-4}	0.017 (3.00)	$SO_{2_{t-4}}$	0.005 (1.84)
Monday	4.544 (1.73)	.14	NO_{t-5}	0.015 (2.94)	$SO_{2_{t-5}}$	0.001 (.32)
Tuesday	3.391 (1.31)	.14	NO_{t-6}	0.010 (1.98)	$SO_{2_{t-6}}$	−0.004 (−.73)
Thursday	0.177 (.07)	.14	NO_{t-7}	0.004 (.51)		
Friday	−1.537 (−.58)	.14	NO_{t-8}	−0.005 (−.39)		
Saturday	3.378 (1.27)	.14				
$R^2 = .291$						

[a] The figures in parentheses are t statistics.
[b] Mean of dependent variable is 116.18.

Almon technique using second-degree polynomials in estimating the lag structure of the two pollutants.[23] The climate variables and the day-of-the-week variables are essentially unchanged from the previous results. The magnitude and significance of the pollution variables display inter-

23. This particular regression was chosen because it had the minimum estimated standard error of disturbances adjusted for the degrees of freedom.

esting characteristics. The maximum effect of NO on daily deaths occurs at a lag of three days. This could have implications as to the physiological mechanism involved. The significant SO_2 weights (t to $t - 4$)[24] imply a different physical response. The maximum effect on deaths takes place simultaneously with peaks in SO_2 pollution levels and then drops off. (Figures 1 and 2 illustrate the two underlying structures.)

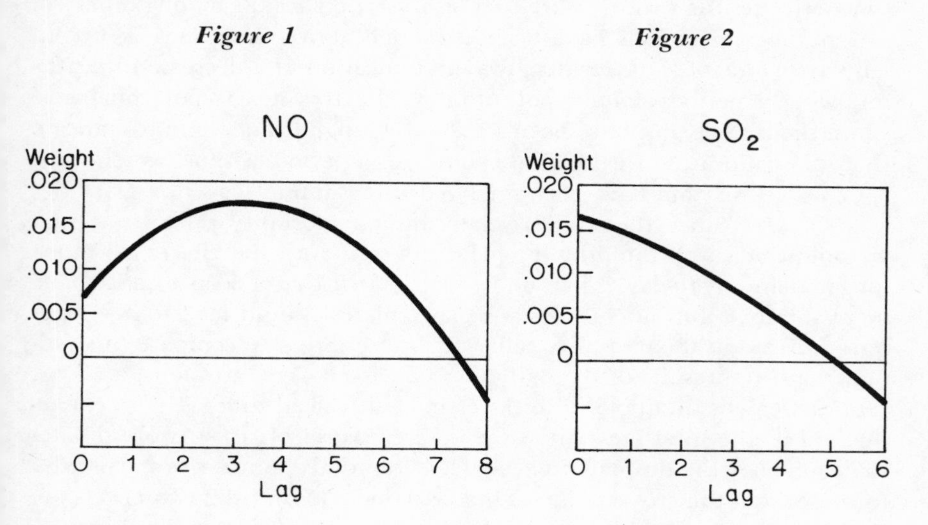

Are Deaths Merely Shifted by a Few Days?

One of the persistent questions is whether the estimated effect of air pollution on health has any policy significance. Surely an individual would have to be extraordinarily ill (or marginally viable) if an increase in air pollution results in death. Surely such an individual would die within a short time anyway, and so the air pollution episode can have no effect on the annual death rate, but only act to reallocate deaths within a short interval. If so, abating the air pollution episodes would do no more than add a few days of life to the few individuals dying and would have no policy significance.

Note that the statistically significant association between daily pollution and daily mortality is not proof that such an assertion is incorrect.

24. The fact that SO_{2t-5} was no longer important probably indicates that its previous significance was an artifact of multicollinearity among the lagged variables.

The pollution might only be reallocating deaths within a few days. One test of the assertion would be to investigate a long enough lag structure to be able to identify decreases in mortality that might follow a few days after an air pollution episode. Another test would be to dampen short-term effects by considering longer intervals (such as the monthly periods used by Hodgson). A more powerful test would be a comparison of time-series results within a city with cross-section results among cities.

A number of models have been estimated above using lags of five to ten days. None of these results give any indication of a decrease in deaths following a period of high pollution. While this test is not conclusive (since the lag may not have been sufficiently long), it does tend to indicate that the reallocation (if it exists) is not over a period as short as ten days.

Cross-section data have been analyzed using annual data for 117 cities in 1960 [24]. Since the studies relate annual mortality rates to annual air pollution levels (holding other factors constant), the effects are much longer than day-to-day. That analysis estimated that a 50 per cent abatement in air pollution (sulfates and particulates) would lead to a 4.5 per cent decrease in the mortality rate.[25] It is interesting to compare this estimate with the results of the daily analysis; for daily regressions, a 50 per cent abatement in air pollution is estimated to lead to a 5.4 per cent reduction in the total mortality rate. The closeness of these two estimates indicates that the estimated effects from the daily time-series are similar to those from the cross-section. Thus we conclude that the two effects are similar and that the increase in deaths when air pollution levels rise is not merely a shifting of these deaths by a few days.

Summary and Conclusions

In this paper we have taken up the question of whether daily mortality is affected by daily air pollution levels. Our method of investigation has consisted of examining some of the relevant literature on the subject as well as performing our own statistical analysis. Taken together, we believe that the evidence supports the conclusion that in large cities there are significant effects of air pollution on the daily death rate.

Our statistical analysis examined the effect of five pollutants in five cities. In only one of the five cities investigated, Chicago, was a significant relation displayed between pollutants and daily mortality when climate

25. See footnote 10.

and weekly effects where controlled. We conjectured that the effect was due to the relatively large number of daily deaths and to the high pollution level. We also investigated a number of other formulations, but found the simple linear one to be at least as good as any alternative we tried.

Only two pollutants, SO_2 and NO, were statistically significant, with SO_2 much more important. Careful investigation of the lag structure revealed that high SO_2 levels had a large immediate effect that gradually diminished over time. For NO, the effect was delayed by several days, reached its peak, and then diminished.

In examining the question of whether the deaths associated with pollution are merely displaced by a few days, we found our estimate corresponded closely to those from cross-section data. Thus we concluded that displacing deaths by a few days is not an important consideration in our estimate.

These results have a number of policy implications if one accepts the association we estimated as a causal one. The daily results strengthen this conclusion and indicate that the short-term effects are important. Authorities should continue to take steps to abate pollution during episodes. But the linear form also indicates that they should worry about air pollution during "clean" periods as well.

References

1. Ashe, W. F., "Exposure to High Concentrations of Air Pollution (1)—Health Effects of Acute Episodes," Proceedings of the National Conference on Air Pollution, Washington, D. C., *Public Health Service Publication No. 654,* Nov. 18–20, 1958, p. 188 (1959).
2. Becker, W. H., et al., "The Effect on Health of the 1966 Eastern Seaboard Air Pollution Episode," *Archives of Environmental Health* 16 (1968): 414.
3. Bokonjić, R., and N. Zec, "Strokes and the Weather: A Quantitative Statistical Study," *Journal of Neurological Sciences* 6 (1968): 483.
4. Boyd, J. T., "Climate, Air Pollution, and Mortality," *British Journal of Preventative and Social Medicine* 14 (1960): 123.
5. Brant, J. W. A., "Human Cardiovascular Diseases and Atmospheric Air Pollution in Los Angeles, California," *International Journal of Air Pollution* 9 (1965): 219.
6. ——— and S. R. G. Hill, "Human Respiratory Diseases and Atmospheric Air Pollution in Los Angeles, California," *International Journal of Air Pollution* 8 (1964): 259.

7. Eckhardt, R. B., "Variations Affecting Death Rate," *Archives of Environmental Health* (Letters) 17 (1968): 837.
8. Ellsaesser, H. W., "Air Pollution in Los Angeles," *Science* (Letters) 173 (1971): 576.
9. Firket, M., "Sur les causes des accidents servenus dans la valleé de la Meuse, lors des brouillards de décembre 1930," *Bulletin de Académie Royale Medicale de Belgique* 11 (1931): 683.
10. Gaver, K. M., "A Note on the Polynomial Technique for Estimating Distributed Lags," unpublished (1971).
11. Georgii, H., "The Effects of Air Pollution on Urban Climates," *World Meteorological Organization Symposium on Urban Climates and Building Climatology, Brussels* (Oct. 1968).
12. Glasser, M., et al., "Mortality and Morbidity during a Period of High Levels of Air Pollution," *Archives of Environmental Health* 15 (1967): 684.
13. Glasser, M., and Greenburg, L., "Air Pollution, Mortality, and Weather," *Archives of Environmental Health* 22 (1971): 334.
14. Greenburg, L., et al., "Air Pollution and Morbidity in New York City," *Journal of the American Medical Association* 182 (1962): 161.
15. ———, "Intermittent Air Pollution Episode in New York City, 1962," *Public Health Reports 78* (1963): 1061.
16. ———, "Report of an Air Pollution Incident in New York City, November 1953," *Public Health Reports* 77 (1962) p. 7.
17. Hechter, H. H. and J. R. Goldsmith, "Air Pollution and Daily Mortality" *American Journal of Medical Sciences* 241 (1961): 581.
18. Hexter, A. C. and J. R. Goldsmith, "Carbon Monoxide: Association of Community Air Pollution with Mortality," *Science* 172 (1971): 265.
19. ———, "Air Pollution in Los Angeles," *Science* (Letters) 173 (1971): 576.
20. Hodgson, T. A., Jr., "The Effect of Air Pollution on Mortality in New York City, A Statistical Study," unpublished.
21. Hodgson, T. A., Jr., "Short-Term Effects of Air Pollution on Mortality in New York City," *Environmental Science and Technology* 4 (1970): 589.
22. Lave, L. B. and E. P. Seskin, "Air Pollution and Human Health," *Science* 169 (1970): 723.
23. ———, "Health and Air Pollution: The Effects of Occupation Mix," *Swedish Journal of Economics* 73 (1971): 76.
24. ———, "Air Pollution, Climate, and Home Heating: Their Effects on U. S. Mortality," *American Journal of Public Health* 62 (1972): 909.
25. Lewis, H. R., *With Every Breath You Take* (New York: Crown, 1965), p. 66–67, 181–182.
26. McCarroll, J., "Measurements of Morbidity and Mortality Related to Air Pollution," *Journal of Air Pollution Control* 17 (1967): 203.
27. McCarroll, J. and W. Bradley, "Excess Mortality as an Indicator of Health Effects of Air Pollution," *American Journal of Public Health* 56 (1966): 1933.
28. Mills, C. S., "Respiratory and Cardiac Deaths in Los Angeles Smogs" *American Journal of Medical Sciences* 233 (1957): 379.

29. *Mortality and Morbidity During the London Fog of December 1952,* Reports on Public Health and Related Subjects 95 (Her Majesty's Stationary Office: London Ministry of Health, 1954).
30. Mosher, J. C., et al., "Air Pollution in Los Angeles," *Science* (Letters) 173 (1971): 576.
31. Rogot, E. and W. C. Blackwelder, "Associations of Cardiovascular Mortality with Weather in Memphis, Tennessee," *Public Health Reports 85* (1970), p. 25.
32. Schrenk, H. H., Heimann, H., Clayton, G. D., et al., "Air Pollution in Donora, Pa.," *Public Health Service, Bulletin 306,* Federal Security Agency, Division of Industrial Hygiene (1949).
33. Silverman, L., "The Determinants of Total Emergency Admissions to Pittsburgh Hospitals," unpublished (1972).

COMMENT

Thomas A. Hodgson, Jr., Social Security Administration

Although I do not accept the Lave and Seskin critique of the literature without objections, I shall not raise them here, as they do not bear upon the authors' analysis. Lave and Seskin have made an important contribution in their paper. For the first time, to my knowledge, a lagged structure is identified for air pollutants that does not exhibit a hit and miss pattern of significant coefficients with reasonable signs; a pattern of many earlier studies which did not inspire confidence that the estimated relations revealed anything at all about the time-dependent structure relating mortality to air pollution. The smooth functional forms taken by the weights of nitric oxide and sulfur dioxide in the final Lave and Seskin model are encouraging evidence that they have uncovered at least a glimpse, if not more, of the underlying structure. The different paths taken by the two sets of weights may, in fact, result from different physiological responses to nitric oxide and sulfur dioxide, or, possibly, to other guilty agents for which nitric oxide and sulfur dioxide are indices. These initial results indicate it may be possible to describe the time-dependent physiological responses induced by individual air pollutants even when the biological mechanisms that are responsible remain a mystery. Further study on the dynamic aspects of the relation between exposure and response, both experimentally in the laboratory and epidemiologically, is called for. Increased knowledge of the time paths of responses may provide clues helpful in discovering the nature of the biological mechanism, which, in turn, will improve the specification of subsequent epidemiological and statistical models.

My reservations on the Lave and Seskin study concern specification of variables. They do not, however, raise serious questions regarding the principal results.

First, rainfall can reduce concentrations by washing pollutants from the atmosphere. I am not aware, however, of evidence of an independent effect of rainfall upon those categories of mortality generally considered responsive to air pollution which warrants including rainfall as an explanatory variable.

Second, there is substantial evidence that mortality is positively related to thermal stress. Thermal stress is relieved by loss of body heat through evaporative cooling and convection from the body.[1] The efficacy of these cooling processes to dispense body heat is proportional to the ventilation rate, which, in turn, depends on wind speed.[2] Moisture content of the air also affects the amount of body heat lost through evaporation, while solar radiation is a primary source of radiant heat gained by the body.[3] Since wind speed, moisture, and solar radiation all influence the effect of heat upon the body,[4] they are candidates for inclusion in the model. In practice, however, they often show up as insignificant explanatory variables. This may be because wind speed, humidity, and solar radiation exert their deleterious effects in conjunction with temperature, with the variation in mortality due to heat reflected in the temperature variable. A complicating factor in specifying climatic variables is the large amount of time spent indoors, in artificial environmental conditions that may not be reflected from day to day and place to place by ambient air temperature, moisture content, solar radiation, and wind speed.

The use of mean temperature, however, does not adequately take into account the detrimental effects of both extremely low and extremely high temperatures. While more research, at least epidemiologically, has examined the effect of extreme heat on mortality, it is known that coldness affects bodily functions.[5]

Thirdly, the day of the week in itself certainly does not affect mortality. If the correlation is not spurious, but a proxy for some other explanatory variable or variables, the question is, what are these variables? My inclination is not to include days of the week without a greater a priori justification than a previously observed statistical relation between emergency hospital admissions and day of the week unless the coefficients of other variables with a stronger a priori justification are severely and adversely affected by excluding days of the week. Emergency rooms in many hospitals are routinely used as a substitute for a family physician. These visits may be related to days of the week in a way that deaths are not.

Fourthly, lack of significant coefficients for those variables formed by the product of four consecutive daily levels of a pollutant does not preclude the possibility of an episodic effect that is generated by only two, or three, consecutive days of high pollution. Including additional days can obscure the effect induced by a shorter period, depending on the serial correlation between consecutive levels of a pollutant. This is probably not the case in the Lave and Seskin analysis given the rather high serial correlations of nitric oxide and sulfur dioxide and the small t statistics of the episodic variables. On the other hand, attainment of statistically significant coefficients for these variables in the face of serial correlation of consecutive values would raise the question of which of those periods included are actually responsible for the observed effect. The episodic effect should be more systematically investigated, and perhaps the authors' footnote on having experimented with other variables without observing an episodic effect bears witness to their having done so.

Although available evidence supports the hypothesis that deaths are not merely shifted by a few days as a result of increased air pollution, I do not believe the Lave and Seskin argument is conclusive in itself. It is not clear to me whether the estimated reduction in mortality following a fifty per cent reduction in pollution according to the daily regressions is based upon all of the five cities considered by the authors, or just Chicago. In any case, we do not know if this small sample is representative of the larger set of 117 cities. Lave and Seskin have shown that the results obtained for Chicago cannot be replicated for Denver and St. Louis. Although every city contains individuals at risk of death when exposed to air pollution, the populations at risk differ among cities with respect to racial and ethnic composition, medical care, diet, income, and other factors affecting health status, including the nature of pollution which threatens them. Accordingly, caution is urged when extrapolating results of daily regressions in one or several cities to a large number of cities. Some hasty calculations (which can be redone to obtain more precise values) from my work on daily mortality and air pollution in New York City indicate a fifty per cent reduction in air pollution would have lead to approximately a nine per cent decline in total deaths and fifteen per cent fewer deaths from heart and respiratory diseases. The data in my analysis are from the period November 1962 through May 1965 and coincide with much of the data utilized by Lave and Seskin. Thus, if the differences between time-series and cross-section results are to be taken as a measure of deaths that were merely shifted a few days, the percentage is five times larger for New York City. The Lave and Seskin daily regressions do not closely reflect the New York City experience.

We should not be surprised, however, at differing figures among individual cities. Nevertheless, Lave and Seskin have raised an interesting point. Certainly, some deaths are merely shifted a few days, and more detailed comparison between time-series analyses within cities and a cross-section analysis among cities may help us understand this aspect of the nature and magnitude of the short-term response of mortality to air pollution.

I believe Lave and Seskin rightfully conclude "there are significant effects of air pollution on the daily death rate," and they have contributed new support for this contention. Numerous investigators have sought to examine the relationship between air pollution and human health. Diverse populations, time periods, indices of air pollution and other explanatory variables, and methods of analysis have been employed. It is difficult to make comparisons among some studies, and lack of a consistent approach has prevented the close replication of results, which has long been an essential ingredient in epidemiological proof of causation. On the other hand, the frequent indictment of air pollution indicates contaminated air is deleterious to human health. Our knowledge of this phenomenon is meager on many counts; for example, the biological mechanism through which community air pollution exerts its effect on morbidity and mortality and those components of the ambient air that are in fact guilty. Investigators have precious little data or a priori knowledge as a basis for seeking answers; but I believe epidemiologic study such as presented here is important, provides insight, and must be continued.

References

1. Clarke, John F., "Some Effects of the Urban Structure on Heat Mortality," *Environmental Research* 5 (1972): 97.
2. Clarke, "Effects of the Urban Structure," pp. 97, 99.
3. Clarke, "Effects of the Urban Structure," p. 98.
4. Clarke, John F., "Some Climatological Aspects of Heat Waves in the Contigous United States," *Environmental Research* 5 (1972): 84.
5. Hodgson, Thomas A., "Short-Term Effects of Air Pollution on Mortality in New York City," *Environmental Science and Technology* 4, no. 7 (July 1970): 589–597.

"People or Ducks?" Who Decides?

John Jackson, Harvard University

Introduction

Most environmental policy debates concern people's conflicting priorities, as embodied in the question of people or ducks, and the problem of who is to resolve the conflicts. The debate usually pits those who want to use our natural resources to support our technology against those who wish to preserve our resources in their more natural state. Conflict arises because both groups cannot use the environment as they wish. We must decide collectively how to use our natural resources and who will pay for technological advances. Such conflicts and the need for a collective decision are not unique to the environmental area. Decisions on land use planning, transportation, education and national economic policy, to name a few, all have the same characteristics. The choice of a particular policy in any of these areas may benefit one group more than another and possibly harm a third. Thus trade-offs between "people and ducks" and the need for collective decisions are characteristic of nearly all public policies.

It is the responsibility of our political process to insure that these different values and interests are taken into account when determining public policies. Problems arising from environmental concerns and from attempts to govern metropolitan areas are forcing reassessments of the

NOTE: This research was supported by grants to Harvard University from Resources for the Future and the Ford Foundation. The author wishes to thank the many officials in Minneapolis and St. Paul who were so helpful and James Q. Wilson and E. T. Haefele for their comments and suggestions on earlier drafts. The content of the paper, of course, is the sole responsibility of the author.

performance of our political institutions in meeting their responsibility. We now routinely create new agencies and alter existing ones to deal with the conflicts inherent in these public policy matters. In some cases organizations are being created to deal with particular problems, while in others existing agencies are being combined into single superagencies. Similarly, there is considerable debate and experimentation with decentralization and local control over the administration of public services at the same time that several areas are moving toward highly centralized metropolitan governments. Finally, there are the perennial questions about which officials should be appointed and which ones should be elected, and whether elections should be by district or at-large. With these debates and changes going on, it is important to understand which interests get considered, and how decisions differ with different institutional arrangements. The only way to gain this understanding is by looking at specific decisions, the associated interests, and the performance of various institutions. This is a case study of one such decision.

The question of people or ducks has been troubling the people of Minneapolis and St. Paul since 1968 when the Metropolitan Airports Commission (MAC) decided to locate a new jetport north of the Twin Cities and adjacent to a large game preserve and bird sanctuary, and the Metropolitan Council (hereafter the Council) vetoed the Airport Commission's decision. The conflict created by these vetoes still exists, in spite of several attempts to resolve it by gubernatorial committees, threatened intervention by the state legislature, and the efforts of a joint Airport Commission and Metropolitan Council committee formed to study the alternatives. This paper is an attempt to explain how two agencies serving the same region reached such different decisions. The explanation illustrates some fairly simple notions about the consideration given economic and environmental impacts of such a decision by specialized executive agencies, multipurpose policy-making bodies, and elected and appointed officials. Given that there are a number of conflicting interests associated with selecting a site for a new airport, the important question is how each agency and each recruitment process weighs these interests, and how the outcomes differ under alternative decision structures.

The Organizations and Their Anticipated Behavior

The central participants in the airport decision are the MAC commissioners, the professional staff of the MAC, the members of the Council, and Northwest Airlines. The latter organization, in addition to being a major airline serving the Twin Cities, started operations there, has its corporate headquarters in Minneapolis, is a major employer in the Twin Cities area, and has acted as the spokesman for the other airlines during the controversy.

There are two important distinctions between the MAC and the Council. These differences are replicated in many of the governmental bodies involved in setting environmental policy and in governing metropolitan areas. The first concerns the scope of the responsibilities of the two agencies and the second is in the qualifications for membership and the methods used to select the members of the two bodies. The MAC, as its name implies, is a single function agency whose sole responsibility is operating the airport system for the metropolitan area. Its responsibilities are largely administrative and its decisions have been made predominately by the professional staff. The Council, on the other hand, is a multi-function agency established in 1967 to oversee and coordinate planning and development in the entire seven-county area. It has varying degrees of control over all existing independent agencies such as the Sewer Board, the MAC, and the Transit Authority. The Council also has the responsibility for planning some metropolitan land use activities, such as highways, parks, and public buildings. The Council's responsibilities then are solely of a policy-making variety encompassing many issues at the metropolitan level. Decisions are made by the councilors through debate and by roll call votes as in most legislatures. The other difference is in the background and selection of the members of the MAC and the Council. Five of the nine MAC members hold elective office in either Minneapolis or St. Paul and all members but the Chairman come from the two central cities, while the members of the Council are appointed by the Governor on the basis of districts which encompass the entire seven county metropolitan area. Finally the MAC staff members are chosen by the Commissioners for their prior experience and expertise in the aviation field.

The Metropolitan Airports Commission

The MAC was created by the Minnesota state legislature in 1943 to end the rivalry between Minneapolis and St. Paul for the most elaborate airport facilities. In part this rivalry concerned which city's airport would be the base of operations for Northwest Airlines, and in part it stemmed from the natural rivalry of the two adjacent cities. At that time, the MAC was given jurisdiction over both cities' airports and any airports, present or future, which the MAC might want to acquire within a 25 mile radius of the two city halls. This has now been extended to a 35 mile radius.

The MAC was also instructed to provide at least one major commercial airport located as nearly equidistant to the two city halls as possible. The MAC designated Wold-Chamberlain field, which had been the Minneapolis airport, as the major commercial airport and proceeded to develop it into a major international air terminal in the succeeding thirty years. The MAC also administers five smaller airports scattered around the periphery of the metropolitan area.

The Commission is made up of nine members, most of whom are elected to municipal offices in one of the two cities. These include the mayors of both Minneapolis and St. Paul, one member of the Minneapolis city council and one member of the Minneapolis Park Board, and two city councilors from St. Paul.[1] Each city also has one "citizen commissioner." These citizen commissioners are appointed by the mayor and city council in each city. The ninth member of the Commission is appointed by the governor and serves as chairman. The Chairman must not be a resident of either city or of a contiguous county, meaning that he comes from outside the Twin Cities metropolitan area. There is no suburban representation on the Commission even though five of the six airports under MAC control and over 60 per cent of the metropolitan population in 1970 reside outside the two cities.

The Commission appoints a staff headed by an Executive Director who serves as the chief executive and operating officer of the Commission. The legislation that created the MAC states that the Executive Director should have business experience, preferably in the aviation field.[2] Both men who have held the position since 1943 have satisfied this requirement as have the other members of the staff.[3] The Commission has kept its administrative staff small. As of August 1970, it consisted of 17 people, including secretarial and clerical employees. This does not include about 200 other individuals employed in operating the airports. The Commission has relied to a great extent on outside consultants and attorneys as needed.[4]

The Commission has three main sources of revenue; charges for the use of MAC property (e.g., terminal facility rentals, landing fees, etc.),

1. The Minneapolis Park board member is included because Wold-Chamberlain is built in Minneapolis park land and prior to the MAC the airport was administered by the Park Board.

2. Minnesota Statue 360,106 (4) (1969).

3. Short biographies of several of the staff members are included in an article, "Jet Noise and Its Impact," *Airport Services Management* Vol. 12, No. 10 (October 1971): 16–26.

4. Donald V. Harper, "The Minneapolis-St. Paul Metropolitan Airports Commission," *Minnesota Law Review* Vol. 55, No. 3 (January 1971): 392.

federal and state aid; and support for its operations and maintenance from the cities of Minneapolis and St. Paul of up to one mill on all assessed valuation. The two cities can also be taxed to pay for the prinicpal and interest on the first $20,000,000 of airport bonds. After that the bonds are supposed to be self-liquidating and the cities are responsible only if the self-liquidating feature fails as the result of a default by a contractor, such as Northwest Airlines. In practice, the levy for operations has ranged between 0.05 mills and 0.465 mills, and there has been no assessment for operations since 1961. The assessment for the payment of principal and interest on MAC bonds has ranged from 0.460 to 2.445 mills and no assessment has been made since 1969. Thus the Commission is now financially independent of the two central cities.[5] The income from the charges and rentals at the MAC airports accounts for most of the Commission's income for both operating and capital investment purposes. As we will see, this is a crucial factor in all Commission decisions.

The Metropolitan Council

The Council was created by the Minnesota state legislature in 1967 to coordinate the planning and development in the seven counties comprising the metropolitan area of Minneapolis and St. Paul.[6] The legislature instructed the Council to prepare a comprehensive development guide for the metropolitan area which would state goals, standards, and programs for both public and private economic development in the metropolitan area. A Development Guide Committee was formed for this purpose and now has the responsibility for making recommendations to the whole Council about proposed construction or development plans by the independent boards, commissions, and municipal governments in the metropolitan area. The Council has the authority to suspend proposals that it decides do not conform to its development guide. This is the major way the Council has of influencing the decisions of the Metropolitan Airports Commission, the Metropolitan Transit Commission, the Metropolitan Sewer Board and other metropolitan agencies and governments. With the exception of the Sewer Board and Transit Commission, all the agencies have independent operating and financial authority and their budgets are outside Council control.

The Council's additional power over the other agencies and governments is limited and requires federal cooperation. Under existing legisla-

5. Harper, "Minneapolis-St. Paul Metropolitan Airports," pp. 434–437.
6. Minnesota Statutes, Ch. 473 B (1967).

tion, local proposals for federal aid for specific projects, like an airport or a housing project, must be reviewed by a local planning agency, in this case the Council. If the Council opposes the proposal, the local agency may still submit the proposal, but with a statement that it was opposed by the planning agency. In the case of such an appeal, it is up to the federal agency to decide if it wants to uphold or override the local planning agency. Although local agencies have seemed unwilling to challenge the Council on such questions, the Council's real power comes from its ability to suspend new construction plans of the different metropolitan organizations. This authority is not inconsiderable.

There are fourteen Councilors, each representing a single-member district. These districts are aggregates of several state legislative districts and are relatively equal in population. There is also a chairman of the Council who does not come from any of the districts. All councilors are appointed by the state governor with the advice and consent of the state senate. Although supposedly nonpartisan, these appointments have reflected the political affiliation of the governor. Councilors are appointed to six-year overlapping terms so that four or five councilors are up for reappointment every two years. The Chairman serves at the pleasure of the governor. There is continual debate over having the Councilors elected rather than appointed. This change would have to come from the state legislature which has previously opposed such an arrangement.

The expected behavior of the MAC and the Council

In this section we will analyze how these two agencies came to different conclusions about the location of the airport and use this analysis to examine how different governmental structures perform at resolving the conflicts inherent in an airport location question. This analysis is based on some simple notions about organizational, electoral, and legislative behavior. The case itself is used to examine and enrich these notions, which can then be used to suggest ways of improving the structure and performance of environmental and metropolitan policymaking agencies.

The easiest participants to describe are the MAC commissioners. The elected members are usually more concerned with the jobs to which they were elected. The operations of the airport are only an ancillary function and certainly are not as important as other municipal decisions in determining their chances for reelection. For many years the mayor of St. Paul did not attend Commission meetings, but simply sent a representative. The citizen commissioners are usually active in local businesses or labor unions and are much more concerned with the operations of these organizations than they are with MAC decisions. The commissioners normally ignore questions about the airport and concentrate on their other responsibilities. They will do so unless there is obvious concern or discontent with the operations of the airport system among their constituents. If such a concern appears, and particularly if the concern is shared by a majority of the Commissioners' constituents, we should expect the elected officials to react quite strongly and visibly. The most obvious issues over which such discontent might be raised are the costs of the airports if they begin to raise local tax rates perceptibly, poor and inconvenient service to airport users, rising noise levels, and attempts to expand the airport within the cities of Minneapolis and St. Paul. The elected commissioners are probably more sensitive than the appointed members to these concerns if they arise. It seems obvious to expect the elected commissioners to be less sensitive to questions and problems raised by people outside the two central cities, who of course are not in their constituencies. Thus objections to a decision to expand one of the outlying airports or to a new airport location outside the two cities are not likely to evoke much response from the commissioners. In the absence of aroused constituents, the commissioners will be fairly complacent about MAC operations and decisions.

A complacent attitude and approach on the part of the MAC commissioners aids the MAC staff members in fulfilling their objectives. These men are professional aviators and airport administrators. Their likely objective will be to insulate themselves from as much external pressure as possible so as to pursue the more limited goals of building and running airports. These goals undoubtedly include accolades from professional groups in the aviation field. The staff's performance criteria would include such measures as the volume of air traffic handled, the growth in this traffic, their safety record, the absence of delays attributable to airport operations, and similar efficiency measures. Satisfying these professional objectives requires providing modern and adequate terminal and air space for expected traffic levels and minimizing externally imposed restrictions on operations. Obtaining the autonomy to pursue their goals means avoiding things like tax increases, congestion delays, and noise problems which are likely to cause the commissioners to become concerned about the staff decisions and operations. Preventing or quickly solving problems which do arouse the commissioners' concern will increase the staff's autonomy and independence and permit them to pursue their own personal and professional goals. In selecting a site for a new airport, the MAC staff will be primarily concerned with their professional concerns of available terminal and air space and the technical feasibilities and costs of alternative locations. In the process of making a decision, they will try to limit outside involvement as much as possible so as to maintain their autonomy and their ability to follow the technical objectives. Focusing on these more narrow technical criteria and preventing external involvement means that the MAC staff will completely avoid the larger conflicts which surround the question of a new airport.

The objectives and behavior of the Council are the hardest to describe. The usual model of such a legislative type policy organization assumes that the members' responsibility is to insure that the interests and concerns of the people in their districts are represented and have a chance to influence the decisions of the agency. This responsibility is fulfilled by having councilors support proposals favored by their constituents and form coalitions with other councilors to promote programs which their constituents consider most important to the metropolitan area. Explanations of such a council's decisions are based on knowledge of the concerns of the people in each district on different issues and the coalitions and negotiations likely to take place among the members. For example, councilors from districts where the people are concerned about the environmental or economic impacts of an airport location can be counted upon to raise these issues and to make sure that the final decisions reflect these concerns. The cost to the councilors for not doing this effectively is heat from their constituents and possible loss of the job.

Such a metropolitan-wide legislative organization would have a much better chance than an administrative agency like the MAC and its staff of resolving or at least of considering the conflicts inherent in issues such as a decision to build a new airport. The sole responsibility of the representatives in such a legislative body would be to mediate the differences involved in public matters. They would not be rewarded professionally for promoting aviation and their futures as councilors would not be primarily dependent upon their performance in another job. At the same time, since they would be elected on a district basis and because the perceived benefits and costs of the airport are not spread uniformly throughout the metropolitan area, it is almost certain that the questions of the distribution of these benefits and costs will be raised, debated, and influential in the final decision of such a council.

Unfortunately this description of a metropolitan legislature and its implication for public policy decisions is not completely appropriate here. The legislative model relies on the assumption that councilors' preferences reflect the preferences of their constituents, which may not be a bad assumption if the members are elected. However, the Council members are not elected, but are appointed by the governor. Consequently the councilors are likely to be voting their own personal views which are derived from their social and economic associations within the metropolitan area and perhaps those of the appointing individual. Only the councilors who are quite personally associated with the interests of their districts by virtue of residential associations or business contacts will act according to the representational model. Even with this qualification on the behavior of the Council, its decisions are more likely to represent concern for the conflicts associated with the airport location decision than are the decisions of the MAC. Because of its young age, the members as a group were quite concerned with establishing the Council as a strong metropolitan-wide policy making agency. They were quite conscious of their responsibilities to guide and plan metropolitan development in a positive fashion. They were also making a determined effort to establish their control over the decisions of the independent agencies which affect this development.

History of the Airport Decision

The history of the airport location issue began in 1967 when the MAC staff determined from FAA and Air Transportation Association data that Wold-Chamberlain, the existing commercial airport, would be inadequate by 1980.[7] This prompted a staff study on possible alternatives to meet the expected demand. The search immediately focused on six possible sites, three north of the Twin Cities and three to the south. The original purpose of the Commission, its charter, and its composition as well as existing geography and airport locations precluded a site in either the east or the west. The Commission had already done considerable work toward a future northern site. In 1950 a secondary airport for use by general aviation was established in Anoka County, to the north of the Twin Cities. Additional land was acquired and the airport expanded in 1961. Part of the justification for the acquisition of these lands was that when a second major airport was needed, this land would be available for further expansion.[8] Unfortunately, in the succeeding years residential

7. This historical account draws heavily on an article by Donald V. Harper, "The Airport Location Problem: The Case of Minneapolis-St. Paul," *ICC Practitioners Journal* (May–June 1971): 550–582.

8. Minneapolis-St. Paul Metropolitan Airports Commission, "In the Matter of the Expansion and Development of the Commission's Airport System by the Expansion of Anoka County-James Field or Acquisition of Lands and Development of a New Airport in Anoka County, Findings, Conclusions, and Order," February 24, 1969, p. 8. (Hereafter referred to as MAC, "Findings.")

development and the location of a series of television antennae made expansion of the Anoka airport unacceptable to the Commission.

Because of the potential of urban development and the distance from both downtown areas, the Commission focused on a northern site called Ham Lake. This is a 15,000 acre tract about 20 miles from each downtown area and 25 miles from Wold-Chamberlain. An additional advantage of this site according to the MAC was that it would not interfere with existing or any proposed operations at Wold-Chamberlain.[9] The Commission hoped to continue to use the existing airport for commercial aviation, as well as the maintenance and repair facilities which the airlines were currently leasing. Their initial statements indicated a feeling that both airports would be needed by the late 1990's, thus making it prudent to maintain Wold-Chamberlain and keep it in use.[10] The Ham Lake site also had the advantage that most of the land under consideration was either marginal swampland or peat bogs or was being used only for farming or sod farming. Consequently it could be assembled at a relatively low cost. At the same time, due to the marginal nature of much of the land, the Commission felt that the likelihood of future development was less than at other potential sites.[11] Finally, the northeast boundary of the airport site was the Carlos Avery Wildlife Management Area, a protected wildlife preserve. The advantage to the MAC was that no development could take place on this site of the airport, and as one person put it, "Birds do not call up complaining about noise."

The decision to go with the Ham Lake site was made largely by the MAC staff. In fact one person suggested that the Commissioners may have been kept out of the decision until the staff had picked Ham Lake, but this cannot be verified.[12] In any event, the first public notification came when the MAC announced that it was going to hold public hearings, as required by law, on a proposal to take the necessary land at the Ham Lake site. Five hearings were held between April 1968 and December 1968, two for public testimony and cross-examination of the MAC staff and the others for the MAC staff and consultants to outline the proposal and specific hearings for other governmental agencies and for the airlines.[13] The opposition to the proposal came from several groups.

9. MAC, "Findings," p. 35.
10. MAC, "Findings," pp. 37–38.
11. MAC, "Findings," p. 33.
12. Harper, "The Airport Location Problem," pp. 558–562. This possibility was also mentioned by several people interviewed in the course of this study.
13. Minneapolis-St. Paul Metropolitan Airports Commission, "Chronological Summary MAC Consideration of Location of New Major Airport," Sept. 20, 1971. Hereafter referred to as MAC, "Chronological Summary."

One vocal opponent was Northwest Airlines, supposedly acting as the spokesman for all the airlines serving the Twin Cities. Their opposition was over costs and competition.[14] Northwest is tied into expensive leases for the existing terminal, overhaul, and storage facilities at Wold-Chamberlain. It feared that building a new airport would raise its costs considerably. Given MAC's usual financial objective to be self-supporting through user charges for airport space, this was a reasonable concern. In addition, Northwest pointed out that virtually all its traffic was to the south of the Twin Cities so that a northern site would measurably add to their operating costs independent of the MAC's charges. At the same time, Northwest argued that all their employees were now located on the southside in the proximity of the existing field and that a northern location would work a severe hardship on them.[15]

There was an additional unstated reason for Northwest's opposition—the potential of additional competition. A member of the Commission's staff said that the most profitable run for airlines serving the Twin Cities is the Chicago trip. This is how Northwest got its start and it now handles about 70 per cent of the traffic on that run. Under the MAC plan which would have kept Wold-Chamberlain open, Northwest feared that it would then be possible for a third level carrier to obtain authorization for a downtown-to-downtown run between Wold-Chamberlain and Midway in Chicago. If this happened, it would seriously undercut Northwest's advantage on the Minneapolis to Chicago run. It was also apparent that for financial reasons, the Airport Commission was not likely to agree to close Wold-Chamberlain. As a result, Northwest was basically opposed to any second airport, and all the more opposed if the new facility was going north.

The second source of opposition was the environmentalists who feared what the new airport might do to the quality of life in the Carlos Avery Refuge and to the wáter table and water quality in the area.[16] The water question was of particular concern to some people because much of St. Paul gets its water from sources which would be affected by the airport if the environmentalists' projections were correct. The environmentalists were also concerned about the effect of increased noise on the wildlife and on people's ability to enjoy the preserve. There were also several second-

14. Harper, "The Airport Location Problem," p. 562.

15. Northwest Airlines has continually cited these reasons for their opposition to a new major airport and particularly to a northern site. These points formed the basis of an August 21, 1972 letter from the President of NWA, Donald Nyrob, to one of the Council members prior to the most recent Council vote.

16. Harper, "The Airport Location Problem," pp. 562–563.

ary questions; the effect of increased air pollution on the local environment, the possibility of birds being drawn into the jet engines, and the likelihood of more frequent ground fog associated with the swampland and high water table.

The third and least organized opposition group consisted of southside businessmen and commercial interests.[17] Most of the past growth in the Metropolitan area has been in the southerly direction. This can be attributed in part to the airport, although there were other contributing factors such as the construction of Metropolitan Stadium in Bloomington, the opening of Interstate 494, and the presence and growth of Control Data Corporation. One consequence of this development was the growth of a large hospitality industry and other firms dependent on proximity to air transportation. These businesses faced the loss of considerable revenue or increased operating costs if a new airport were located at a northern site and the volume of traffic at Wold-Chamberlain reduced. This opposition of course was partially offset by the support generated among northern business interests and downtown interests that had been losing business to the newer hospitality and retail concerns south of the city. These downtown interests hoped that a northern site would mean increased business for them as they would now be the most proximate service area.[18]

In spite of this opposition, the Commission voted unanimously on February 24, 1969 to proceed with land acquisition at the Ham Lake site. Although the Commission indicated that it looked at the five other sites, these are dismissed rather quickly in their report, called a "Findings, Conclusions, and Order." [19] In this document, the Commission put forth its reasons for favoring the Ham Lake site as follows.

1. Sufficient available land (15,560 acres).
2. This land is suitable for airport construction in terms of slope and soil conditions and is comparatively cheap due to its unsuitability for alternative uses.
3. Noise problems will be minimal because of the presence of the Carlos Avery area to the northeast, the reduced potential for further development, and the ability to implement the proper zoning controls over potentially usable land.

17. Harper, "The Airport Location Problem," p. 563.

18. There was even one proposal to establish a direct nonstop rapid transit link between the airport and downtown. This would further encourage downtown growth.

19. See footnote 8.

4. There would be minimal conflict with the airspace at Wold-Chamberlain and other general aviation fields.
5. The possibility of bird strikes was small and no worse than other areas in the vicinity of the Mississippi River.
6. "The reduction in value of the game refuge is at this point speculative and can only be determined after the airport has been in operation." [20]

The Commission report did not address the water quality and level questions, the problems which might be encountered in constructing runways and terminal facilities on the peat bog (although the Commission claimed that, "the predominant soil is fine sand which makes an excellent material for subgrade and subbase under pavement. . ."),[21] nor with the airlines' contention that they would not operate out of two airports and preferred a southern site.

The next step in the decision process was approval by the Metropolitan Council. The Council had sixty days to decide whether it would concur with the Commission, send the proposal back to the Commission for further study, or reject it outright. In the event the Council decided on this last alternative, the Commission could then go to the state legislature asking them to resolve the dispute.

On April 24, 1969 the Council voted to suspend the Airport Commission's Order to begin land acquisition at the Ham Lake site.[22] It contended that the Commission had not adequately considered other sites and that there were sufficient potential problems with the Ham Lake site to warrant a more detailed examination of alternative sites. The Council recognized those advantages to the Ham Lake site recommended by the Commission. However, it felt that the potential problems were such that more study of the specific questions was needed to make the appropriate comparative analysis between Ham Lake and other sites. Ham Lake might finally prove to be the most desirable site, but the Council was not yet satisfied of that on the basis of the MAC analysis and the public hearings. The Council was primarily concerned about the potential impact on the water level and quality, the future use and enjoyment of the Carlos Avery Refuge, the accessibility of a northern site to current airport users who were assumed to reside on the southern side of the two

20. MAC, "Findings," p. 34.
21. MAC, "Findings," p. 34.
22. Metropolitan Council, Minutes of the April 24, 1968 Council Meeting.

cities, and the Commission's view that two major airports were needed. The vote of the Council was ten to four in favor of suspending the Commission's order.[23]

At this point, Northwest Airlines entered (or was dragged into) the discussion.[24] In February, Northwest had proposed a 25 to 30 million dollar expansion of its facilities at Wold-Chamberlain to accommodate the new 747 jet planes it had on order. Less than two weeks after the Council vetoed the Ham Lake proposal, the MAC announced that it was not approving the Northwest expansion. The Commission said that it could not approve the expansion at Wold-Chamberlain until the question of the new airport had been settled. The airline accused the Commission of using this decision to try to coerce the airline into supporting the Ham Lake proposal. The airline in turn threatened to build the new facilities as well as move its existing operation to Seattle if the second airport were built on a northern site and if Wold-Chamberlain were left open. Since the airline employed about 6,000 people in the metropolitan area, this was not a trivial threat from the communities' standpoint. The threat, of course, brought charges that the airline was trying to blackmail and coerce the Commission into dropping the Ham Lake site and pressuring the Council into continuing to veto it if the Commission submitted the proposal again.

On May 19 the Commission voted to resubmit the Ham Lake proposal. New airport zoning ordinances had been passed by the state which gave the Council the right to regulate land use for three miles outside the boundaries of a new airport and up to five miles in certain directions if that was deemed necessary. The MAC had also conducted a new airspace survey which it hoped would satisfy the Council. This plan was submitted to the Council on May 29 and the Council again had sixty days to respond. Neither of these changes seemed to reduce the opposition to the Ham Lake site, however. The MAC was also under considerable pressure to change its mind on the Northwest expansion proposal. This was coming largely from commercial interests, politicial leaders, and the news media. On June 2, the MAC announced that it would approve the Northwest expansion and the use of 747s if the airline agreed to pay off the MAC bonds in ten rather than thirty years. Ten years was the MAC esti-

23. Harper, "The Airport Location Problem," pp. 567–568 and references cited therein.

24. Harper, "The Airport Location Problem," pp. 567–568, recounts this part of the controversy in great detail.

mate of the time it would take to get the new airport operational. On June 4, Northwest announced that it was going ahead with an $8,000,000 expansion at Seattle to accommodate storage and training facilities for the 747s. The airline implied that this was part of the 25 to 30 million dollar expansion earlier planned for Wold-Chamberlain. Northwest also stated that the remaining part of the expansion at Wold-Chamberlain was still contingent upon the airport location decision. In a special meeting called for later that afternoon (June 4), the Commission agreed to go ahead with the 747 facilities at Wold-Chamberlain and to ask the Council to stop further consideration of the new Ham Lake order in view of Northwest's opposition.

The President of Northwest, Donald W. Nyrop, sent a telegram to the mayors of Minneapolis and St. Paul saying that Northwest was withdrawing its expansion requests and stated that before they were considered again the MAC would have to withdraw the Ham Lake proposal, settle the one or two airport decision, presumably in favor of one, and come to an agreement on how any new expansion or new facilities were to be financed. Governor LeVander of Minnesota, a Republican, then called for a meeting including himself, Nyrop, and the Commission, and Council Chairmen. The result of this meeting was that the Commission agreed to withdraw the proposal and to build the new facilities at Wold-Chamberlain. At subsequent meetings between the staffs of the Commission, the Council, and the airlines, a statement was issued that called for further study of the airspace conflict between Wold-Chamberlain and a southern airport, of the fog problems at both sites, and an origins and destination study of airport users. These ignored the more basic questions of the potential environmental problems and the future use of Wold-Chamberlain.

While these studies were being conducted debate continued, although at a lower level. The debate became focused on two questions. The first issue was that of one versus two airports and the future of Wold-Chamberlain. The second was the related problem of the financing of a new airport. Put quite simply, the MAC wanted to keep Wold-Chamberlain open because of the large fixed investment in facilities there and the considerable revenue generated by leasing the facilities to the airlines. If Wold-Chamberlain were to be closed down, the airlines would be freed from their obligation to pay the leases on the facilities. The MAC would then be responsible for the debt on the existing facilities as well as the debt incurred in building the new airport. The taxpayers of Minneapolis and St. Paul would have to carry this debt through the property taxes.

The Citizens League in a special study contended that this was the major reason for the Commission's attitude toward Wold-Chamberlain.[25] The Commission also contended that it was senseless to close Wold-Chamberlain. They maintained that it was a well equipped airport and that at some point in the future the demand for airspace in the Twin Cities would be such that the second airport would be needed. Consequently it would be shortsighted not to maintain Wold-Chamberlain, let alone shut it down.

The airlines were taking the opposite view, particularly Northwest who had the largest investment in facilities at the existing airport. Northwest maintained that they would not operate out of two sites and were reluctant at best to move to a northern site even if Wold-Chamberlain were closed. The airline pointed out that if Wold-Chamberlain were closed, the justifications for a northern site over a southern one on the basis of potential airspace conflicts were no longer supportable. Northwest then argued that since most of the flights came from the southeast or southwest and since a vast majority of employees and passengers lived on the southern side of the Twin Cities, a southern site made much more sense.

While this debate was going on, there was mounting pressure from people residing in the vicinity of Wold-Chamberlain to do something about the noise problem created by the increased use of jets.[26] Their demands to reduce the noise levels resulted in the first critical involvement in airport business by elected officials from either of the Twin Cities. Minneapolis has a ward-based city council. The citizens of South Minneapolis, the area adjacent to the airport, went to their councilor, Mrs. Gladys Brooks, asking if she could get the MAC and the airlines to reduce the noise levels and limit the noise to certain hours. Mrs. Brooks sent letters and pleas to the MAC and the airlines requesting their cooperation and introduced several city ordinances which would have restricted activity at the airport. She succeeded in stopping training flights and the testing of jet engines at night, in limiting the use of certain runways,[27] and in establishing a noise abatement council, MASAC, which had representatives from the MAC, the airlines, and citizens groups. These restrictions and the citizen interference were seen as a severe inconvenience by the MAC.

25. Citizens League, *New Airports for the 70's and After* (Minneapolis: Citizens League, 1969).

26. Stan Olson, "MASAC's Role," *Airport Services Management* Vol. 12, No. 10 (October 1971): pp. 24–26, and interview with Councilwoman Gladys Brooks, Aug. 23, 1972.

27. William K. Matheson, "Operational Procedures," *Airport Services Management* Vol. 12, No. 10 (October 1971): pp. 19–20.

These noise problems succeeded in generating additional incentives and pressures to get on with a new facility and to close Wold-Chamberlain or at least limit it to general aviation. During this time, the Council reiterated its previous position that the new airport should be a replacement for Wold-Chamberlain at least as far as commercial traffic was concerned. In December 1969 the three parties agreed that a potential site south of the two cities located at Rosemount-Farmington would be evaluated and compared to the Ham Lake site in the studies currently under way.

During this period of the controversy the Federal government passed and began to implement two pieces of legislation which specifically affected the deliberations in the Twin Cities.[28] The first was the Environmental Quality Act of 1969 which stated that all federally funded projects had to consider the environmental impact of the proposal and avoid detrimental impacts whenever possible. The second was the Airport and Airway Development Act of 1970. This latter bill supplements the first by stating that the Secretary of Transportation must approve all new sites or improvements before federal aid may be provided and that such approval may be forthcoming only after consultation with the Secretaries of HEW and the Interior to determine the effects on natural resources. Supposedly the Secretary of Transportation may not approve projects which have an adverse effect on the environment if a feasible alternative exists. If no such alternative exists, the Secretary must insure that all steps will be taken to minimize such effects. The implications of these bills for the deliberations in the Twin Cities are fairly obvious. Both the Commission and the Council initiated environmental impact studies by outside consultants of both the Ham Lake and Rosemount-Farmington sites.

During the spring and summer of 1970 the reports of the different consultants on airspace, meteorological, accessibility, and environmental problems were completed and reported to the Commission. One of the consultants, R. Dixon Speas Associates, summarized these reports in a document prepared for the Airports Commission in October 1970.[29] In this summary the Ham Lake site is preferable to, or at least not inferior on most grounds to the Rosemount-Farmington site on all criteria except

28. Robert Jorvig, "Synopsis of Airport Planning, 1968–1971," Metropolitan Council Memorandum, August 11, 1971, p. 2, and Harper, "The Airport Location Problem," pp. 574–576.

29. R. Dixon Speas Associates, "Summary Report of Studies and Documents Related to Twin Cities Metropolitan Area Second Major Airport Site Selection," October, 1970. Hereafter referred to as Speas, "Summary Report."

for a slight difference in potential accessibility.[30] According to the origin and destination survey and the projections made by the consultant for the year 1985, the average trip length to Ham Lake would be 48 minutes, compared to 42 minutes at Rosemount-Farmington. In terms of airspace Ham Lake was clearly preferable because it would not interfere with operations at Wold-Chamberlain. According to the report, a new airport at the southern site could only operate at 70 to 80 per cent of the capacity at Ham Lake for similar facilities because of the airspace conflicts. The meteorological consultant said that both sites were equivalent from the standpoint of meteorological interference. The main question here was the potential incidence of ground fog. The consultants and a physics professor from the University of Minnesota argued that there was no problem in this regard and in fact the problem might be less than at Wold-Chamberlain.

On the environmental question, they concluded that the Rosemount-Farmington site would require diversion of the Vermillion River, a small feeder of the Mississippi River, but this did not involve any ecological hazard because, "This river has not in the past received much attention from conservationists and recreationists. . . ." [31] With reference to the Ham Lake site, the consultant claimed that the new airport "will not necessarily adversely affect the ecological balance in the game refuge." [32] It was further suggested that a cooperative refuge/airport system could enhance and protect Carlos Avery.

> A refuge/airport system could be designed and negotiated, which would provide a new concept in airport development and a more reasonable approach to otherwise conflicting needs of transportation and environment.[33]

Upon receipt of these reports, the Commission met and again voted approval of the Ham Lake site. On November 9, 1970 the Commission issued a "Revised Findings, Conclusions, and Orders" specifying that land be acquired at the Ham Lake site for the construction of the new airport. The Commission stated in its conclusions as follows.

> This conclusion (to acquire the Ham Lake site) is essentially a reaffirmation of the conclusion arrived at through the Commission's "Findings, Conclu-

30. Speas, "Summary Report," pp. 6–9.
31. Speas, "Summary Report," p. 9.
32. Speas, "Summary Report," p. 9.
33. Speas, "Summary Report," p. 9.

> sions, and Orders" of February 24, 1969 as to the location of the new major airport within the Anoka County lands studied and the subject of public hearings, revised solely to meet the requirements to make the same compatible with its environs.[34]

The Commission spent about 15 pages comparing the two sites and justifying its selection of Ham Lake.[35] These justifications pretty much incorporated the consultants findings, as summarized in the R. Dixon Speas Associates report. The accessibility of the southern site was minimized, ". . . accessibility to the two sites would be approximately equal (favoring the southern site by five and one-half minutes)," [36] while pointing out that the northern site was closer to the terminal handling most of the air freight which moved through the Twin Cities in 1968. The final aspects of the comparisons were the cost and financing estimates. At each point, the Commission maintained that the southern site would be more expensive. Land acquisition costs would be twice as much at Rosemount-Farmington, site preparation would be 10 per cent more, and an additional 37 million dollars in highway construction would be needed to provide adequate access to the airport. The conclusion of these findings was that the Ham Lake site could be financed without recourse to revenues from the cities of Minneapolis and St. Paul. The Rosemount-Farmington site, however, would require revenue from these sources.[37]

The Council wasted no time in beginning consideration of the Commission's Order. The Council's referral committee held public hearings on November 24, 1970. The Council staff prepared a long synopsis of the current issues, the findings of the different consultants with respect to Ham Lake, and made a series of recommendations to the Council.[38] For the most part the Council staff agreed with the findings and conclusions of the Commission. There were two major, highly significant, differences. One was that the Council staff said that the new airport should be the only commercial airport and that if Wold-Chamberlain stayed open it should only serve general aviation.[39] The second was that the Council staff

34. Minneapolis-St. Paul Metropolitan Airports Commission, "Revised Findings, Conclusions, and Orders," November 9, 1970. Hereafter referred to as MAC, "Revised Findings."

35. MAC, "Revised Findings," pp. 33–47.

36. MAC, "Revised Findings," p. 43.

37. MAC, "Revised Findings," pp. 44–47.

38. Metropolitan Council, "Report of the Staff (Ham Lake Referral)," December 2, 1970. Hereafter referred to as MC, "Report of the Staff."

39. MC, "Report of the Staff," p. 14.

was not as sanguine as the Commission about the environmental compatibility of the airport and the natural resources adjacent to the Ham Lake site.[40] The Council staff claimed that,

> It is clear that noise from the aircraft would have an adverse effect on man's enjoyment of observing the wildlife in Carlos Avery and for this reason would effect man's use of Carlos Avery.[41]

and in the next paragraph,

> . . . it is clear that certain types of urban development will have an adverse effect on the fish, waterfowl, fowl and small game habitat in this area.[42]

The staff ends up by concluding though that because of the Airport Zoning Act

> . . . the staff concludes that the airport can be built and operated compatible with natural resources by careful airport design and through operation of the Airport Zoning Act.[43]

On December 10, 1970 the referral committee met to make its recommendation to the Council. In this recommendation, the committee agreed that a new airport was needed, that it should be the only airport until at least the year 2,000, and that the Ham Lake site could not be developed without some permanent damage to the environment of the Ham Lake area.[44] The Council followed the recommendation of its referral committee by a 9–5 vote later on the same day. This amounted to a switch of one vote from the decision made in April, 1969 when the vote was 10–4. The reason given for the switch was that one Councilor had become convinced that the new airport at Ham Lake did not pose a threat to the quality of the water supply for the city of St. Paul.[45]

A map of the metropolitan area showing the Council districts, the incumbent Councilors at the time of the two airport votes, and the airport locations is given in figure 1. The Councilors and their districts voting for the Ham Lake proposal on the first vote were: Craig (3), Pennock (5),

40. MC, "Report of the Staff," pp. 49–59.
41. MC, "Report of the Staff," p. 52.
42. MC, "Report of the Staff," p. 52.
43. MC, "Report of the Staff," p. 59.
44. Metropolitan Council, "Report of the Referral Committee," December 10, 1970.
45. Statement of Mr. James Dorr to the Metropolitan Council, August 24, 1972.

Figure 1

How the Council Represents the Area

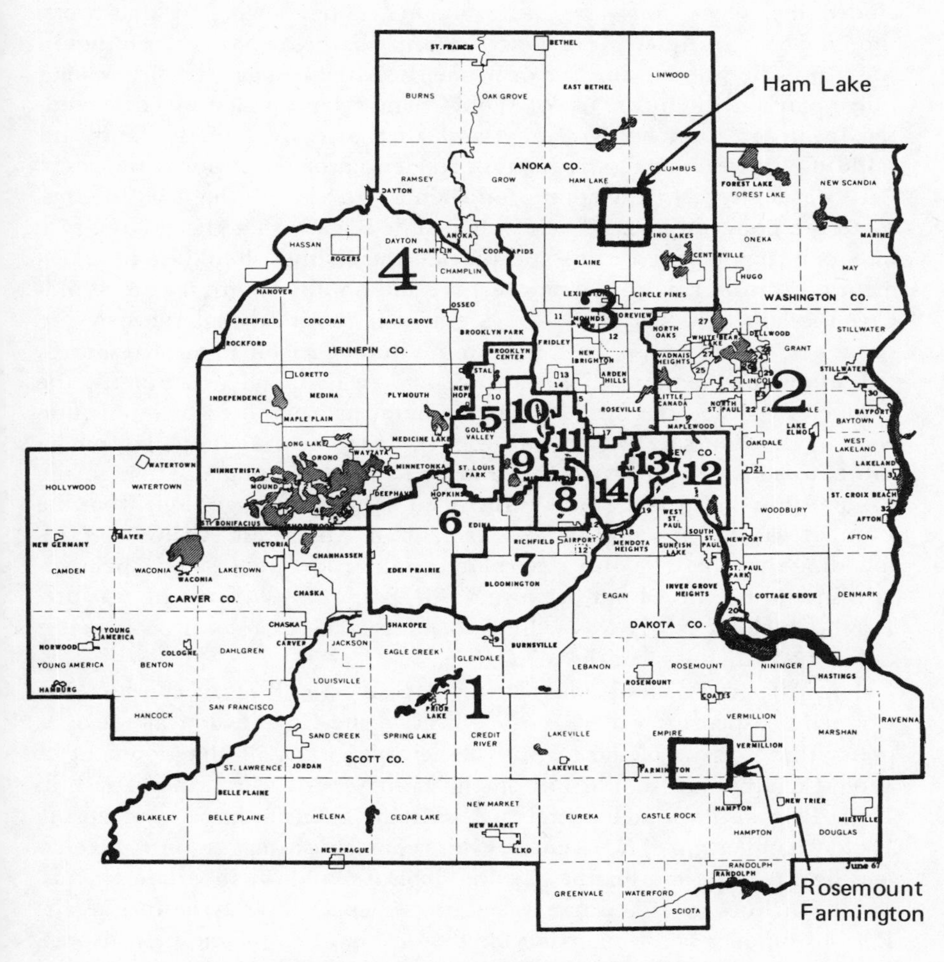

The councilmen and their districts are as follows:
Chairman—James L. Hetland, Jr., Minneapolis.

1. Marvin F. Borgelt, West St. Paul.
2. Milton L. Knoll, Jr., White Bear Lake.
3. Joseph A. Craig, Coon Rapids.
4. Donald Dayton, Wayzata.
5. George T. Pennock, Golden Valley.
6. Dennis Dunne, Edina.
7. Clayton L. LeFevere, Richfield.
8. Glenn G. C. Olson, Minneapolis.
9. E. Peter Gillette, Jr., Minneapolis.
10. James L. Dorr, Minneapolis.
11. George W. Martens, Minneapolis.
12. The Rev. Norbert Johnson, St. Paul.
13. Mrs. James L. Taylor, St. Paul.
14. Joseph A. Maun, St. Paul.

Dunne (6), and Maun (14). The Counselor to change his vote was Dorr (10).

When the Council transmitted its decision to the Commission, it included five suggestions for future action.[46] They suggested that work should begin on selecting a new site and that Rosemount-Farmington was a feasible site, but not the only one. It was also suggested that a long range plan for the future use of Wold-Chamberlain be developed, including future aviation needs, if any, and alternative uses of the land and buildings for commercial, industrial, residential or recreational purposes. This was desirable, the Commission claimed, in view of the need to eliminate the noise problem. The Council stated again that there should be only one commercial airport and that all the airlines should be in agreement on transferring to the new airport and on any future uses of Wold-Chamberlain. The Council also reiterated its position that the new site should be developed so as to have minimum effect on natural resources and allow for maximum joint airport and recreational use. Finally, the Council suggested the formation of a committee composed of representatives of the Council, the Commission, the airlines, and the environmentalists to consider potential sites and to make a recommendation to the Commission and to the Council. Included with these suggestions from the Council was a letter from the President of Northwest Airlines recommending that there be only one commercial airport, that Wold-Chamberlain should be phased out between 1978–82, that a southern site was preferred, and that NWA was willing to undertake the role as coordinator between the airlines and the MAC.[47]

The Airports Commission met on January 18, 1971 to consider the Council's suspension order. At this meeing, the Commission affirmed its policy that no part of the cost of the new airport or of abandonment of Wold-Chamberlain should fall on the taxpayers of Minneapolis and St. Paul. They also initiated a study assessing the financial impacts of closing Wold-Chamberlain. The essence of this report was a suggestion to accelerate the payoff period for the existing debt at Wold-Chamberlain by raising the airlines rents and usage fees and concurrently relying on the airlines to support bond financing for the development of the new airport. Predictably, the airlines strongly opposed this suggestion in a position paper sent to the Commission in March, 1971 and signed by six of the

46. Letter from Mr. James L. Hetland, Jr., Chairman of the Metropolitan Council, to Mr. Lawrence M. Hall, Chairman of the Metropolitan Airports Commission, December 23, 1970.

47. MAC, "Chronological Summary," p. 5.

eight airlines serving Wold-Chamberlain. In this paper, the airlines also stated that it was the job of the local authorities to decide on a new site and the airlines were not going to assume the power of decision in this area.[48]

No further formal action was taken during the remainder of the legislative session which ended in the spring of 1971. Several bills had been filed dealing with the airport question specifically and with the Airport Commission and the Metropolitan Council generally. Since both of these local bodies were created by the state legislature, the legislature has the right to alter their powers, procedures and authorities whenever it wishes. Although nothing came of these bills, their presence was enough to halt any action by the Commission.

Later in 1971 the Commission and the Council agreed to establish the Joint Committee suggested by the Council in its rejection letter in December, 1970. The only difference between the Joint Committee as established and the Council suggestion was that it did not contain airline or environmental representatives. It was constituted of eight members, four from the Commission and four from the Council. This Joint Committee had no formal powers, as people involved in the decision were quick to point out.[49] It was established as an ad hoc advisory body to both the Council and the Commission and neither body was in any way bound by any recommendations of the Joint Committee.

The advent of this Joint Committee instituted an important procedural change. The recommendation of this committee was not for a specific site, but for a "search area," and this recommendation then went directly to the Council for consideration. The Council then considered this recommendation in its airport systems Development Guide chapter on airport systems. In writing and approving this chapter, the Council is in a position of choosing between several areas. The decision puts the Council in the position of playing a positive decision-making role for the first time and presumably gives the Council more control over the decision. The Airport Commission does not have to go along with the Council's decision and could continue to pursue the Ham Lake site or any other alternative. However, it would be doing this in spite of a clear signal from the Council about what location is acceptable.

The Joint Committee agreed to consider four different sites. Two of these were Ham Lake and Rosemount-Farmington. The two new ones

48. MAC, "Chronological Summary," pp. 5–6.
49. Interview with Mr. Raymond G. Glumack, Director of Operations, Metropolitan Airports Commission, July 17, 1972.

were located slightly further from the downtown areas, one in the north and the other in the south and labeled the north and south search areas.

The Joint Committee met on July 24, 1972 to vote on its recommendation. This was one of the most interesting meetings of the prolonged controversy. The Committee took four votes. The first was a unanimous vote eliminating Ham Lake. Rosemount-Farmington was eliminated by a 7–1 count on the second vote. The dissenter was Marvin Borgelt, one of the Council members on the Committee. His Council district covered the entire southern tier of the metropolitan area, including the Rosemount-Farmington site. The third vote was on a motion to recommend the new southern search area. This motion did not pass on a 4–4 tie vote. The final motion was for the new north search area and this passed 4–3 with one abstention.

The interesting votes were cast by Martin Companion, a St. Paul AFL-CIO labor leader and the citizen member of the MAC appointed from St. Paul. He abstained on the vote for the north search area. Companion was a close personal and political associate of David Roe, the leader of the state AFL-CIO and former Minneapolis citizen commissioner on the MAC, who was supporting a northern site. A major concern of the unions in both cities and of the St. Paul politicians was the question of union jurisdiction at the new airport site. The Minneapolis unions had jurisdiction over any northern site while a southern site would belong to the St. Paul groups. Several people mentioned this as one of the economic considerations which had played a part in the controversy even though it was pointed out on several occasions that the MAC had always found a way to spread contracts fairly evenly between representatives of the two cities. However the unions in both cities and some politicians were concerned about this matter. Companion had previously stated however that he favored a northern site, but had not said why. At the time of the Joint Committee votes Companion had arranged with George Pennock, the Committee Chairman, to vote last on the final two motions. Pennock was a supporter of the northern site and had voted for Ham Lake both times it was considered by the Council. When it was Companion's turn to vote on the southern search area, he voted for the southern site saying he had received a telegram threatening his job if he did not vote for a southern site. This made the vote 4–4 and Pennock ruled that the motion had not passed. Companion also voted last, but then abstained on the motion to recommend the northern site. This resulted in the 4–3 vote with one abstention by which the motion passed. These two votes led to considerable bitterness among the supporters of the southern site who claimed that Pennock was not supposed to vote except in the case of a tie. They felt

Pennock had arranged this tie by letting Companion vote last so the vote was 3–3 when it was his turn to vote. Although Companion did not reveal who sent the telegram, there was considerable agreement among observers at the time that it came from the St. Pauls unions. This seems to be a case of an appointed official following his personal interests rather than those of a supposed constituency since all the elected St. Paul members voted for a southern site.[50]

The Joint Committee's recommendation was received by the Metropolitan Council and referred to the Development Guide Committee for inclusion in the airport systems chapter of the Guide. The Council convened on August 24, 1972 to consider and take a preliminary vote on this chapter. The only section to receive any attention was the discussion of the location for a new airport. Within the Council, debate centered on the economic questions. Environmental questions were no longer important. Several councilors who voted against Ham Lake said they now were supporting the new northern site because it did not have adverse environmental impacts. The major economic concern was the economic loss to the Bloomington-Richfield area specifically and the southern area generally. This concern was countered by an argument that the northern site would "redress the economic imbalance in the metropolitan area." Several councilors emphasized that they felt the current circumstances did not warrant proceeding with the new facility, but they were only voting for the Development Guide chapter stating where a future airport should be located. Finally, several councilors felt it was important for the Council to use this opportunity to go on record making a positive statement as to where a new airport should be to indicate they were not merely an obstructionist agency. The Council then voted 9–5 to accept the draft of the airport systems chapter specifying the northern site. The councilors casting no votes and their districts were: Borgelt (1) and Hoffman (7) from the southern areas, Johnson (12) and Reed (13) from St. Paul, and Martens (11) who was opposed to any new airport.

This will not terminate the controversy however. The Council must still hold public hearings on the airport chapter and then take a final vote. The expectations were however that the recommendation for a northern search area will stand. Then it is up to the Airports Commission to decide if they want to proceed with plans for a new airport and if they do, if they want to locate it in the northern search area. There is considerable speculation that the Commission may still come up with a plan of

50. This account was provided by Mr. David Rubin, Airports Program Manager for the Metropolitan Council, in a telephone conversation August, 5, 1972.

its own. The Executive Director of the MAC has publicly stated that he found a proposal by a St. Paul city councilor to expand Wold-Chamberlain interesting and thought it had possibilities. Furthermore, three of the St. Paul members of the Commission have indicated that they might vote against any northern site at this point. Under the rules establishing the Commission, this will be enough to block a northern site. If this happens, several people predicted that the state legislature may step in and restructure the Commission.

Influences on the Commission and Council Decisions

It should be quite clear at this point that the single function agency, the MAC, gave different consideration to the issues involved in the airport location question than did the multipurpose body, the Council. Furthermore, there were indications that the appointed officials responded to different concerns and interests than did the elected officials. The question now is what have we learned about the ability of different governmental structures to handle the conflicts inherent in public policy decisions from the way these two organizations considered the issues in the airport location question. Unfortunately, given that we have only one case and several institutional differences, it is difficult to answer that question. It is not clear how much of the difference between the Commission and the Council decisions may be attributed to the fact that one is a specialized administrative agency and the other a multifunction policy body; that the Commissioners are elected officials, whereas the Councilors are appointed. However, it is possible to make some observations about these structural differences.

Single function and multifunction agencies

The overwhelming majority of the MAC decisions were the decisions of the MAC staff. The individual Commissioners seemed unconcerned about the operations and decisions of the Commission and gave the staff considerable discretion on these matters. At least until the airport location and noise questions arose, the Commissioners were continually accused of being a rubber stamp for the staff.[51] This lack of concern can be attrib-

51. Harper, "The Minneapolis-St. Paul Metropolitan Airports Commission," pp. 381, 393 and 398.

uted both to the structuring of the Commission itself and to the actions of the staff who did a good job of avoiding problems which could arouse the Commissioners' constituents. The staff seemed solely concerned with airport operations and with avoiding problems such as tax increases which might cause the Commissioners to become involved in airport decisions. These two general objectives manifested themselves in the form of three criteria which the staff imposed on a new airport location. The MAC staff wanted to remain financially self-supporting, minimize potential air space conflicts, and avoid future urban development in the vicinity of the airport.

Staying off the tax rolls of the two major cities substantially reduced the likelihood of involvement by the Commissioners. Undoubtedly the mayors and councilors from the Twin Cities were more than happy not to have the airport operation adding to their property tax rate. The taxing arrangement of the MAC made some sense in 1943 when the Commission was established and the two cities had over 80 per cent of the metropolitan population. By 1970 this rate had fallen below 40 per cent, so that these citizens would be certain to object to financing the airport for the metropolitan area. The elected officials from each city, who comprise most of the Commission, can see this inequity and are likely to be strong supporters of the "zero-tax support" position. Thus, as long as the operations of the airport could be kept off the cities' tax rolls the Commissioners would be less motivated to participate in running the airports and the staff would have more autonomy. Maintaining "zero-tax support" necessitated keeping Wold-Chamberlain open and dictated the two airport solution favored by the MAC.

Once the MAC staff decided that Wold-Chamberlain would remain open, considerations of air space conflicts became very important. The MAC staff was continually receiving professional accolades for its efficient operations.[52] Widely separated air spaces, which permit easier methods of aircraft handling and the minimization of traffic delays, are certainly preferred from the standpoint of maintaining their professional reputation. Air space considerations virtually dictated a northern site, once it was decided to keep Wold-Chamberlain in operation. Any southern site which avoided a conflict with Wold-Chamberlain would have been too far from either city to be practical. However, at one of the Joint Committee meetings an FAA official said it is possible to take a three dimen-

52. Harper, "The Minneapolis-St. Paul Metropolitan Airports Commission," p. 381, and John F. Judge, "The Minneapolis-St. Paul Dilemma," *Air Line Pilot* Vol. 41, No. 3 (March 1972), pp. 6–9.

sional view of the problem. This in effect means layering the approach routes rather than following the two dimensional view the Commission was taking by laying out rectangles on maps to define airspace.[53] This three dimensional concept provides more flexibility and airspace, but is difficult to control. It would have meant that Rosemount-Farmington might be compatible with Wold-Chamberlain. The staff was not enthusiastic about the idea and had either rejected it or had not considered it earlier.

The final consideration to which the Commission seemed to give considerable weight was the potential for future urban development. Preventing potential development near the airport minimized future difficulties from noise and pollution and the resulting demands that restrictions be placed on aircraft and airport operations. This consideration of land usage favored the Ham Lake site with the refuge to the northeast and its abundance of undevelopable wetlands, peat bogs, and marshland on the remaining sides. On this criterion, Ham Lake was preferable to any other site, north or south. This land also was available at the lowest price, but the low potential for future development and the reduced airspace problem, given that Wold-Chamberlain was to stay open, were the prime considerations in the Commission decision.

There is no question that the attributes the Commission was concerned about—the minimization of airspace conflicts and the lower potential for a noise problem—are to be valued in an airport site. But what are these advantages worth? Locating the airport next to the refuge involved some trade-off between the advantages of reduced airspace conflicts and noise pollution and the costs of uncertain risk to wildlife and some damage to the value of the refuge, even though people could reasonably disagree about the magnitude of the potential damage. The staff of the MAC minimized both the amount and extent of these damages and were clearly willing to make the trade in favor of reduced airspace and noise pollution problems.[54]

The objectives and decision criteria of the Metropolitan Council are not so easy to define. It would be easy to claim that the Council had the reverse view of the trade-off made by the Airports Commission since they

53. July 17, 1972 meeting of the Joint Committee.

54. One MAC staffer pointed out the relatively small amount of land involved by showing what area the refuge would cover if placed adjacent to the existing airport. In this case the refuge would extend well beyond areas currently experiencing any inconvenience from air traffic. His conclusion was that the airport would not seriously affect the preserve. Interview with Mr. William Olson, Project Engineer for the Metropolitan Airports Commission, July 17, 1972.

approved a northern site after rejecting Ham Lake. However, this is probably not the whole case. To be sure, the Council and its staff did consider the risk the Ham Lake site held for the Carlos Avery and for water quality in the area. The switch from a 9–5 vote against Ham Lake to a 9–5 vote for the new northern search area can be partly attributed to the reduced environmental risks associated with the new location. The veracity of these statements was supported by the transportation planner for the Council. In July he said several councilors had opposed Ham Lake for environmental reasons and predicted they would vote for a northern site. In addition, Mr. Dorr's stated reason for changing his vote on the Ham Lake site the second time it came up for Council consideration was that he had been convinced by various consultants that it would not endanger water quality in the area. The Council also stated that possible harm to the environment was one of the reasons for suspending the Commission's proposal each time it did so. Consequently it seems fair to conclude that the Council was giving more weight to the environmental risks than was the Airport Commission.

There were several other considerations in the Council decisions, however, which complicate the analysis. The most important of these was the question of the distribution of the economic benefits expected to be associated with the airport. The Councilors' votes in part reflected these economic impacts, particularly after the environmental questions were settled. The lengthy arguments by Mr. Hoffman against the northern search area, his and Mr. Borgelt's votes against this proposal, and the favorable votes of the councilors from northern and western areas reflect these considerations. Councilors' votes did not always reflect the expected interests of their districts on this matter, however. One councilor from a southern district, Dennis Dunne, voted for the Ham Lake site and the northern search area. He was the President of the Minneapolis Chamber of Commerce, and a suggestion was made that he was voting the Chamber's interests. The Chamber hoped that a northern site would help the downtown area regain some of the business it had lost to southern establishments since the expansion of the airport, the construction of the stadium, and the growth of new manufacturing firms and their associated commercial interests and population. The Minneapolis Chamber of Commerce had come out strongly in favor of the Ham Lake site when it was first proposed.

The second dimension to the geographic issue was the split between Minneapolis and St. Paul. In addition to the important question of union jurisdiction, it was generally agreed that Minneapolis stood to gain more from a northern site and St. Paul more from a southern one. This was particularly true if the new northern search area was selected, since it is further west than Ham Lake. The implications of these city differences were apparent in the three Council votes. Two of the three St. Paul members of the Council voted against both Ham Lake and the proposed northern search area. The only St. Paul councilor not to do so had been the St. Paul citizen representative on the Airports Commission until late 1967, when he was appointed to the Metropolitan Council. The Minneapolis councilors had opposed Ham Lake, but when the votes were cast for the new northern search area, three of the four Minneapolis councilors voted yes. The no vote was cast by George Martens who said he opposed building any new airport and thought the money could be put to better use elsewhere.

The Airport Commissioners did not become concerned about the Minneapolis versus St. Paul question until after the Joint Committee decision to go north, even though the two cities have equal representation on the Commission. The new concern on the part of the St. Paul commissioners is mostly attributable to a change in the membership. During the course of the controversy the citizens of St. Paul elected a much more active mayor and city council. Although these elections cannot be related strictly to the airport issue, they changed the complexion on the Commission.

There is one piece of evidence to suggest that the Council was functioning as a general legislative body trying to consider the overall distribution of the economic benefits associated with different projects within the metropolitan area. During Council deliberations on a proposal by the Metropolitan Transit Authority to build a rail rapid transit system, one of the councilors suggested deferring the vote until the airport decision came up. He then proposed to one of the other councilors that he would vote for the rapid transit if the other member would support a northern site. The offer was declined for any number of possible reasons, such as the trade would not have changed the outcome of either vote. Whatever the reason, it shows that the Council tried to relate the consequences of one decision to an outcome on a different issue, particularly with respect to the distributional question.

The final consideration which seemed to affect the Council's decision was a consideration of Northwest Airlines interests. As was pointed out, the airline was vigorously opposed to both the two-airport concept and a northern site. The Council, although it eventually adopted a northern site, consistently backed Northwest's demand for a single airport. This seems to be more important to the airlines than having a southern site. It is likely that the Council is more sensitive than the MAC to Northwest's threats to move their operations elsewhere. If Northwest did go through with this plan it could affect the local economy in a significant fashion, particularly on the southside where most of the 6,000 employees live. This may not be the only reason the Council supported the airlines, but it is in a better position than the MAC to see the economic consequences of a move by Northwest.

On balance it becomes difficult to say precisely what weight the Council gave the various factors in considering and rejecting the Ham Lake proposal and in selecting the new northern search area. They appeared to be more concerned with the environmental effects, the economic impact of the activity associated with the airport, and possibly the interest of a large local business which would be significantly affected by the location decision. It should be clear that the Council considered a much broader set of issues than did the Airport Commission and its staff who were almost solely concerned with their cash flow, airspace conflicts, and the absence of existing or potential urban development.

Elected and appointed officials

There are noticeable differences between the behavior of the elected and the appointed participants in the airport location controversy. The differences are not apparent until late in the case because the only elected officials are the MAC Commissioners who played a minor role in the early decisions to locate a new airport at Ham Lake. The Commissioners' apathy toward Commission decisions suggests that unless officials are elected directly to the decision-making agency and held responsible for agency decisions in the eyes of the voters, they are not likely to be active participants.

The first elected official to take an active role in MAC affairs was Gladys Brooks, the Minneapolis city councilwoman, whose ward adjoins the airport. In response to her constituents' complaints about the noise level at Wold-Chamberlain she was able to get the MAC to agree to participate in a sound abatement council and to restrict operations at Wold-Chamberlain. She subsequently requested and was appointed by the Minneapolis city council to a seat on the MAC and served on the Joint Committee. She has been a proponent of the new airport because it would alleviate the noise problem in her ward and voted for the southern site on the Joint Committee because it is more convenient to her district.[55]

55. These statements were made in an interview with Mrs. Gladys Brooks, August 23, 1972.

The other elected official to take an active role in the case was Leonard Levine, a young St. Paul city commissioner who was appointed to the MAC in 1971. Levine proposed expanding the current airport by constructing a set of detached runways in Eagan Township, across the Minnesota River from Wold-Chamberlain. He came from the Highland Park area of St. Paul, the part of that city closest to the airport. He was a leader of the Highland Park Community Council which had been protesting the increased air traffic noise. Levine first ran for one of the St. Paul city commission positions, which are elected at-large, on a platform advocating, among other things, a reduction in airport noise levels. Levine opposed all four potential sites for a new airport. He opposed Ham Lake because of its potential impact on St. Paul water quality. The reasons he gave for opposing the new sites, both north and south, and for his proposal to expand Wold-Chamberlain was that the new sites were too far from the cities, that people would not use the new airport, and that it would be another Dulles, which he pointed out was operating at a deficit.

There are several representational explanations for Levine's positions. The most important fact is that Levine must run at-large both to hold his seat or to move to higher office. One reason for his opposition to a northern site is simply that St. Paul would benefit more from a southern site. The St. Paul AFL-CIO of course was opposed to a northern site where the Minneapolis unions have jurisdiction and union support is important for any Democrat running at-large in St. Paul. Another reason is simply the cost question. The MAC staff and their financial consultant estimated that if the Council enforced the one airport requirement the new airport would require tax support which would only come from the two central cities. It is reasonable to expect that the people of St. Paul will object to paying for the new airport since most of the metropolitan population resides in the suburbs. This financial consideration may have been the motivation behind Levine's detached runway proposal, which would be considerably cheaper. Of course this would not alleviate the noise problem at Wold-Chamberlain and the citizens in its vicinity were strongly opposed to Levine's proposal, as were the people of Eagan Township—who of course do not vote in St. Paul. Several people said Levine was in trouble with his constituents in Highland Park because of the proposal and his opposition to a new airport. Presumably to counter this problem Levine had the MAC authorize a poll in areas around the airport on reactions to the noise problem. The results of the survey purport to show that people preferred to have a convenient airport, even at the cost of the noise problem, and that people felt that aircraft noise was no worse now than it was a few years ago. However Levine's critics were quick to point out that the survey was taken while Northwest Airlines was still shut down by a pilots' strike which had been going on for several months.

The individual councilors and appointed commissioners did not exhibit the sort of constituency orientation shown by the elected officials. The discussion of the Council's consideration of the economic impacts of the airport location suggested that some of the councilors' votes, such as Hoffman, Borgelt, and the northern members, were consistent with their districts' interests. However there were also some which possibly were not, such as Dunne's which was more consistent with the interests of the Chamber of Commerce. The one versus two airport issue and the question of Northwest's interests also suggests that some councilors may have been influenced by personal considerations. Several of the councilors have personal economic interests which would be adversely affected if Northwest were to move most of its operations to another city. One councilor is an officer in a local bank, one founded the metropolitan area's largest department store, and another is a local realtor. If Northwest were to substantially reduce the size of its local payroll and transfer jobs out of the area it would seriously affect the prosperity of all three concerns In addition, one of these people was a personal friend of the president of Northwest Airlines. The realtor voted no on all three votes and stated his opposition to any new airport. The other two men both voted against the Ham Lake site and for the northern search area. However both said they were opposed to a new airport at the present time but were voting yes because they felt the future of the Council required a positive vote. One of the people interviewed who is close to both the Council and the airport decision said these economic and personal effects were strong enough to make the councilors at least sympathetic to Northwest's interests, if not to vote on those interests on several occasions.

The most interesting behavior from the representational standpoint was Companion's votes at the last meeting of the Joint Committee which recommended the north search area. His arrangement with Pennock and his abstention on the vote for the north site were better expressions of personal interest than of supposed constituency concerns. Companion supported the north site in spite of considerable pressure from an important St. Paul constituent group, the AFL-CIO, who supposedly threatened him with the usual pressure reserved for elected officials. However in Companion's case the threat was not effective, possibly because of his tie to the state AFL-CIO leader. The fact that the elected St. Paul Commissioner on the Joint Committee voted for the southern site and that all three elected St. Paul Commissioners have subsequently stated that they will not support a northern site should indicate the extent to which Companion deviated from the position preferred by his nominal constituency.

The votes of some of the councilors and Companion's behavior on the Joint Committee indicate that personal preferences, private loyalties, and personal objectives can play an important role in the decisions of appointed officials. Personal considerations are not absent from elected officials' decisions although they seemed less important for the few elected officials involved in the case. The major difference however is that voters, if they are so motivated, can take sanctions against elected officials who put their personal views above constituency concerns. In the case of an appointed official, the question of sanctions is left up to the appointing individual or body, or to private organizations such as the AFL-CIO or the Chamber of Commerce. These may be of limited value if sanctions are felt to be a useful or necessary control.

The decisions of the Commission and the Council and the behavior of the individuals associated with both organizations support most of the expectations about the effects of institutional structure on public decisions. The single function, executive type agency, the MAC, took a much narrower view of the problem of locating an airport and gave much greater weight to the issues of financing and operating the new facility than did the Metropolitan Council, the general purpose, multifunction body. Secondly, the elected officials, such as Brooks, seemed more sensitive to constituency concerns and less susceptible to private pressures and personal considerations than did the appointed officials, like Companion and Dunne, although this could only be examined in a limited fashion here. These differences should be important when it comes to designing and altering governmental institutions.

Who Decides and What Difference Does It Make?

The characteristic of public policy decisions which sets them apart from private decisions and which presents a problem for the design of governmental structures is the need to accommodate the legitimate but conflicting interests and values associated with public activities. Conflicts arise in part because of the collective nature of the decisions. We can use the environment to support technological advances, such as air travel, or we can use it to support wildlife and recreation. However we cannot do both, at least to the degree that some people wish. Consequently there were conflicts in the Twin Cities over differences in the values attached to such things as additional airport capacity, the reduction of aircraft noise in residential areas, the presence of a large bird sanctuary and the possibility of harm to it caused by low flying planes, and avoidance of potential damage to water quality if the swamp areas were drained and the water runoff from the airport allowed to drain into the water shed.

A second source of conflict in most public decisions, and which was apparent in the airport location decision, is that attempts to improve the level, quality, or distribution of one public service invariably have significant impacts on other public activities. In this case the initial decision was to expand the capacity of the air terminal and to reduce the level of noise pollution in populated areas. The consequences of that decision were changes in the distribution of income in the region, as it is effected by the level of economic activity, decreased enjoyment of a large wildlife sanctuary, and major alterations in the accessibility of a major transportation and workplace center for many people.

The need for a collective decision on matters of land use and environmental policy and the consequences of the airport location decision for issues outside the field of air transportation were at the root of the conflicts in the decisions by the MAC and the Council. The behavior of the MAC and the Council and of the members of these organizations suggests that a multifunctional legislative type body whose purpose is to make choices among different activities and to control the various operating agencies can do a better job of mediating the conflicts generated by these different activities and their consequences than the special function agencies. An important characteristic of the Council in this regard is that it has no operating responsibilities and is not in the position of trying to sell people on the advantages of one particular activity, such as more sewers, rail transit, or an expanded airport. The Council's sole function is to insure that the decision process is sensitive to all the conflicting interests associated with each activity. Mr. James Hetland, the first Chairman of the Metropolitan Council, underscored this responsibility by saying that the Council will be much more effective once it recognizes that it is not supposed to completely agree with the different operating agencies, but is supposed to be in conflict with them on the levels of service, where activities will be located, and how they will be provided. He said the Council's job is to make these decisions and that it necessarily involves disagreeing with the special agencies because each organization has different functions to perform.[56]

The experiences of the Metropolitan Council also suggest that a body

56. Interview with Mr. James L. Hetland, Jr., Vice-President, First National Bank of Minneapolis, August 23, 1972.

structured to include all the legitimate conflicts of interests involved in public policy decisions and which has no administrative functions of its own makes more appropriate use of its staff and the information they provide than the executive agencies. The Council staff did not try to promote or even suggest a particular decision during the controversy. It presented its evaluations of the MAC consultant's reports and its own analysis of the various consequences of each alternative on matters which concerned the Council, such as aviation needs and economic and environmental impacts. In each of its presentations the Council staff was subjected to considerable cross-examination by the councilors. This cross-examination usually brought out additional questions and considerations relevant to the decision which the staff had not included, even though they had tried to anticipate councilors' concerns. Consequently the Council not only had control of its staff but was able to draw a considerable amount of analysis out of it.

The better experiences of the Council with respect to its staff may be accounted for by several explanations, even though we may assume that the general objectives of the Council staff were similar to those of the MAC staff in that both presumably had a strong professional orientation and wanted to maintain their autonomy. In the first place fulfilling the professional objectives of the Council staff did not require the promotion of any particular public policies such as rail transit or additional air space. Consequently the staff had less of a vested interest in the final decisions of the Council from a professional standpoint. Secondly, because the various interests involved in the decision were represented on the Council, although possibly not in the most desired fashion, the staff was subject to serious cross-examination which would have revealed biases on their part. At the same time, because there were councilors with strong interests on both sides of the issue, the best way for the staff to keep its independence in the long run was to avoid taking sides or getting drawn into the controversy in terms of promoting particular outcomes. In fact one of the staff members of the Council involved in both the airport and the transit issues said that his strategy was to stick with the analysis and avoid the appearance of taking sides. The most likely circumstance in which this strategy would break down is if one person or group came to dominate the council decisions and thus to "own" the staff. However in a multifunction body with many cross-cutting issues and interests it is unlikely that such a dominant group could emerge.

It is important that a multifunctional body be charged with the task of mediating the public policy conflicts and making sure that all the interests associated with a decision have an opportunity to influence that decision.

The most obvious reason for this requirement is that much of the conflict results because decisions in one area affect activities in other areas and special function agencies cannot take account of these relationships. Because the Airports Commission does not have any responsibility for environmental matters or for regional income distribution there is no way it can adequately arbitrate between those interests who want to get a new airport and those who want to protect the bird sanctuary or those who are concerned about the possible decline of economic activity in part of the region. A multifunction organization whose sole responsibility is to mediate the conflicts inherent in public policy decisions and whose members are tied to the participants in the conflict by some means of institutional design, such as by district constituencies, stands a better chance of at least debating the programmatic conflicts.

A second important reason for having decisions made by a multipurpose body is its ability to execute side payments among groups who have opposing interests. There were no apparent examples of this taking place in this case, however. The closest the Council came to doing this was the councilor who supposedly suggested that his vote on the rail transit question could be tied to promises to vote for the northern search area. In part the absence of visible side-payments or log-rolls may have resulted from the fact that each area wanted to get the airport rather than prevent its presence. In most cities the contrary is true; each area is fighting to block the construction of the airport in their vicinity even though they may agree in theory that a new facility is needed. We could envision a situation where the citizens in south Minneapolis, Highland Park, and Eagan Township, for example, were trying to force the construction of a new airport to get rid of the noise problem and airport users were demanding a new facility because of congested conditions at the current one but wanted to maintain the convenience of the existing location. Each time the local authority proposed a possible site the citizens of that area would get sufficiently aroused to block the proposal either through public hearings, the state legislature, or court proceedings. A multifunctional body such as the Metropolitan Council, if they had control of the various functional agencies, could then arrange a set of side-payments or log-rolls whereby any area willing to accept the airport could be promised additional positively valued services, such as mass transit, parks, etc. These types of side-payments in kind are about the only compensations possible in the public sector because it is not possible to explicitly reduce the tax rate in one area simply because it agrees to accept certain costs in the form of undesirable activities. These compensations can only be accomplished by an organization which has control over the operations of the various specific departments or agencies. In this regard, Mr. Hetland also said that for the Council to become effective it had to have the ability to get budgetary control of the operating agencies, such as the Sewer Board, the Transit Commission, etc. so that the Council had some control over how the operations were conducted and which areas received what kind of services.[57]

The requirements for membership and the procedure used to select members are also important in how well different interests will be considered. One of the characteristics which made the Council more effective than the Commission was the fact that Council members represented specific geographic districts throughout the metropolitan area, even though the Councilors were not accountable to their districts and some did deviate from district concerns. Since the interests affected by most public decisions are not uniformly distributed over a metropolitan region, a geographic pattern of representation means the various groups with an interest in the decision are likely to be explicitly included in the deliberations. This is certainly not the case with the MAC or with many single-function agencies where the governing board, the commissioners, or the directors are appointed for their expertise, political connections, etc. and without regard to any systematic representational criteria. The question of apportionment is clearly important and application of the one-man, one-vote criteria would surely be a necessity. A Council of Governments in which Eagan Township or the other suburban communities had an equal voice with Minneapolis or St. Paul would distort the representational gains made by moving to a Metropolitan Council type legislature. The Council solves this problem by making Council districts coterminous with state legislative districts which by law should be apportioned equally.

The procedure used to select the members of the legislative body does have an important bearing on how various interests will influence public decisions. Elected officials seem to be more concerned with the interests and concerns of their immediate electoral constituency and less influenced by outside private organizations and personal considerations than appointed officials. Of course elected officials are not immune from these latter influences, but they presumably give relatively less weight to them, particularly if the elected office constitutes a full time occupation. Though not particularly surprising, this is an important observation when combined with the previous observation that membership on the decision

57. Interview with Mr. James L. Hetland, Jr., Vice-President, First National Bank of Minneapolis, August 23, 1972.

making body should be district based. The logic for requiring elected officials is that they are more likely to insure that various interests are represented in the decision process and have an opportunity to express themselves in an influential way. Numerous people, including several incumbent councilors, in discussing the future of the Council, said it was a necessary and hopefully an imminent step to make the council an elected body. Their reasoning was precisely the argument presented here. Since the councilors' function is to insure that the various metropolitan interests are taken into account in planning and development decisions, it is necessary to have them elected by the people who hold those interests.

The implications of this case and the conclusions of this paper are quite straightforward. The best way to insure adequate consideration of the legitimate interests associated with decisions in the public sector is by a multipurpose legislative type body which has jurisdiction and control over the various functional agencies or departments and whose members are elected from equal sized districts throughout the area affected by the decisions.

COMMENT

Richard E. Quandt, Princeton University

The problem of locating an airport provides an almost ideal framework for discussing some important questions that often arise in large public investment projects. These include some very general questions such as (a) how one should quantify, for purposes of a cost-benefit analysis, certain items that are generally regarded as defying quantification and (b) whether certain requirements, frequently of an environmental nature, should be regarded as inviolable constraints upon the airport location decision or whether they should be quantified and entered into the objective function. On a more specific level airport location problems force us simultaneously to deal with policy variables (land-use and regional planning targets), endogenous variables (traffic generation, modal split, air traffic control, surface access, etc.) and exogenous variables (meteorological and geographic data). John Jackson's fascinating and at the same time somewhat depressing case history of the Minneapolis airport location problem focuses on one particularly important aspect, namely on the nature of the body (or bodies) responsible for making the location decision and on the possible consequences for the ultimate decision of the composition and aims of such bodies. His basic finding is that a multifunction legislative body with elected members representing all affected constituencies is the most suitable type of body because (a) it can do a better job of mediating conflicts that arise in connection with the airport location decision, (b) it can better execute side payments and provide compensation to those whose interests may be injured by the decision, and (c) it can be expected to be more responsive to the interests of the electorate.

These are reasonable conclusions and it is not my aim to challenge them. Rather I should like to consider further (a) some of the particular problems involved in locating a new airport, (b) the entire locational decision-making process with specific reference to what is probably the most systematic and massive effort in this regard, namely the work of the Roskill Commission on locating the third London airport.

It may be difficult to measure precisely when a process designed to produce an airport location decision was in fact initiated. In Minneapolis this process started no later than 1968 and, as far as I understand, has not terminated yet; nor is there any promise of speedy termination in the future. The mind boggles at the thought of how long the fourth New York jetport has been under discussion. In contrast, the appointment of the Roskill Commission and its terms of reference were announced in the House of Commons in May 1968; the Commission completed its inquiry and produced its report by late 1970 and a decision was made by the Government in April 1971. In spite of the fact that there had been discussions and white papers concerning a third London airport prior to 1968, the Commission appears to have been justified in its rather understated view that "The story of the fourth airport in New York is, we believe, longer than the story of the third London airport." [1]

What is the source of the difference? In Minneapolis at least the story appears to be one of a sequence of paralyzing strokes and counterstrokes by the Metropolitan Council or the Metropolitan Airports Commission, or by Northwest Airlines, or representatives of South Minneapolis interests or representatives of St. Paul interests, etc. It is bad enough to have the Commission and the Council, with their significantly different mandates and functional aims, both involved in the decision making process; this is compounded by blackmail by private business interests and by the conflict of interests that arises from the fact that any Minneapolis airport, as indeed many others including the New York airports,

1. Commission on the Third London Airport, "Report," HMSO, 1971, p. 6. Hereafter referred to as Commission, "Report."

would lie near the boundaries of several municipalities or jurisdictions each of which must concur in any decision.

The contrast between the decision making structure in Minneapolis and the Commission on the Third London Airport is enormous. The members of the Commission were appointed by the President of the Board of Trade for the single task of recommending a site, if any, for the third London airport. They themselves had no vested interest in the decision to be made; they were neither to be reelected by any constituency nor reappointed by any authority to a body with continuing existence. Their recommendation was to be made directly to the Government. Among their members we find a judge (the Chairman), three professors (of transport, economics, and aircraft design, respectively), a partner in an engineering firm, an inspector from the Department of the Environment and the deputy chairman of a large business firm. The terms of reference given to the Commission specifically instructed its members to consider questions of planning, noise, agriculture and environment, air traffic control, surface access, and defense and to employ cost-benefit analysis. The Commission felt, and nobody would accuse any of the participants in the Minneapolis situation of harboring similar anxieties, that "Not the least of the tasks facing the Commission upon its appointment was the need to establish public confidence that its work would be impartial, unbiased and entirely uninfluenced by anything which had gone before." [2]

The work of the Commission proceeded in five stages. In the first stage public and commercial organizations were invited to present evidence on problems of regional planning, noise, surface access, etc. and extensive hearings were held. Next local hearings were held at potential sites and the Commission welcomed the formation of "resistance groups" in the various localities. In the third stage the Commission's Research Team engaged in detailed investigations of all aspects of the airport location problem from traffic forecasts to questions of locating particular runways, etc. This was followed by discussion among outside experts and the Research Team. The final stage consisted of one more set of public hearings. The outcome of all these deliberations was an impressive tableau in which the Commission listed for each of the four potential sites on the short list (Cublington, Thurleigh, Nuthampstead, Foulness) some 20 categories of associated cost.[3] Not all costs were explicitly entered

2. Commission, "Report," p. 10.

3. One may fault the procedure because of its omission of computing explicit benefits. Given the forecast that existing airport capacity was inadequate to meet future

here. Neither the destruction of the breeding grounds of the brown bellied Brent goose that would result from locating the airport at Foulness nor the destruction of Stewkley Church, said to be the finest example of Norman architecture in England, that might occur if Cublington were selected were explicitly quantified. Yet the Commission attempted to determine all costs, whether these were actually to be incurred as in the case of airport construction costs or whether they were purely notional as the cost figures associated with noise pollution. In all this work the Commission attempted to view the matter in a general equilibrium framework. Examples of this range from purely technical matters to rather broad policy questions. Among the former we find that the construction of the airport at Foulness (but not at the other sites) would have necessitated an expansion of the relatively minor airport at Luton; the increase in noise at Luton was appropriately attributed to Foulness. Among the latter we find the attempt to coordinate with the South East Joint Planning Team, an independent government body responsible for developing and implementing regional plans for the South East region. They attempted to obtain at least qualitative indications of undesirability even in those matters where they abandoned the hope of quantification: it turned out that Stewkley Church could be moved whereas the Brent geese appeared to be resistant to persuasion to change their habitat.

The sum of all costs was a minimum for the Cublington site and this is the site that the Commission recommended.

Although the comprehensiveness and the dispatch with which the Research Team and the Commission dealt with the task at hand are impressive, these characteristics do not insure that all analyses and, indeed, the final decision itself are correct. It must be recorded that the work of the Research Team, its methodology, its data and its assumptions have come under serious fire from a number of scholars, engineers and consultants.[4] Nor does a decision-making structure such as the Commission's

---

demand, the benefits from providing additional capacity were in fact postulated rather than demonstrated empirically. This aspect of the Roskill procedure is particularly subject to criticism in the light of the opinions that emerged during the latter part of the Commission's work to the effect that existing airport capacity at Heathrow, Gatwick, Luton and Stansted, with possible additions, was adequate to meet the foreseeable demand until sometime in the 1980's. See "Who Needs Wing," *Economist* (December 26, 1970): 65–67.

4. See for example, J. Parry Lewis, "Misused Techniques of Planning: The Forecasts of Roskill," *Regional Studies* 5 (1971): 145–155; F. A. Sharman, "The Third London Airport," *Regional Studies* 5 (1971): 135–143; N. Lichfield, "Cost-Benefit Analysis in Planning: A Critique of the Roskill Commission," *Regional Studies* 5 (1971): 157–183; E. J. Mishan, "What is Wrong with Roskill," *Journal of Transport Economics and Policy* 4 (1970): 221–234 and others.

guarantee that the criteria of decision making are uniformly accepted: one need only point to Buchanan's dissent from the majority opinion in which he (a) rejects the compelling nature of an explicit cost-benefit calculus and (b) tends to treat environmental consequences as absolute constraints rather than part of the objective function. Be that as it may, the majority of the Commission accepted the principle of decision making expressed in the following:

> We believe therefore that, following our refusal to accept the existence of any absolute constraint upon the choice of site, the right answer in the interests of the nation rests in a choice of site which, however damaging to some, affords on a balanced judgment of advantages and disadvantages the best opportunity of benefiting the nation as a whole.[5]

It seems to me that the principal differences between the cases of London and Minneapolis are three-fold. (1) The Roskill Commission explicitly adopted an overall cost-benefit calculation and the aggregate of miscellaneous participants in the Minneapolis case have not. That is not to say that the examination of issues in the case of Minneapolis did not, item by item, consider the same issues as the Roskill Commission did, such as noise, environmental factors, surface access, air traffic control, regional impact, etc. It is just that there seems to be no evidence that (a) all these items were costed out according to some reasonably objective and uniform methodology and (b) that all costs were aggregated for purposes of making a final decision. (2) A second and not unrelated difference is that in the Roskill case the principals had no vested interests, did not represent constituencies with veto powers and could be regarded as repositories of a social welfare function as much as this could ever be hoped. These two differences are responsible for the fact that a recommendation was made speedily and for the belief that it had been arrived at fairly and rationally.[6] (3) The ultimate decision-making body in the Roskill case was the national government, satisfying Jackson's requirement that the decision be made by a multifunction elected body, but one much broader in scope than the particular decision at hand. Although elected, many of the participants in the final decision

5. Commission, "Report," p. 137.
6. Not all observers are fully agreed on this latter point. The Commission has been severely criticized by some for being neither a fair nor an entirely rational decision-making body.

could not have had as specific a vested interest in it as did the members of the Minneapolis Metropolitan Council.

In contrast, the Minneapolis case appears to be characterized by a decision-making process in which (a) those responsible for making the decision had a strong vested interest, (b) they could effectively block action, and (c) it never became necessary to evaluate the aggregate cost-benefit. The only major advantage of such a procedure is that no decision need be made until and unless an appropriate process of compensation is agreed upon. Although the Roskill Commission was fully aware of the need to compensate the losers from any rearrangement in the social fabric, the structure of the Commission did not ensure that such compensation would actually be paid; rather this matter was left up to the Government and Parliament. These considerations must have influenced the Government to reject in April, 1971 the site recommended by the Commission and to choose Foulness instead. The principal reason for this seems to have been the persuasive nature of the arguments put forth by Buchanan who felt that the environmental damage at Foulness was less than at the other sites. Added to this was the apparently growing concern that some of the intangibles, the cost associated with noise pollution, the defacing of the Buckinghamshire countryside, etc., had not been correctly evaluated in the cost-benefit analysis. Since no adequate mechanism of compensation could be devised, the Government preferred the site with the lesser environmental damage and possibly lesser cost to the nontraveling public. In the Minneapolis case, and certainly under Jackson's preferred system, the guarantee of compensation can at least in principle be made part of the price for agreeing to a decision at all. But the need for compensation ought ideally to be ascertained by some objective criterion by objective investigators; it ought to be fixed in magnitude by a broadly constituted legislature and ought not depend on the relative bargaining strengths of a very small number of people. Consequently, although one must agree with Jackson that among the systems that are at least partially present in the Minneapolis case, a multifunction body of elected representatives for decision making is preferable to other alternatives, I find it difficult to ignore the advantages of Roskill.

INTERNATIONAL AND COMPARATIVE ANALYSIS

On the Economics of Transnational Environmental Externalities

Ralph C. d'Arge, University of California, Riverside

As man socialized and the human population clustered into geographically, culturally, and politically defined nation states, reaping the material benefits of specialization and efficiency, human interaction increased both within and among these states. However, because of the sheer magnitude of the environment and its assimilative capacity in relation to human populations and industrial production, in the historical past there was little, if any, interaction between states that involved a purely nonhuman or environmental element. A nation's waterborne wastes were easily assimilated by river, estuary, and coastal waters before such wastes could interfere with other nations' activities. No state produced and consumed chemical compounds that were not easily neutralized by a seemingly infinite natural environment. In the past few decades, such a perception of the natural environment has increasingly changed to the realization that the planet's environmental assimilative capacity, even augmented by substantial investment, cannot sustain an endlessly growing population or material consumption per capita. Nations are coming to realize that not only are they economically and politically interdependent but also environmentally dependent with no well-defined international markets or political mechanisms for efficient regulation.

Economic science has analyzed problems of nonmarket interdependence for a very long time, applying the concept of externalities since Pigou. In essence, externalities are social interdependencies not taken into account

NOTE: I have benefited from the comments of A. V. Kneese, W. J. Baumol, and L. Westphal on an earlier draft.

by formal markets or by agreements between the affected individuals or nations. Thus, externalities embrace a spectrum of problems from the neighbor's noisy and disturbing stereo to the commitment of all nations today toward rapid economic development which may preclude choices between material consumption and aesthetic enjoyment of future nations. In either case, the affected party's, i.e. the disturbed neighbor's or future generations', preferences are not adequately being considered when a decision is made.

The classical economic solution to such externality problems was to "internalize" them by either developing a well-defined market for the "spillovers" or controlling them through collective provision of regulations. Neither of these possibilities appears easily amenable to the problem of transfrontier externalities in general, and environmental externalities in particular. First, environmental externalities have arisen because most dimensions of the natural environment on a regional or global scale are resources without rigidly defined or enforceable ownership rights. The oceans, stratosphere, and electromagnetic spectrum are typical examples. These resources are viewed as being commonly owned or not owned at all. A nation which agreed to a particular pattern of ownership of these resources could potentially lose some of its implicitly controlled resources and thereby national wealth.[1] As long as international entitlements are obscure, any nation can lay implicit claim to the common property resource exceeding any equitable share it may presume to receive if entitlement were made explicit. This is not to say that once some other nation impinges on a nation's perceived implicit entitlement, that it will not find a negotiated settlement and thereby explicit entitlement to be superior to an implicit one. However, the impinging nation, in negotiating, must revise downward its own perceived ownership of the common property resource. In consequence, proceeding from a situation of implicit entitlements of common property resources to explicit regulation and thereby ownership means that some (or all) nations must revise their expectations of national wealth, stemming from the resources that each implicitly believes it controls.

A second aspect of major importance arises from the concept of na-

1. Christy draws a very useful distinction between the production of wealth and distribution or ownership of wealth with regard to ocean fisheries. The first concept involves issues of access and free use while the second involves specification of shares. The discussion in this paper will be centered on distributional as opposed to use issues. See F. T. Christy, Jr., "Fisheries: Common Property, Open Access, and the Common Heritage," *Pacem in Maribus,* edited by Elizabeth M. Borghese (New York: Dodd, Mead & Company, 1972), Ch. 6.

tional sovereignty. Not unlike consumer sovereignty as conceptualized by economists, national sovereignty implies the idea that governments, acting in their own interest will, omitting deviations in power or information implying political or economic monopoly, achieve the greatest welfare for all by independently pursuing autonomous goals and interacting competitively through international markets. The belief in national sovereignty as an ideal is so imbedded that it is impractical to presume it will be easily given up.

Coupling the concepts of national sovereignty in decisions and the idea that implicit, as opposed to explicit, entitlement of international common property resources yields a greater perceived wealth for nations, is suggestive that resolution of transnational externality problems will generally need to be embedded within the following restrictions:

1. No nation will easily accept international agreement on entitlement of significant common property resources without compensating payments to retain its perception of national wealth. In consequence, the classical answer to externality problems of internalizing the decision-making process for the resource is not easily transferable to transnational problems. A new overriding element of distributional gains and losses must be simultaneously included in efficiency consideration.

2. Undirectional transnational externalities, if they are of substantial importance to the emitter country as a method of waste disposal, will in general be resolved by some form of a compensation system where compensation flows *from* the receptor country. The nonliability case (or "victims must pay" principle) will generally be dominant.[2] This is to be contrasted with the reciprocal environmental externality case where compensation may flow in either or both directions.

3. International court settlements of transnational externalities are not likely to yield satisfactory results. There appear to be three almost insurmountable problems. First, how are damages to be measured and damage payments assessed? The receptor country's social values may be strikingly different than the emitter country's. In consequence, there may not be a social welfare index that is applicable or acceptable by both. If the damage is entirely confined to hindering production in the receptor country, then international trading prices, at least at the margin, offer a measure of welfare loss. However, if the impact is on individual citizens with no market prices representing their losses, then a measure of welfare loss is not available

2. The appearance of reciprocity may negate this statement, particularly in those cases where an external diseconomy in one direction between countries is offset by an external economy in the opposite direction.

except through direct examination and questioning. A very striking example of this is the preservation of Nubian monuments behind the Aswan Dam where the value is *only* to "all of humanity." In addition, there may be uncertainty as to the magnitude of loss unless the externality is allowed to continue to the point of maximum damages, i.e., threshold levels of fish populations. And, with irreversible biological or other processes, this becomes a dangerous, if not intolerable, experiment. Second, given the sovereign rights of nations, no nation can be forced to pay environmental damages. The trade-off here is in terms of loss of international prestige and goodwill or increasing the possibility of conflict versus monetary payments based on possibly misrepresented public preferences of the receptor nation. Third, there is basically the "chicken and egg" problem of historical precedent most dramatized by airports and noise pollution. An airport is built drawing in people that then are affected by airport noise. As Coase cogently stated the problem, who is responsible? Who is the polluter in the "polluter must pay principle"? As environmental problems increase in severity and potential damages induced among countries rise, it seems that assigning responsibility will become increasingly difficult. In this context, there is also the problem of assigning damages when more than one nation's waste residuals contribute to total damages. If the different nations' residuals are synergistic or if damages are nonlinear relationships of waste intensity, then there is no easy method of determining how much responsibility each nation should take even with the "polluter must pay all costs and remaining damages" principle. Thus, it can be anticipated that international courts or international commissions will have difficulty in arbitration even if such institutions were given partial regulatory powers.

To conclude this rather negative introduction, transfrontier environmental externalities are not likely to be resolved by international organizations without some form of compensation to both countries—an unlikely case unless a transnational external economy can be discovered to offset the inherent costs of a transnational external diseconomy.[3] In this paper, I attempt to develop a rather abbreviated taxonomic discussion of models on transnational environmental externalities and analyze them in the context of bilateral and multilateral negotiation. The role of international tribunals, courts, or management commissions is not analyzed

3. It should be recognized that a great many offsetting negotiations which are indirect, i.e. without direct monetary compensation, occur currently. The rule is preferential nation status, military gifts, U. N. vote commitments, economic aid unrelated to the conflict "a priori," often are a form of compensation. Direct monetary compensation is only one form of retribution within the context of intergovernmental arbitrage and agreement.

explicitly except as an occasional point of reference for "ideal" efficient utilization and distribution of global environmental resources.

Wealth Effects of Transnational Externalities in Production

In order to clarify taxonomically transnational externalities, we shall employ the traditional concepts of production and utility relationships in economics. In so doing, externalities will be divided into four categories.

1. The first category is externalities generated in production processes that affect production costs and processes in other nations. Examples include industrial water pollution that requires treatment before industrial use in a downriver nation or the salt pollution of the Colorado River by U. S. agricultural tailings' water that reduces farm productivity in Mexico.
2. The second category is externalities generated in production in one country that do not affect production but degrade the environment of citizens in other countries. An example might be the acid rains in Scandinavia emanating from the Ruhr industrial complex.
3. The third category is externalities generated by acts of consumption or final use of goods which affect production costs, i.e. rude tourists.
4. The fourth category is externalities generated by acts of consumption which influence environmental quality in other countries, i.e. municipal wastes affecting recreational use of Lake Erie. It appears relevant that for transfrontier environmental externalities the first two cases could be considered the dominant ones currently. However, as urban sprawl continues and populations centralize and enlarge, pollution from acts of consumption might become increasingly relevant as a transnational problem. Municipal wastes in the Great Lakes and elsewhere offer support for this allegation.

In addition to classifying externalities on the basis of source such as production processes or acts of consumption, they can also be usefully classified by how each enters utility and production relationships of citizens in nations. Definitionally, a truly global externality problem is one in which the externality enters some production function, utility function, or both in every nation. Alternatively, a regional externality problem connotes that the externality only enters a prescribed subset of all nations' utility and or production functions.

Using the neoclassical concept of utility and production function, the various types of transnational environmental externalities can be catalogued. Let $F^i(x, k, y)$ denote the production function for country i where x denotes a vector of outputs, k a vector of resource inputs, and y a vector of environmental externalities that influence production, but cannot be controlled by autonomous decisions of firms in nation i. The vector y may include such components as water quality in a river, lake, or estuary jointly used by several nations, or a common airshed. Thus, $F^i(x, k, y)$ represents the traditional economic concept of a production frontier for country i, where given its domestic resources k, externality components y which, prior to regulation, it does not control, and levels of production for N–1 of the N products it produces, the function yields the maximum of the Nth product the country can produce. Omitting considerations of environmental externalities within a nation for the moment, i.e. presuming all y's are externally determined for the nation, then F^i can be viewed as a wealth measure for the ith country. As some or all of the components of y are changed, then so must the actual (and perceived) wealth of the ith country. In figure 1, a simple diagram is given where the country produces two goods, and a component of y is changed which influences the production of one of those goods, x_2. Note that y^* represents an external diseconomy since maximum production of x_2 is reduced for any predetermined production level of x_1. If both products of country i are affected by the transnational external diseconomy, then the curve in figure 1 would move more uniformly inward.[4]

A glance at figure 1 easily confirms several expectations on international trade patterns. Countries affected by uncompensated transnational externalities in production will tend to produce more of those commodities less influenced by external diseconomies if firms in these countries respond to international prices. If a country is relatively small and the transnational externality is also relatively small so that compensation that

4. In the international trade literature, resource endowments traditionally are not presumed to move internationally. Thus, with this assumption transnational environmental externalities will shift the production curve inward, but generally not alter its convexity properties regardless of how pervasive the external diseconomy. This can be easily demonstrated. Since an external diseconomy in production may reduce the productivity of inputs, as output shifts from one commodity to another, one can expect diminishing marginal productivities of inputs to continue at an accelerated rate, as production of the second commodity is increased. Consequently, convexity will generally be retained. Recently, Baumol, Starrett, and others have pointed out that the highly pervasive externalities in production may destroy the convexity of production curve for commodities if resources are *transferable* in production. Transnational externalities do not present this problem so long as resources are immobile. This is suggestive that multinational corporations that contribute to international resource mobility may at some point contribute to nonconvexity, making it more difficult to identify efficient intracountry environmental control strategies. See W. J. Baumol and David Bradford, "Detrimental Externalities and Non-Convexity of the Production Set," *Economica* (May 1972); and D. Starrett, "On a Fundamental Non-Convexity in the Theory of Externalities," Harvard Institute of Economic Research, *Discussion Paper 115,* 1970.

Figure 1

Effect of Transnational Externality on a Country's Production Possibilities

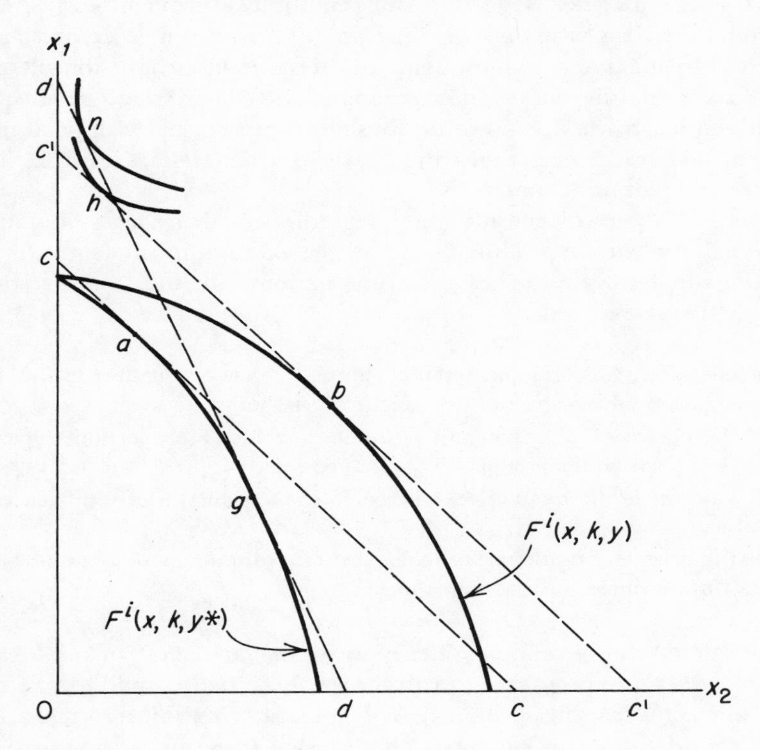

adjusts trade patterns has no perceptible influence on international trade prices, then a clear measure of loss in wealth of the country can be obtained. If the country produced at point *b* prior to the transnational externality or in its absence, but with the externality it produces at *a*, then income loss measured by international prices equals the distance between *c* and *c′*. However, if compensation or removal of the externality changes international prices, then the direct linkage between measured income losses and welfare losses is obviously broken. As an extreme example, let us presume that compensation or payments for resolving the transnational externality in production of commodity x_2 shifts international prices from *dd* to *c′c′* in figure 1. The country thereby shifts its production from *g* to *b*. But now it shifts its consumption point from *n* to *h*, thus exporting more x_2 for less x_1. The country has been made *worse off* from the removal of or compensation for the transnational externality because of international price effects. This is again obviously an extreme case, but it demonstrates the problem of determining welfare loss when international prices are affected by the compensation adjustment for external diseconomies. It also underscores the point that even a receptor country may lose by making a commitment to international arbitration of transfrontier externalities unless international trade effects are given adequate consideration. Normally, however, we should expect that with good information, bilateral negotiation would lead to a net welfare gain for both emitter and receptor countries.

A second aspect of consideration regarding wealth effects of transnational externalities in production is how the externality enters production functions in both emitter and receptor nations. Obvious alternatives in the receptor case include:

1. effects on all or some resource inputs, either through reducing their marginal productivity, total productivity, or both;
2. the choice of processes and techniques or factor proportions separable from impacts on the productivity of inputs;
3. adjustment in qualitative aspects of the product which influences its value;
4. the need for additional technology or other inputs to alter inputs before they can be used, such as water purification.

Note that 1 and 4 relate explicitly to particular factors of production, while 2 affects the production process directly. Finally, under 3, not only must one consider cost-minimizing control strategies, but also market demand considerations to measure welfare losses. In terms of the functional relationship F^i discussed earlier, the four impacts can be translated as follows.

Table 1 contains a symbolic categorization of the various impacts of external diseconomies on production. For any particular transnational environmental externality, each impact in isolation or any combination may be operative. However, the policy implications for each in measuring wealth or income losses can be substantially different. For example, in assessing the extent of receptor costs of the externality, if the only impact is a shift downward in productivity of all inputs, then costs are easily measured applying international prices. This is also the case when all that is required is some additional residuals processing. Alternatively, in the case of product change, costs must be determined as the difference in net rent between production of the qualitatively different commodities,

TABLE 1

Impact of External Diseconomies on Production Processes

Impact of Externality	*Symbolic Representation*	*Comments*
Productivity of inputs	$F_{y^*k} < 0$[a]	Marginal productivity of selected inputs decreased
	$F_{y^*x} < 0$	Total productivity of all inputs decreased uniformly
Process or technique substitution	F replaced by F′	Transformation process of inputs to outputs altered separable of effects on input productivity
Change in qualitative aspects of output	$F(x^*, k, y^*)$ with $x^* \neq x$	New vector of outputs x with different qualitative characteristics or dimensions
Additional technological adjustment of other inputs	$F(x, k^*, y^*)$ with $L^* = F(x, k, y^*)$	Production function is altered to embody additional processes

a. F_{y^*k} represents the derivative of the production function with respect to externality y^* divided by the derivative of the production function with respect to input k.

i.e. beer produced in Olympia, Washington, or in St. Louis with variations due to water quality. Such an assessment requires explicit knowledge of consumer preferences in the receptor country and countries it exports to. Likewise, unless there are accurate measures of the cost of process change or change in technique, it will be most difficult to assess receptor nation costs from this source of impact. A simple dynamic example with technological change will suffice to illustrate complexities here. A nation downstream is affected by increasing deterioration in water quality. In consequence, firms there respond by selecting a less water-intensive production process. Embodied technical change in this process leads to greater efficiency in production and efficiencies in production compensate partially or offset completely inefficiencies due to water quality deterioration. Where is the loss to the nation? It surely includes technology conversion costs, but an unanticipated external economy has otherwise offset external diseconomies associated with poorer water quality. This example is cited only to point out the difficulties inherent in measuring the wealth or income loss of the receptor nation. Conceptually, the rule to follow would be the so-called "with and without" principle encountered in benefit-cost analysis. Thus, an appropriate static measuring device of wealth-income loss is output valued at international prices with and without the transnational externality. However, if there are underlying dynamic processes such as learning embodied in external diseconomies, such static measurements will be disputed in bilateral negotiation, particularly with the adoption of the "polluter must pay" or absolute liability principles.

In the emitter country, the same four cases of impacts on imputs (in a positive sense) and shifts in processes are possible because of the country's ability to use environmental assimilative capacity and cause downstream damages without compensation. When the transnational externality is strictly related to production sectors in the several countries, it is significant to ask whether the impact on factor use will be completely symmetrical. In other words, if the marginal productivity of a factor is reduced downstream, is it carried upstream for the same factor or is the upstream country's outward shift in $F^i(x, k, y^*)$ induced by different causes? If, for example, the upstream country uses the river for upstream dumping rather than incurring process changes (while the downstream effect is to reduce the marginal productivity of a single factor downstream), then assessment of damages is relatively straightforward with international prices constant. However, if there are qualitative changes in the upstream countries' product (which is competitive or complementary with the downstream countries' product that also undergoes qualitative changes), then again demand conditions must be understood to assess damages and measure the involuntary transfer in wealth.

Take two countries with equal factor endowments and resource size facing competitive international prices that are constant, i.e. each country is small in relation to global or regional markets. The upstream country increases factor productivity per se by dispersing waste heat into the streams rather than using cooling towers requiring 10 per cent of its productive factors, while the downstream country experiences the need for cooling towers in order to use river water for production of commodities which requires, by assumption, 8 per cent of its factors of production. Since in neither case do productivity of individual factors or international prices shift, the composition of output in the two countries would not change. All that occurs, is that the emitter country receives a 10 per cent increase in factor income and the receptor an 8 per cent decrease in factor income while the world as a whole undergoes a net 2 per cent rise in the amount of commodities produced by the two countries. If the externality can only be corrected by stopping the heat emission upriver, then it should not be. Allowing the externality to continue is economically efficient, but distributionally adverse to the receptor country. Since there generally will be no mechanism to require the emitter to provide compensation, i.e. except by threat of receptor country or altruistic goals of the emitter country, no transfer payment will occur. A "polluter pays" principle similar to the one adopted by Organization for Economic Cooperation and Development (OECD) member countries where the polluter must pay for controls and not pay compensation for damages would lead to an inefficient allocation of resources in this case.

What is important to note is that transnational production externalities which are uncompensated induce distortions of production relationships in *both* emitter and receptor countries. Perceived wealth of both change as a result of the involuntary character of externalities. However, it is very likely that the impact downstream, because of different factor endowments, production technologies, factor mixes, and comparative advantage in commodities, will be different than the upstream country's impact. A guess might be that in most cases, the emitter country's firms save in costs of additional technology, i.e. recycling and in-factor productivity per se, while the receptor country firms tend to lose in terms of required process or technical changes in production and qualitative aspects of the commodities produced. In consequence, the usual implicit assumption that upstream and downstream control costs are similar in magnitude is not likely to be valid.

To summarize this rather scattered discussion on the wealth effects of transnational externalities in production: (1) these types of externalities can be viewed as involuntary transfers of a nation's wealth; (2) if the externality influences competitive international prices, then its resolution by the OECD "polluter pays" principle may economically harm the receptor country as well as the emitter country. Likewise, the "victim must pay" principle may economically harm both emitter and receptor countries. However, transnational externalities in production generally yield a potential rise in income in one country and a fall in income in the other. In certain instances, where control costs or efficiencies are different between countries, it may pay to retain the externality and, if necessary, use lump-sum transfers to achieve equity.

Wealth Effects from Transnational Externalities on Consumers

Thus far, the discussion has been developed from a perspective of externalities in production, but a few words are added here to describe the impact of externalities in final use or consumption as distinct from production. In order to develop the idea of a wealth transfer resulting from transnational externalities in consumption, I apply a simple graphical technique first used by Dolbear.[5] Let there be two countries with collective utility functions as follows:

$$\begin{aligned} &U^1(x_1, y_2), \\ &U^2(x_2, y_2), \end{aligned} \tag{1}$$

where x_1 and x_2 are the quantities of a private good consumed in countries 1 and 2, and y_2 is the quantity of a private good consumed by 2, but yielding an external diseconomy to 1.

With first and second derivatives denoted by subscript, we assume:

$$\begin{array}{lll} U_x^1 > 0, & U_y^1 < 0, & \\ U_{xx}^1 < 0, & U_{yy}^1 < 0, & U_{xx}^1 U_{yy}^1 > (U_{xy}^1)^2; \\ U_x^2 > 0, & U_y^2 > 0, & \\ U_{xx}^2 < 0, & U_{yy}^2 < 0, & U_{xx}^2 U_{yy}^2 > (U_{xy}^2)^2. \end{array}$$

Note that the usual strict concavity assumptions are specified for preferences toward the two private goods, x and y. The significant difference, however, is that individuals in country 1 are presumed to undergo increasing marginal disutility if the externality is intensified.

Next, a budget constraint for both countries is postulated:

$$\begin{aligned} px_1 &= M_1 + B, \\ px_2 + ry_2 &= M_2 - B; \end{aligned} \tag{2}$$

or

$$p(x_1 + x_2) + ry_2 = M_1 + M_2; \tag{2'}$$

where p and r are the given international prices for private goods x and y, and M_1 denotes the initial income of country 1. Negotiation for external effects between 1 and 2 is allowed through payment of a bribe, B with $B \gtreqless 0$, which cancels out of equation 2′. For simplicity, we shall assume $p = 1$ by suitable redefinition of quantity units of x, and that the externality and consumption of y_2 are strictly joint products.

In figure 2, indifference maps are depicted along with initial (before

5. F. T. Dolbear, Jr., "On the Theory of Optimum Externality," *American Economic Review* (March 1967).

the externality appears) endowments.[6] Country 2's indifference map is identified with the origin 0. However, country 1's indifference map is not. Since 1 cannot exchange y_2 in an international market, its budget constraint would be a sloped line commencing from point *a*, i.e. the country's exchange price for y_2 is initially zero; and the country will, given a choice, be at a corner solution. As point *a* moves to the left, country 1 has a higher budget. But, in order to represent 1 and 2's combined budget in x–y space, the budget line and indifference map of 1 must be rotated in order to be parallel with the slope of the combined budget line such that the sum of 1 and 2's budget does not exceed their combined budget. Thus, the indifference map for country 1 is rotated in x–y space so that x_1 is always measured from the aggregate budget line from right to left in figure 2.

Initial consumption is at point *a* where both 1 and 2 are consuming x but neither is "consuming" y_2. Next, country 2 "purchases" the externality yielding commodity and moves to point *b* where the budget line aa' is tangent to the indifference curve 22. Clearly, this point is not Pareto optimal in a global sense, since residents of country 1 suffer a loss of utility and are moved *involuntarily* from indifference curve 1′1′ to 11. By negotiation, they could achieve the contract curve *cc*, which implies a reduction in intensity of the externality. It should be reemphasized that country 1 has control only over consumption of product x and cannot either purchase or sell y_2, except by negotiating with 2, thereby ruling out the development of a competitive market for the externality. Also, the contract curve defined as *cc* is *not* one where marginal rates of substitution for x and y are equated with relative prices defined as $1/r$, although the budget line of country 2, *adjusted* for fees or payments, must pass through this tangency.

What we now wish to introduce are simplified versions of the "third party liability" (TP) and "victim must pay" (VP) principles. First, the TP rule is presumed to require that country 2 compensate country 1 for all disutilities caused by the externality.[7] With negotiation and the TP

6. Collective indifference maps for each country are presumed to exist which effectively assumes away difficulties of interpersonal comparisons within countries.

7. Note that country 2 by assumption can only regulate the externality's impact by reducing consumption of y_2 or paying country 1 some x to tolerate the external diseconomy. Thus, country 2 cannot provide controls and country 1 implicitly cannot undertake defensive expenditures internally to reduce the intensity of the externality. Recently, the OECD member countries agreed to implement a "polluter pays" principle which in effect requires the polluting firm to pay all control costs and the receptor firm or consumer to pay all residual damages. Also, transfrontier pollution was explicitly excluded from this agreement. For clarity, we adopt the following terminology: polluter being responsible for all control costs and residual damages is denoted as the

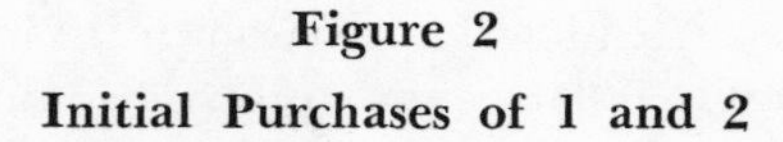

Figure 2
Initial Purchases of 1 and 2

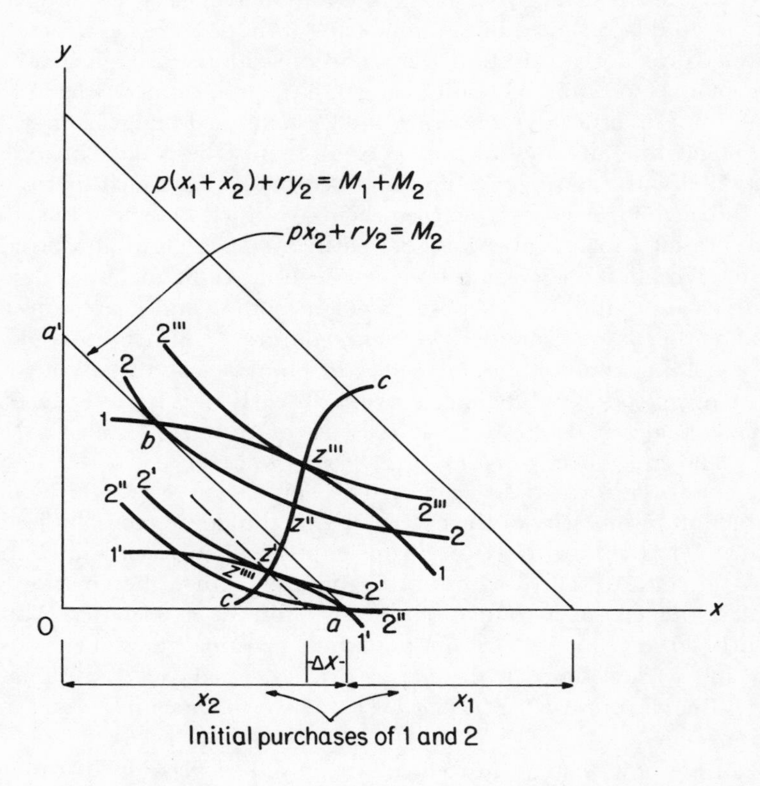

rule specified, the point z' on the contract curve in figure 2 is achieved. Note that if the TP rule is specified, damages cannot occur or must be paid in kind, i.e. reduced noise only, then point *a* will be achieved once again but at a loss to at least one, if not both countries. Also, a TP rule which specified that the damaged country must be made better off while the liable country cannot be better off, will result in negotiation to the contract curve. The ultimate point achieved on the contract curve in this

---

"third party principle" where third party connotes the implicit need for an outside enforcement body or court in order to establish exclusive right for the receptor nation; polluter pays only the control costs and no residual damages is referred to as the "polluter pays" principle. See Organization for Economic Cooperation and Development, *Recommendation of the Council Concerning International Economic Aspects of Environmental Policies,* Paris (May 26, 1972).

case will be to the left of point z' in figure 2 at point z'''', i.e. x_1 will be higher while y and x_2 will be lower than for the first PP rule.

We next turn to the case of VP rule. By VP rule, we mean that there is a rule which specifies that country 2 by inadvertently creating an external diseconomy is not required to compensate country 1 for damages. Also, there are assumed to be no provisions contained in this rule which would impede private negotiation between affected countries. Clearly, in this case a negotiated settlement would involve a payment from country 1 to country 2. Costless negotiation would result in a distribution of x and y on the contract curve between z' and z''' in figure 2. Between z' and z''' on the contract curve, both countries 2 and 1 are made no worse off and perhaps better off than at point b.

A TP rule, as specified here leads to a negotiated settlement on the contract curve cc on or to the left of point z', while a VP rule leads to a negotiated settlement on or to the right of point z''. Thus, the distribution of wealth between countries is altered by adoption of a TP or VP rule which changes the consumption patterns for x and y. Since negotiation leads to attainment of the contract curve regardless of rules, Coase is correct in asserting short-run efficiency.[8] The above derivation can be taken as a proof of the "Coase proposition" amended as follows: regardless of whether a TP or VP rule is adopted, "perturbations" arising from nonmarket forces can be resolved through negotiations of the countries affected, and provided transactions costs are zero *after* perturbation, Pareto efficiency is attained in the short run and in the long run if consumers and factors of production are immobile internationally.[9] In Appendix I, a simple model of international production is analyzed which demonstrates that under several rather general assumptions neither the TP nor VP rules will achieve global production efficiency without other internationally agreed upon secondary controls. The assumptions include certain types of positive transactions costs and that countries are essentially price takers. A second assertion which becomes obvious in the diagrammatic analysis is that with more or less continuous perturbations and negotiation, a VP rule will have "wealth" effects toward making some countries worse off, that is, if one assumes that, ordinarily, the emitter country is made better off by "creating" external diseconomies. Not unlike gambling, there will be some countries that emerge as winners and others as

8. R. Coase, "The Problem of Social Cost," *Journal of Law and Economics* (August 1960).

9. See Appendix I for a discussion of the possibilities when transaction costs are positive.

losers. Since the perturbations are unforeseen in this world, it would appear Paretian in spirit to adopt a TP rule.

Some Brief Notes on Bilateral Negotiation and Game Theory

Whether the "third party must pay," "victim must pay," or some intermediate principle is implicitly or explicitly adopted, there are potential gains from trade via bilateral negotiation. Both the emitter and receptor have incentives to undertake negotiation.[10] Thus, in terms of game analysis, unadjusted externalities without threats of conflict are essentially the form of cooperative or Nash types of games. Both countries can potentially gain by cooperation although each may decide that it is in its best interest not to cooperate, or pretend not to and thereby possibly achieve a greater expected threat payoff or value. With externalities generally, it can be expected that *side payments* will be made in order to achieve cooperation. Thus, the classical Nash cooperative solution without side payments appears to be inappropriate. Side payments might enter, for example, if the upstream and downstream control costs were different.

In order to illustrate some basic problems involved in negotiation for transnational externalities, we shall resort to a simple example. Let it be assumed that there are two countries 1 and 2 where 2 is the emitter and 1 the receptor. Downstream damages are 10 prior to control and 2 with control, with upstream gains due to savings in pollution control of 4. Downstream control costs are assumed to be 6. There are two decisions for each country that are not mutually exclusive, namely whether to join the coalition and at what level to participate, i.e., how much control to provide. We shall concentrate on the decision whether to cooperate or not presuming there is *no* double layered process of irrevocable agreement to negotiate which is followed by negotiations. Also, it is assumed for the moment that both announce their intention to negotiate simultaneously with offers. Finally, neither country is assumed to know precisely the other's control costs and the emitter country does not accurately know the level of downstream damages.

Given the previous assumptions and a set of expectational values of each country, the maximum bid and expected bid of the receptor country and minimum acceptable offer and expected offer for the emitter country

10. That is, assuming the externality is continuous over time, has inadvertently arisen, and the emitter, if he does not pay, must cease the activity-generating externalities under the "third party must pay" rule.

can be established. To make things simple, we shall set the following probabilities.

1. Country 2's government believes the probability of downstream damages being avoided by control of 10 is .2 and the probability of damages being avoided of 2 is .8. Thus the expected value of damages avoided by controls is simply: $E = .2(10) + .8(2)$.
2. Country 2's government believes the probability of the downstream country's control costs being 4 is .5, and the probability of these costs being being 6 is .5.
3. Country 1's government believes the probability of the upstream country's control costs being 4 is .5, and the probability of them being 6 is .5.

Presuming the "victim must pay" principle is in effect, the maximum bid that 1 would make to 2 would be initial damages less damages after control which is 8. The minimum acceptable offer for country 2, of course, is 4. These amounts are in effect the "threat payoffs" of the Nash type of game for coalitions. Thus, there are clearly gains from negotiation if negotiation can be *initiated*. However, note that "expected rent" beyond payment for control by country 2 from country 1 *prior* to negotiation is negative, i.e. rent defined as 2's expected value of downstream damages reduced is .2(10) + .8(2) less 2's expected value of 1's control costs .5(6) + .5(4). Thus, 2 may not undertake negotiations due to the fact that it expects 1's offer will be less than its control costs.[11] Alternatively, 1 will decide to attempt negotiation with 2 because its expectation of gain would be 8 − .5(4) − .5(6) which exceeds its gain of 2 from instituting control unilaterally.

In figure 3, the possible bargaining positions and "threat payoffs" for this problem are depicted. Any point on *SS′* is Pareto efficient in that all gains from negotiations on the transfrontier diseconomy between countries 1 and 2 are exhausted. Which point on *SS′* is ultimately selected via negotiation depends on the rules established for negotiation, i.e., once-and-for-all bid by the receptor, a sequence of simultaneous bids and demands, or sequential bids and demands. Also, the point depends on the exact specifications of the VP or TP rules. For example, if a TP principle is adopted which specifies the emitter country cannot be any better off than before the emergence of the transfrontier externality, point *S* will

11. It should be pointed out that this grossly simplifies expectation problems in that we omit consideration of country 2 calculating country 1's expectations and likewise country 1 calculating country 2's expectations on country 1, *ad infinitum*. Here we only consider country 1 and 2's expectations directly.

Figure 3

Efficiency Locus for Transfrontier Externality Between Two Countries

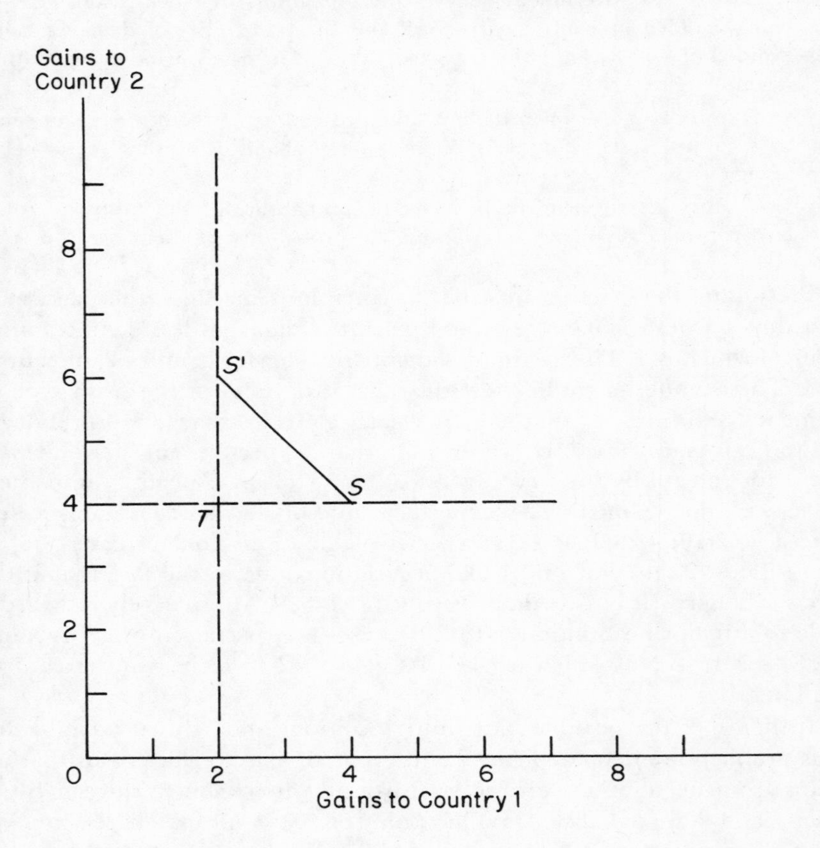

be the negotiated solution. Alternatively, if under the VP principle the receptor country cannot be made better off, then S' will be selected. Finally, if transactions-negotiation costs are introduced, then it is conceptually possible for country 2 to decide not to negotiate under either the TP or VP principles, since the country *a priori* perceives that it would be better off without it. (See Appendix I for an elaboration on this point.)

This most simple example of a cooperative game between two countries on transfrontier externalities underscores several points. First, the bidding process itself, i.e. who bids first and how binding is the bid, may significantly alter the outcome. In the example above, country 2 will not establish negotiations or make the first bid because expected gains are negative. Second, precise information on damages and control costs by both countries will reduce the risk associated with bidding too high or setting payments too low. Under the TP principle, this risk factor is intensified since the emitter country now confronts uncertainty with respect to both receptor damages and downstream control costs. Alternatively, under the VP principle, the country making the payment, i.e. receptor, confronts uncertainty only with regard to the emitter's control costs. Such a reduction of uncertainty on information does not occur in OECDs "polluter pays" principle where the emitter is uncertain on damages *and* downstream control costs and the receptor is uncertain as to upstream control costs.

Conclusions

The major conclusions of this paper are as follows.

1. Transnational externalities in production or consumption are involuntary transfers of perceived wealth among emitter and receptor countries. Given the general acceptance of the national sovereignty principle, such externalities of a significant magnitude will in general only be resolvable via bilateral negotiation with side payments, usually from receptor to emitter.
2. Unadjusted transnational externalities, if their resolution through compensation or other means, shifts international prices, may make a single emitter *or receptor* nation better *or worse* off. Thus, the link between wealth-income and welfare of a country is broken. In order for the receptor nation to be made worse off, international prices must be markedly affected and willingness to pay for the commodity, influenced by the externality, must be strong if its relative price increases as a result of compensation or weak if its relative price decreases resulting from compensation.
3. Transnational externalities, unlike domestic externalities in production, will generally not cause a shift from concavity to convexity of domestic production functions. This is due to the usual underlying assumption in the classical trade literature that resources are mobile nationally but not internationally.
4. A "victim must pay" principle or a "third party" principle will lead to an inefficient long-run allocation of resources between countries,

provided resources are mobile between them and certain assumptions on positive transaction costs are valid.[12] These assumptions are that firms only respond to realized profits in international location decisions and are price takers. To achieve global efficiency in resource allocation, some other types of controls need to be implemented such as restrictions on international location. The same inefficiencies might arise for transnational consumption externalities unless consumers are not internationally mobile.

5. Transactions costs in negotiating for externalities are shown to markedly affect the expected outcomes of bilateral negotiation. If such costs are high, countries are generally risk averters, and these costs must be borne by the receptor country under the "third party" principle, then it can be expected that pollution will be greater than is desirable from a global viewpoint. There appears to be a basic asymmetry in incentives for negotiation of externalities between the "victim must pay" and "third party" principles, when transactions costs are positive. Under the VP rule, both emitter and receptor have incentives to negotiate and thereby incur transactions costs. Alternatively, with the TP rule, only the receptor has such incentives since the emitter is always better off by not negotiating.[13]
6. Unidirectional transnational externalities were shown to be as conceptualizable as a special case of cooperative games under the "victim must pay" principle. It was argued that rules on negotiation may substantially alter gains between countries undertaking bilateral negotiation for transfrontier externalities and thereby influence international efficiency in the utilization of resources.

12. See Appendix I for a semi-rigorous proof.
13. See Appendix I.

Appendix

A NOTE ON TRANSFRONTIER EXTERNALITIES, TRANSACTIONS COSTS, AND LONG-RUN INTERNATIONAL ADJUSTMENT IN FIRMS AND FACTORS OF PRODUCTION

Ralph C. d'Arge, University of California, Riverside
William Schulze, University of New Mexico

Transactions costs (or information, negotiation, and policing costs) may affect negotiations in a multitude of ways depending on their source. These include: uncertainty and information gaps (or costs); known or unknown contracting or negotiation costs; costs associated with organizing and sustaining negotiation between countries including dissemination of information; and enforcement costs for existing contracts or treaties. Of these different types of transactions costs we shall concentrate briefly on two types—those associated with confronting an uncertain prospect of a future perturbation (externality) and those costs associated with negotiation once the externality has occurred. What we wish to establish is that differences in risk aversion between emitters and receptors will cause adjustments in the allocation of resources and negotiated level of externality depending on whether the "third party" principle or "victim pays" principle prevails.

There appear to be four cases regarding transactions costs for trans-

NOTE: A model not too dissimilar to the one applied here emphasizing intracountry location is developed by W. J. Baumol in "On Taxation and the Control of Externalities," *American Economic Review* (June 1972). This is an abbreviated and transformed section of a paper by the same authors entitled, "Coase Proposition, Wealth Effects, and Long-Run Equilibrium," Working Paper No. 19, *Program in Environmental Economics,* University of California, Riverside (April 1972).

frontier externalities: (a) where transactions costs are always or nearly zero between emitter and receptor countries; (b) where transactions costs are positive and significant both before and after emergence of an externality; (c) where transactions costs are positive before the externality appears but zero thereafter; and (d) where transactions costs are zero before the externality appears but positive and perceptively significant thereafter. Case (a) can easily be disposed of as one which rules out the existence of externalities that are not a priori resolved by market negotiations. The Coase proposition is a special case of (c). Case (d) appears to be unreasonable. Finally, case (b) is the important one taxonomically for analyzing "real world" problems. An important subset of cases under case (b) arises where transactions costs are different for the two parties either independent of or dependent upon the prevailing rule for liability, i.e., either the "third party," "polluter pays," or "victim pays" principle are operative or not.

A semirealistic case is one where, under complete liability, the receptor country incurs negotiation costs and, under complete nonliability, the emitter country must pay such costs.[1] Thus, we simply assume those who potentially gain are assumed to initiate negotiation and underwrite the cost of negotiation. Let $\Psi_{TP}{}^*$, $\Psi_{VP}{}^*$, Ψ_{TP}, and Ψ_{VP} denote the level of externality-generating activity, for the cases of zero transactions costs and *TP*, zero transactions costs and *VP*, positive transactions costs with *TP*, and positive transactions costs with *VP* rules, respectively. Given the assumption that the contract curve is upward sloping, and not too dissimilar marginal utilities of income or marginal disutilities of payments for negotiation exist between countries, then it can be asserted $\Psi_{TP}{}^* \leqq \Psi_{TP}$ and $\Psi_{VP}{}^* \leqq \Psi_{VP}$. That is, positive negotiation costs, regardless of *VP* or *TP* rules, will impede negotiation so that the optimal level of externality-generating activity *with* zero negotiation costs is not achieved. Given the assumptions above, then $\Psi_{TP}{}^* \leqq \Psi_{VP}{}^*$. This implies $\Psi_{TP}{}^* \leqq \Psi_{VP}{}^* \leqq \Psi_{VP}$, but does not imply $\Psi_{TP} \leqq \Psi_{VP}$. Thus, costly negotiation where one country incurs these costs may lead to a case where the TP rule results in a higher level of externality-generating activity than the VP rule. This outcome can be induced by differences in marginal utility of income between the emitter and receptor countries as well as a large number of other assumptions on initial endowments or preference maps. The important point here is that it cannot be a priori determined that complete liability or the "third party" principle will reduce external diseconomies by a greater amount

1. However, that component of information and negotiation attributable to court proceedings may be awarded to the receptor after settlement.

than no such principle when negotiation costs are introduced and must be paid by the country initiating negotiation. If negotiation costs are different for the two countries, the outcome is even less clear. It has often been contended that emitters must have lower negotiation or organization costs than receptors since receptor countries are generally more in number while the emitter is usually viewed as a single (source) country. With this type of differential in negotiation costs, there is still no clear-cut statement that can be made on the inequality between Ψ_{VP} and Ψ_{TP}. It depends again on who incurs the costs of organization and negotiation. Under the TP rule, the receptor country must undertake negotiation costs since there is no incentive for the emitter to do so. However, under the VP rule, there is an incentive for both to undertake negotiation and incur such costs. If the receptor must pay negotiation costs under either legal rule, then without further assumptions the inequality between Ψ_{VP} and Ψ_{TP} cannot be determined. If the receptor country pays negotiation costs with the TP rule but the emitter country pays these costs under the VP rule, then it can be expected that $\Psi_{VP} \leqq \Psi_{TP}$, provided the emitter has lower negotiation costs. What is important from the above statements is that the "third party" principle (TP) or lack of it with negotiation costs for allocative efficiency requires an additional rule specifying who incurs these costs. Without such a rule, negotiation may be completely stopped and thereby yield inefficiencies.

Thus far, we have not introduced uncertainty explicitly into the discussion even though the characterization of a world with externalities hints at analyzing externalities as unexpected, or at least, uncertain events. If externalities can be identified as uncertain events where a probability distribution is identified for each type of externality and there are methods to reduce the probability of occurrence to zero or some reasonable level, e.g., construct and operate a tertiary treatment plant on the Rhine, then the externality problem can be analyzed, at least partially, with tools from probabilistic microeconomics. We shall not do that here but suggest some obvious results. First, if the emitter country is more risk averse than the receptor, then its government may require purchase of control dvices under the TP but the receptor may not with its absence.

So far, we have viewed transactions costs as the dominant factor in externality negotiations. Next, we turn to a semiclassical long-run case of competition between firms, in an international context. In so doing, a particular set of transactions costs are preserved, namely, that firms are price takers, but output adjustments shift international prices. Firms are assumed to move internationally in search of the highest profits with no hindrance by governments or entry costs and observing no other signals than current profits. Thus, a "perfect capital" market internationally is implied. We also make the simplifying assumption that each country has a distinct comparative advantage in producing one type of commodity and that is *all* it produces. In addition, to avoid balance of payments and other complications, we presume that each country is in a partial equilibrium world where demand price for their product is prespecified at any level of output. Finally, no third party is presumed to enter and arbitrage externalities such that potential gains from trade are exhausted. The question to be resolved is whether or not negotiations between producing countries' firms exploiting the gains from trade made available by an externality in production will result in an efficient solution not just in the short run but in the long run among countries. Since profitability among countries determines entry (presumed to be costless), and since the adoption of either the TP or VP principles will affect relative profitability in emitter and receptor countries, it is clear that the international location of industry will be affected by which principle is adopted.

We will consider a two-country partial equilibrium model with a diffuse externality such that the output of industry in country 2 adversely affects the production of every firm in country 1 in a like manner. An example might be the release of air pollutants by industry of one country into an airshed with instantaneous horizontal mixing which also contains a second country's industry. The questions would then be to determine how large the industries should be in each country from the viewpoint of international efficiency. Where there are n_1 identical firms each producing output y_1 in country 1 and n_2 identical firms each producing output y_2 in country 2, we can write inverse demand functions (price P_i as a function of country output n_iy_i) for country 1 as $P_1(n_1y_1)$ with $P_1'(n_1y_1) < 0$ and for country 2 as $P_2(n_2y_2)$ with $P_2'(n_2y_2) < 0$, where the prime denotes a first derivative. The total production cost for each firm in country 1 is given as $C_1(y_1) + D_1(n_2y_2)$ where $C_1(y_1)$ is the direct cost of production for each firm in country 1 and $D_1(n_2y_2)$ is the damage incurred by each firm in country 1 as the result of the emissions of country 2. Note that we are presuming damages are separable of other production costs in the receptor country. However, our results concerning the optimality of various policy measures are generally not dependent on this assumption. Total cost for each firm in country 2 is $C_2(y_2)$. We assume C_1', C_2', C_1'', $C_2'' > 0$ and D_1', $D_1'' > 0$ in the relevant regions of production.

The conditions for a global optimum for both countries taken together is obtained by maximizing net benefits (NB) which can be defined as the difference between willingness to pay for the output of both countries and the total cost of production in both countries. Thus, the international optimum in a partial equilibrium framework is a maximum of:

$$NB = \int_0^{n_1y_1} P_1(s_1)\, ds_1 + \int_0^{n_2y_2} P_2(s_2)\, ds_2 \qquad (1)$$
$$- \{n_1[C_1(y_1) + D_1(n_2y_2)] + n_2C_2(y_2)\} \qquad y_1, y_2, n_1, n_2 \geqq 0$$

where s_1 and s_2 are dummy variables of integration in the demand functions for the output of each country. Assuming an interior solution, the first order conditions are:

$$\partial NB/\partial y_1 = n_1(P_1 - C_1') = 0, \qquad (1.1)$$
$$\partial NB/\partial y_2 = n_2(P_2 - C_2' - n_1D_1') = 0, \qquad (1.2)$$
$$\partial NB/\partial n_1 = P_1y_1 - (C_1 + D_1) = 0 = \pi_1^*, \text{ and} \qquad (1.3)$$
$$\partial NB/\partial n = P_2y_2 - (C_2 + n_1D_1'y_2) = 0 = \pi_2^*. \qquad (1.4)$$

The interpretation of (1.1) and (1.2) is quite straightforward and implies, where $n_1, n_2 \neq 0$, that for each firm in country 1 price should be equal to marginal cost (C_1'), and that for each firm in country 2 price should be equal to marginal cost (C_2') plus marginal damages to country 1 (n_1D_1'). These are the usual short-run conditions with a unidirectional external diseconomy between countries. The conditions for a long-run optimum, (1.3) and (1.4), are more interesting since they should correspond to the definition of zero profits for firms in countries 1 and 2 respectively (assuming firms enter or leave countries until profits are zero). Equation (1.3) implies that $\pi_1^* = 0$ is the optimum level of profits where the receptor country bears the full cost of the externality D_1 at the optimum. This result suggests that compensation for damages will distort long-run equilibrium in the receptor country. Equation (1.4) implies that $\pi_2^* = 0$ is the optimum level of profits where the emitter country must bear an additional cost of n_1D_1' per unit of output y_2 produced. This can be interpreted as an optimum long-run Pigovian tax equal to marginal damages on the output of the firms in the emitter country.[2] We note then, that the optimal policy after taxation by an "international tribunal" or agency is to do nothing with respect to the receptor country, allowing it to bear the cost of the externality after the optimal tax on output has been applied to firms in country 2. This will assure the optimum number of firms in each country.

2. A similar result is obtained by Baumol in his *American Economic Review* article. Baumol, "On Taxation."

The relationship between the optimal Pigovian tax case (denoted *), the unadjusted externality case (denoted E), and the TP principle case (denoted P) can be best demonstrated with the aid of figures 1 (a firm in country 1) and 2 (a firm in country 2). In figure 1, the optimal long-run

Figure 1

Costs and Output Adjustments of One Firm in Receptor Country

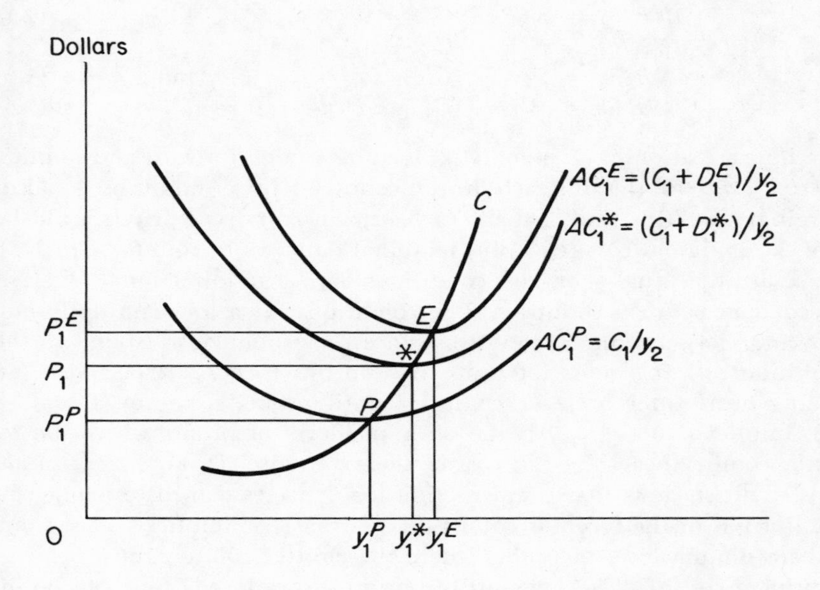

equilibrium point for receptor firms is * at the lowest point of the average total cost curve *including* optimal damages suffered (AC_1*). This point is defined by the intersection of the marginal cost curve (C_1') with the adjusted average total cost curve (AC_1*). As damages (D_1) increase with increasing output of country 2, the average total cost curve of firms in country 1 including damages shifts upward. We presume that where the optimum tax is applied to firms in country 2 and free entry exists for both countries, there will be a convergence to D_1*, the optimal long-run level of damages, and optimal price P_1* and quantity y_1* will result from the long-run equilibrium point (*). Note that this optimum is a basis of comparison since gains from trade are possible between the two countries

Figure 2
Costs and Output Adjustments of One Firm in Emitter Country

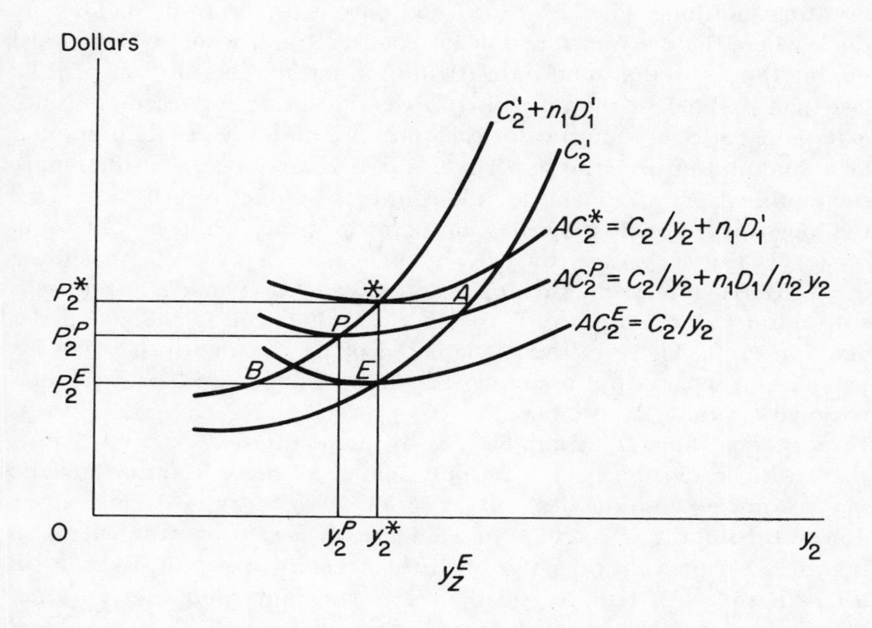

so long as a Pigovian tax is not charged and receipts randomly distributed.

The optimum equilibrium point in figure 2 for firms in country 2 is also denoted *. This can be defined by the intersection of the average total cost curve including the tax (AC_2^*) with marginal social cost ($C_2' + n_1D_1'$). Note that this point corresponds to the zero profit condition for firms in country 2 where AC_2^E is the unadjusted average cost and the area $P_2^E\, E * P_2^*$ is the optimal long-run tax collected from each firm in the emitter country. This implies that if through some mechanism not involving a tax or levy the two countries reach the optimum points * in figures 1 and 2, positive profits equal to the area $P_2^E\, E * P_2^*$ times n_2^* will be obtained. Since positive profits will induce more firms to enter country 2, the optimum point * cannot be a stable equilibrium under free entry. Thus, the Pigovian tax, if achieved through international management agencies, serves to remove these destabilizing profits.

The uncompensated externality case results in a long-run equilibrium at point E in figure 2 for firms in country 1. Here, since damages received (D_1^E) will be greater than optimal (D_1^*), the average cost curve (AC_1^E) will lie above AC_1^*. Since the externality is separable, the marginal cost function (C_1') does not shift, so entry or exit occurs until profits are zero resulting in a long-run price of P_1^E and output per firm of y_1^E. This implies, where the demand function for country 1 as a whole is downward sloping ($p_1' < 0$), that total industry output and number of firms will be less than optimal in the unadjusted externality case for receptors, since externality price and output for each firm are both greater than optimal price and output per firm, respectively. This occurs because at the higher (nonoptimal) price, demand for the country's product is less. In figure 2 the long-run externality equilibrium point for firms in the emitter country occurs where average cost (AC_2^E) equals marginal cost (C_2') resulting in a price (P_2^E) lower than the optimum price (P_2^*). However, firm size is still optimal since output (y_2^E) in this case is identical to the case under taxation (y_2^*). The intuitive explanation of this result which is not dependent on separability is simply that, in spite of the externality, international product is still maximized by producing each unit of y_2 as cheaply as possible. This implies that in the unadjusted externality case, there will be too much total output from the emitter country and too many firms, because the demand curve for the country is assumed to be downward sloping ($P_2' < 0$) even though each firm perceives demand as infinitely elastic. Thus, in the unadjusted externality case, there is an *underallocation* of resources to the receptor country and *overallocation* of resources to the emitter country compared with the international optimum.

In the complete liability or TP case discussed here we assume that firms in country 1 are compensated for damages and that potential entrants into the industry of county 1 are aware that they too will be compensated. Firms in country 2 are responsible for damages done to country 1 and we assume, since the externality is diffuse and the firms are taken as identical, that each must bear the cost of compensation equally. With liability, profits for firms in country 1 and 2 can then be written as:

$$\pi_1 = P_1 y_1 - C_1(y_1) - D_1(n_2 y_2) + [D_1(n_2 y_2)] \tag{2}$$

and

$$\pi_2 = P_2 y_2 - C_2(y_2) - [n_1 D_1(n_2 y_2)/n_2],$$

where the terms in brackets are compensation or liability payments by firms in each country respectively. The first order conditions for maximum profits in each firm, assuming an interior solution, are:

$$\partial \pi_1 / \partial y_1 = P_1 - C_1' = 0 \tag{2.1}$$

and

$$\partial \pi_2 / \partial y_2 = P_2 - C_2' - n_1 D_1' = 0, \tag{2.2}$$

which imply that the conditions for short-run optimality are satisfied. However, if it is assumed that firms enter until profits are zero, compensation to country 1 results in a long-run equilibrium (position P in figure 1), the lowest point on the average cost curve without damages (AC_1^P) for firms in the receptor country. Again, the marginal cost function (C_1') does not shift since damages are separable and the resulting price (P_1^P) and output for each firm (y_1^P) under the TP principle is less than optimum. Thus, both *total* output of country 1 and the number of firms will be too large for the receptor country in the TP principle case since demand for the country's output will be greater at the lower price.

Turning to figure 2, firms in country 2 will reach an equilibrium point (P) in the long run under the TP principle which is the lowest point of the average total cost curve including each firm's share of damages to be paid (AC_2^P). Note that because total damages increase at an increasing rate, marginal damages ($n_1 D_1'$) are greater than average damages ($n_1 D_1 / n_2 y_2$), so AC_2^P lies below AC_2^* and therefore the intersection of the marginal social cost function ($C_2' + n_1 D_1$) and AC_2^P must be below and to the left of the optimum point (*). This is the point of zero profits including liability for damages for the emitter country. Both price (P_2^P) and output (y_2^P) are too low for each firm in country 2. However, total country output will be too high and there will be too many firms in the emitter country under a TP principle since, given the lower price, aggregate demand for the country's output will be too high. Thus, there results a long-run *overallocation* of resources to *both* country 1 and 2. Taxes or other controls are necessary to prevent a misallocation of resources in long-run international production even if there are well-defined rights with a TP principle on the externality.

A TP principle solution *could* be adjusted to the optimum equilibrium point by taxing receptors an amount equal to the damage payments they receive and taxing emitters an amount equal to the difference between average damages and marginal damages, a procedure inefficient as regards international information and enforcement costs to make this bargaining solution appear unattractive.

The complete nonliability case or victim-must-pay principle can best be explained in two steps. First, we will demonstrate that a negotiated solution under a nonliability rule cannot sustain the optimum points (*)

assuming that the number of firms in each country is constrained to be less than or equal to the optimum ($n_1 \leqq n_1{}^*$, $n_2 \leqq n_2{}^*$). As will be seen later, this assumption prevents a free rider problem from upsetting the potential equilibrium point (*) in figures 1 and 2. Assume that firms in both countries are initially at *. Next, observe that * for firm 1 in figure 1 is a point of zero profits. However, in figure 2 it is clear that firms in country 2 could earn profits greater than those obtained under * by moving to point *A*. Thus, for * to be achieved by negotiation for firms in country 2, firms in country 1 must offer to pay a bribe at least equal to the difference between profits at * and profits at *A* to existing firms in country 2. Clearly, firms in country 1 are making zero profits at * in figure 1 and cannot pay any bribe. Thus, the optimum points * in figures 1 and 2 are not feasible under a "victim pays" rule even ignoring the destabilizing effects of entry on coalitions since firms in country 2 must be made at least as well off at market price $P_2{}^*$ as they would be by not adjusting for the externality. Thus, firms in country 2 would be unwilling to remain at *. It is conceivable, with the number of firms fixed by controlling entry through some licensing or national ownership process, to achieve a short-run optimum with a solution somewhere along the marginal social cost function ($C_2{}' + n_1 D_1{}'$) in figure 2. However, the number of firms in country 1 must be fixed ($n_1 \leqq n_1{}^*$) such that profits sufficient to cover bribes to firms in country 2 can be obtained. Clearly, without some taxation policy, even by controlling entry, the long-run optimum solution is not attainable under the VP principle since the receptor country cannot afford to bribe emitters.

If free entry is allowed, potential entrants always have a valid threat of entry in the nonliability case if market prices are above $P_2{}^E$ in country 2 or above the lowest point of the current average cost plus average damages in country 1. It is clearly impossible to bribe potential firms to stay out of a country as long as they could earn positive profits by entering; because of free information regarding current profitability, an indefinitely large number of potential firms would eventually threaten to enter. In figure 2, entry would result in an eventual price of $P_2{}^E$ for firms in country 2 with a short-run optimum position at point *B*, implying negative profits for emitter firms. However, this point cannot be stable because firms in country 1 must earn sufficient profits to compensate firms in country 2 for their losses. But free entry into country 1 will tend to force profits in that country to zero by a free rider process where receptors will enter (given a level of emissions reduced by negotiations between existing firms), join the coalition, but find profits eventually reduced to the point where firms in country 2 can no longer be bribed to reduce output. Potential entrants cannot realize that entry must lower prices, thereby destabilizing existing solutions; nor does the country's government establish entry constraints by assumption. This sketch of events implies that under a VP rule with free entry, negotiated solutions are unstable. One can again imagine a sufficiently complicated set of regulations and/or taxes to allow an optimal solution to be obtained under a VP rule.

Although we have excluded explicit negotiation costs from this analysis, it is difficult to imagine any set of circumstances, in which firms are competitive (free entry is of course a necessary condition for the existence of competitive firms), an externality exists between countries, and long-run optimality is achieved without taxation or other forms of internationally established entry restrictions. It is possible that governments could act in concert to simulate a "Pigovian" solution provided agreement could be achieved on the "third party" principle. In this case, an efficient allocation of global resources could be achieved by the receptor government collecting damage payments from the emitter government and using them for purposes other than compensating the firms or citizens adversely affected.

COMMENT

Larry E. Westphal, Northwestern University

D'Arge's paper contains a provocative argument that is unfortunately obscured by his attempt to cover too much ground in a single paper. This discussion will concentrate on that argument both because of its importance and because much of d'Arge's subsequent analysis stems from it. In reading the paper, one is immediately led to ask what distinguishes transnational environmental externalities from those environmental externalities whose effects are confined within a single nation (hereafter called national environmental externalities). One might argue that there is no *analytical* difference, so that the distinction is unwarranted. But d'Arge's contention is that there is a basic *political* difference that has substantive implications.

His argument rests on two observations. First, national sovereignty makes it impossible to force a nation to pay environmental damages. Second, nations are likely to find it politically expedient to avoid the explicit entitlement or regulation of common property resources, as each nation's perceived ownership in the absence of explicit rules exceeds the share that it could "reasonably" expect under a system of rules. In short, each nation (or at least a sufficient number of nations to block any common agreement) expects that it will be better off without explicit entitlement or regulation, and there is no means to force either upon unwilling nations. Following from these observations is the conclusion that transnational environmental externalities will be dealt with through bilateral negotiations coupled with side payments between receptor and emitter nations. This is in contrast to their resolution through the action of supranational agencies having regulatory control over particular resources or through explicit multinational agreements on the entitlement of or regulating the use of common property resources such as the oceans. D'Arge also draws from these observations the implication that negotiations will follow the "victim must pay" principle under which the receptor nation either (a) does what is possible within its borders to resolve the effects of the externality without the cooperation of or compensation from the emitter, or (b) "bribes" the emitter to take appropriate remedial action.

The original contribution of the paper is this argument: it is an argument that bears scrutiny. d'Arge's implicit view of the resolution of national environmental externalities seems naive and may lead him to overstate the distinction between expected modes of resolving environmental externalities at the national and multinational levels. In the first instance, the observations made regarding national sovereignty and perceived versus "actual" common property rights apply as well to individuals. Without government coercion, individuals (at least the strongest) cannot be forced to pay damages and may be unwilling to establish rules concerning common property rights. And, to varying degrees, governments are dependent upon the will of constituent interest groups and so do not constitute agents that are somehow completely independent of these. Why is it, then, that individuals within a nation find it in their collective interest to subscribe to a system of rules enforced by government coercion, whereas nations do not or will not find it in their interest to come under an analogous set of enforceable rules?

Early in the paper, it appears to be suggested that the answer rests upon the greater severity of distributional gain and loss problems, measurement problems, questions of responsibility assignment, uncertainty, etc. in the case of transnational externalities. But this answer is not acceptable; for each of the observations made on these pages applies equally well to national externalities. Reading between the lines of d'Arge's paper, I suspect that ultimately the answer lies in two peculiar attributes implicitly ascribed to transnational externalities: they are limited in number and they are predominantly unidirectional (i.e., with a single emitter causing damages to a single receptor without the receptor of one form of pollution being the emitter of another form and vice versa). If this is true, we may then imply reluctance on the part of nations to enter into multinational arrangements parallel to those found at the national level by the following argument. It seems reasonable to argue that "sovereign" entities, be they individuals within a nation or nations within the world, decide whether or not to enter into a collective and enforceable agreement regarding externalities on the basis of the expected gains and losses from entering into the collective agreement. Where the probability of being adversely affected by the external effects of others' actions is large enough (and the costs high enough), an individual party will find it in his interest to subscribe to a set of rules even though this limits his own action or imposes costs. Where that probability is small, it is less likely that the individual parties will agree to a common set of rules. It further seems reasonable to argue that an individual party's assesssment of the probability of being adversely affected by externalities depends upon the number of externalities perceived. If they are few, then the expected gain from entering into an agreement will be small. Thus, if the number of significant transnational externalities is limited, it follows that enforceable agreements among nations will not be in the interests of any but those on the receiving end of particular externalities. In particular, such agreements will not be perceived to be in the self-interest of "today's" emitter nations who will assign a very low probability to being receptors at some future point in time.[1] Finally, the unidirectional nature of the externalities implies that no additional gain results from increasing the number of parties to any agreement.

If the above argument is essentially right, then whether d'Arge's expectation of resolution through bilateral negotiations based on the "victim must pay" principle is correct or not is an empirical issue. Are transnational externalities limited in number and predominantly unidirectional? One source of examples with which I am familiar suggests that neither characterization is correct.[2] The paper's almost exclusive concern with the bilateral negotiation of unidirectional externalities therefore appears too restrictive, and it may be misleading.

1. A corollary to the foregoing argument is that agreements are more likely to be mutually acceptable the broader their scope, i.e., the larger the number of externalities (existing and potential) dealt with in the agreement. For example, nations that might block agreements on water pollution within a particular body of water or along a specific water course may find it in their self-interest to subscribe to agreements covering all water and air pollution.

2. See Baumol, William J., *Environmental Protection, International Spillovers and Trade,* Wiksell Lectures 1971 (Stockholm: Almquist & Wiksell, 1971).

To conclude these more general comments, I do not agree with d'Arge's expectation that bilateral negotiation will be the common mode of resolution.[3] Nor does he appear to agree with it completely, for he seems to feel that the probability of acceptance of the "third party" principle is sufficiently great to warrant its detailed analysis. If there can be common agreement on the third party principle as the basis for bilateral negotiation, then why is agreement on, for example, the formation of supranational agencies impossible? This is not to suggest that supranational agencies are the answer to the problem; however it does appear that continuing and simultaneous negotiations involving a number of nations is a more efficacious way of dealing with the problem than is case-by-case bilateral negotiation. As regards the empirical basis for d'Arge's conclusion, two further observations come to mind. First, it is only recently that transnational externalities have become a matter of great concern. As more thought and effort are directed toward this problem one can expect the perceptions of its dimensions to increase along with the number of perceived cases of transnational externalities. Second, some of the most glaring instances of transnational externalities involve more than two parties, at least as receptors (e.g. pollution of the Rhine in Europe).

Why should we be concerned with the mechanisms through which transnational externalities are likely to be resolved? Here I think that d'Arge has a valid and extremely important point: the efficiency with which resources are allocated depends upon the negotiating mechanisms through which transnational externalities are resolved. Equally important, the distribution of gains and losses depends critically on the negotiating rules. If resolution is to be expected through bilateral negotiation, then d'Arge is more than justified in investigating the allocative and distributional consequences of alternative bilateral negotiating schemes, as he does throughout the paper. On the other hand, if bilateral negotiation has as its consequence continued misallocation, then it is important to scrutinize the empirical basis for expecting bilateral negotiation and to seek multilateral alternatives. It may be expected that bilateral negotiation, which implies that each nation *separately* strives for efficient allocation within a set of negotiating rules, will be less efficient globally than arrangements under which system-wide efficiency is sought.[4] D'Arge demonstrates that, under a plausible set of assumptions including interna-

3. In fact there are already some instances of multilateral agreements. See Baumol, *Environmental Protection*.

4. See Baumol, *Environmental Protection*, p. 42, for an additional example to that cited from d'Arge's paper.

tional capital mobility, none of the forms of bilateral negotiation contemplated lead to an efficient allocation of resources.[5]

I now turn to a number of specific comments. With the exception of those parts of the paper which use a game theory approach, the remainder of d'Arge's paper is largely an application of models developed in relation to national externalities to the problem of transnational externalities. The conclusions reached, for the most part, parallel similar conclusions regarding national externalities. One shortcoming of the paper is the complete neglect of trade policy, which in isolated cases may yield a feasible first best solution. To give but one example: assume that country A in producing commodity X transmits an external diseconomy to country B, and assume further that country B is the sole consumer of commodity X, all of which it obtains through trade with country A. If country A cannot be forced to take action resolving the external diseconomy, then country B could levy a tariff on imports of X and thereby achieve a Pareto efficient allocation of resources with respect to its own consumption and production. (In this case the global allocation is also Pareto efficient; see Baumol, *Environmental Protection,* section 13.) The matter becomes more complicated in an "n" country world, but "defensive" trade policy may still yield an improved allocation of resources from the receptor's point of view.

In the second section of the paper, in which the measurement of wealth effects is discussed, it is alleged that removal of the externality or compensation may make the receptor worse off due to a change in world market prices. This is a much more complex issue than is recognized in the discussion, which is based on the assumption, employed throughout, that individual nations are price takers in international trade. If this assumption is removed and, for example, the possibility for an "optimum" tariff is introduced, then it seems unlikely that the receptor can be made worse off through the resolution of an externality. The only case in which such a possibility might be realized arises where the receptor's actions have no effect on world prices, all of the effect coming from the emitter's actions.

The third section of the paper deals with consumption externalities and concludes that bilateral adjustment schemes lead to a Pareto efficient allocation so long as all factors of production are not internationally mobile. This is a consequence of the derived result that transnational externalities will generally not cause a shift from convexity to concavity of

5. D'Arge does not analyze the "polluter pays principle" which might, if suitably implemented, lead to an efficient outcome.

the domestic production transformation frontiers. But this last result is surely a trivial one. It still remains true that the "international" transformation frontier may be concave, implying a globally inefficient allocation. Thus there is still cause for concern. D'Arge's point is simply that, so long as the receptor acts as an externality taker (i.e., perceives the level of the externality as fixed and optimizes subject to that fixed level), the receptor's national transformation frontier will remain convex if it were convex in the absence of the externality. The point being made in criticism is that only in exceptional cases will the level of the externality be completely independent of the receptor's actions.

The game theory discussion, though far from rigorous, is suggestive and is the most interesting and significant part of the paper. The discussion is ambiguous at many points, however. To cite but one example, the concluding sentence under point 5 appears to confuse two issues with regard to negotiation. The first issue is whether any action at all toward the resolution of an externality will be taken on the part of the emitter. Clearly, under the "third party" principle, such action will be forced upon the emitter by the receptor unless transactions costs are both prohibitive and levied upon the receptor. In this sense, negotiation will take place because of the incentives to the receptor to force negotiation (i.e. resolution) on the emitter; clearly the emitter is better off not opening up negotiations through a third party. The second issue concerns whether the resulting resource allocation is efficient. With respect to this issue, it is not clear that the emitter is better off not negotiating. To take a simple example, let us assume that the costs of pollution abatement in the emitter country are 100 while they are 80 in the receptor country. Under the "third party" principle, the emitter must pay for the corrective action, but he may either undertake the action himself or pay the receptor to do so. In the example, assuming there were no uncertainty, the emitter clearly has an incentive to negotiate with the receptor to pay the latter to take the corrective action. The same incentives exist under the "victim must pay" rule if the costs in the receptor country are 100 while they are 80 in the emitter country. Thus the asymmetry is only with respect to the first issue, and not with respect to the second.

D'Arge's argument that pollution may be greater *than desirable* under the "third party" rule with transactions costs being paid by the receptor (point 5) is unconvincing, particularly insofar as it rests on the observation that the receptor might perceive that it would be better off without negotiation. For this to be the case, the expected negotiation costs would have to exceed the known costs of the externality (if these are not completely known, they are presumably better known than the negotiating costs). Unless the uncertainty regarding negotiation costs borders on complete and at least partially avoidable ignorance, it would seem that where expected negotiation costs exceed resolution benefits, the global optimum is no resolution. My point here is that one should not define efficiency on the basis of complete and certain knowledge when in fact there is uncertainty; we need an efficiency criterion that explicitly introduces uncertainty.[6]

One wonders how the results derived from the two country partial equilibrium model would hold up under a more realistic set of assumptions. Rather unrealistic assumptions have been chosen to reduce the discussion to the level where a model applicable in the single country case can be applied. (To illustrate with but one assumption, the unit price of nonmobile factors such as labor is assumed constant, as each firm's production cost function depends only upon its own level of output.) In this argument, as in several others, d'Arge would at least make a stronger case for his conclusion by using a model whose assumptions are compatible with international trade theory rather than merely adapting a model used for the analysis of national externalities.

Finally, insufficient attention is paid to several aspects of transnational externalities that will significantly impact on the negotiating process. First, where there is more than a single receptor, certain forms of negotiation will be subject to the "free rider" problem. This is clearly the case under the "victim must pay" rule if one assumes that the efficient solution is to bribe the emitter to take corrective action, for there is then a problem in assessing payments by the individual receptors. Second, negotiations are likely to be multileveled with various interest groups within each nation being involved as well as the national governments. In many cases the pressure for resolution will come from segments of the receptor's population and their stake in the outcome will require them to participate, at least through discussions with their government; likewise, costs, if imposed, in the emitter nation are likely to fall unequally upon segments of its population. Third, to the extent that the externality also adversely affects groups within the emitter nation, it is not merely a matter of negotiation between two nations with distinctly identifiable interests. Fourth, the "victim must pay" principle may lead to individual

6. It is also unclear why d'Arge insists on assuming that transactions costs must be paid by the receptor under a "third party" rule. If agreement on a "third party" rule is possible, then so too is agreement on the emitter's paying transactions costs. The emitter's lack of incentive to pay these costs appears irrelevant once the principle is adopted, for they can be forced upon him.

nations profiting from threats of directed diseconomies; on this ground alone it is likely that international agreement on an unmodified "victim must pay" negotiating process would not be achieved. Lastly, it may be that in some cases the most effective strategy open to the receptor will be an offensive one, perhaps in areas not directly related to the externality. If the diseconomy is sufficiently great and the receptor is an important export market for the emitter, then in the last resort the receptor may threaten to close his market to the emitter unless the externality is resolved. Or, where a common resource is involved, the "injured party" may threaten to jump on the band wagon of injudicious depletion or he may actually join in to get what he can before the resource is exhausted.

In conclusion, d'Arge's paper leaves open a number of interesting and relevant questions. The paper's virtue is that it asks the questions and poses the issues in a fruitful manner that should clarify the needed research in this heretofore little studied area. D'Arge's search for an efficient mechanism for negotiating the resolution of externalities is valuable and needs to be carried forward, particularly with respect to multilateral negotiation and nonunidirectional externalities, which appear to be far more relevant than bilateral negotiation and unidirectional externalities, respectively.

Management of Environmental Quality: Observations on Recent Experience in the United States and the United Kingdom

Maynard M. Hufschmidt, University of North Carolina at Chapel Hill

Introduction

Comparative analysis of experience on public policy of different countries is hazardous. This is especially true for environmental quality where public and governmental attitudes and values have been changing so rapidly. In 1960, the terms "environment" and "environmental quality" were in the lexicon of few public policy makers, except for specialists in public health. By 1965, scholars and public policymakers were using the terms as a means of organizing analysis and defining problems related to various types of pollution—air, water, and land.[1] And by 1970, environmental quality had become a major issue of national policy in this country and that of most developed nations, including Canada and Japan, as well as Eastern and Western Europe.

The views presented in this paper are the result of observations of this dynamic process at work in Europe and the United States in late 1970 and 1971—the very period when environmental quality policies and pro-

NOTE: This work was supported in part by EPA Research Project R-800901.

1. Lynton K. Caldwell, "Environment: A New Focus for Public Policy?" *Public Administration Review* Vol. 23 (September 1963): pp. 132–139. Allen Kneese and Orris Herfindahl, *Quality of the Environment* (Washington, D.C.: Resources for the Future, Inc., 1965); "Restoring the Quality of Our Environment," Report of the Environmental Pollution Panel, President's Science Advisory Committee, The White House (1965).

cedures were being forged, and when the countries were working together in preparing for the U.N. Stockholm Conference on the Human Environment held in June of 1972. The results presented here are accordingly modest in scope and content. They are akin to a snapshot taken about the middle of 1971—one that shows only general outlines rather than fine-grained detail. Furthermore, it is a view that is affected by the approach of this observer, which is that of public investment analysis.

Outline of the Approach

The approach used to describe the experiences is a straight-forward version of the public investment or benefit-cost model with the important addition of an objective labeled "environmental quality," that is assumed to be distinguished from the "economic efficiency" objective.[2] In summary terms, we include a multivalued *objective function* of the usual form:

$$\max \sum_{t=1}^{T} \frac{E_t(Y_t) + \alpha Q_t(y_t) - M_t(x_t) - K_t(x_t)}{(1+r)^t},$$

where E, M, and K are economic efficiency benefits, operation, maintenance and replacement costs, and capital costs, respectively; Q is environmental quality "benefits," and α is a weighting factor that transforms Q benefits into the metric of E benefits.

Associated with the objective function are the usual relationships of costs to inputs, $M = f(x)$ and $K = f(x)$; benefit to outputs, $E = f(y)$ and $Q = f(y)$; and outputs to inputs, $y = f(x)$, which we also know as the production function.

Because contributions to environmental quality (in addition to those that can be expressed and measured in economic efficiency terms) can at best be expressed in physical, quantitative terms, and because environmental quality weights are hard to come by, and perhaps impossible

2. An earlier formulation of this approach is contained in an unpublished paper of mine entitled "Environmental Planning with Special Relation to Natural Resources," revised, Dec. 1966; further discussion is contained in my paper "Environmental Quality as a Policy & Planning Objective," *Journal of the American Institute of Planners,* Vol. 37, No. 4 (July, 1971). In order to keep the exposition simple, income redistribution, either as an objective to be achieved or as a consequence whose value is to be checked, has been omitted from the approach. In general, income distribution must be taken into account in multiple-objective public investment analysis.

to determine a priori, a more realistic formulation of the objective function for our purposes is:

$$\max \sum_{t=1}^{T} \frac{E_t(y_t) - M_t(x_t) - K_t(x_t)}{(1+r)^t},$$

subject to

$$Q_t \geqq Q_t^*,$$

as well as

$$y = f(x),$$

when Q_t^* (which along with Q_t is almost always a vector) represents a minimum level of environmental quality to be achieved in year t.

We recognize this formulation as consistent with the familiar "standards" strategy commonly used in the United States. But, in terms of the approach outlined here, the crucial issue is the weight that is implicit in the standard or set of standards in any given situation of public investment or resource allocation.

This approach is closely related to the formulations of the problem of environmental quality management by Kneese and other economists concerned with the economics of environmental quality.[3] Obviously, many of the important problems and issues of environmental economics are not made explicit in this simple formulation. Accordingly, the description and comparisons to follow will rely primarily on the barebones approach set forth above.

Overall Policies on Environmental Quality

Nineteen hundred and seventy was a watershed year for environmental policy in both the United States and Western Europe. Beginning in that year, legislation and administrative actions involving both the substance of problems and organization for problem solving were framed in terms of overall environmental quality rather than exclusively, as in the past, in terms of air pollution, water pollution, solid waste management, pesticides damages, and the like. Because this broader policy framework is so

3. A. V. Kneese and B. T. Bower, *Managing Water Quality: Economics, Technology, Institutions* (Baltimore, Johns Hopkins Press for Resources for the Future, Inc. 1968).

new, the administrative workings of government are still busy adjusting to it. It is possible, therefore, to examine the present situation with some assurance that the new policy framework will be stable at least over the next few years. Public and governmental concern for environmental quality is not a passing fad; it is gradually, but steadily, being assimilated into governmental policy and administrative practice in both the United States and Europe.

The United States

After about ten years of working up to the action incrementally, in 1970 the federal government adopted the objective of environmental quality as a major national goal on a par with economic development, employment, and price stability.

The National Environmental Policy Act (NEPA) of 1969 (signed by the President on January 1, 1970) and the newly created Environmental Protection Administration are the two chief instruments of this basic change in policy, but many other administrative, regulatory, and legal actions provide support.[4]

NEPA established the Council on Environmental Quality (CEQ) in the White House in the image of the Council of Economic Advisers. NEPA also required that an "environmental impact statement" be prepared before any federal action that significantly affects the environment is taken.[5] In a landmark decision on the Calvert Cliffs case, a Federal District Court ruled that federal agencies, in this case the Atomic Energy Commission (AEC), must comply with this requirement in a substantive way.[6] Significantly, the Executive Branch decided not to appeal this decision, and AEC changed its procedures to conform with the Court's interpretation of the law.[7]

In a parallel and related development the U.S. Water Resources Council in December 1971 formally proposed that environmental quality be included as a national objective of water resource planning, with status

4. See *Environmental Quality,* Third Annual Report of the Council on Environmental Quality, U.S.G.P.O., August 1972.

5. See Richard N. L. Andrews, "Three Fronts of Federal Environmental Policy," *Journal American Institute of Planners* Vol. 37, No. 4 (July 1971): pp. 258–266; and Richard N. L. Andrews, "Environmental Policy and Administrative Change: The National Environmental Policy Act of 1969, 1970–1971," unpublished Ph.D. dissertation, University of North Carolina, Chapel Hill, 1972.

6. Calvert Cliff's Coordinating Committee v. AEC, 449F.2d (D.C.Cir.1971).

7. AEC revised regulations (10 CFR 50, Appendix D).

equal to the objectives of national economic efficiency and regional development. Intensive public discussion of the Council's proposals over a three-month period revealed major differences of view as to appropriate levels of discount rate, and validity of the regional development objective. However, there was overwhelming approval of the environmental quality objective, and a majority of respondents on the issue expressed the view that the objective should be given top priority.[8]

Environmental quality as a major policy was also institutionalized in 1970 through the establishment of the Environmental Protection Agency via transfer of air, water quality and solid waste management functions, and pesticides and radiation regulation from existing departments and agencies.[9] Administration of the major environmental quality programs was brought together, and a powerful advocate for the environment was established.

In fact, however, the relative weight to be given environmental quality as compared with national and regional economic development in public investment decisions has not been determined by Congress, the Executive, or the Judiciary. Weights are being established implicitly in the process of making decisions on specific projects and programs, in which environmentalists and supporters of economic development along with their allies in the Congress and the administrative agencies take strongly opposing positions.

But it is clear that a series of court decisions, administrative actions, and Congressional followups since 1970 have confirmed that environmental quality is a major national goal, not to be lightly disregarded.[10] Furthermore, via the requirement for environmental impact statements, the issue of environmental quality must be faced in every public investment plan and program directly or indirectly supported by the federal government. In this sense, we can assert that, conceptually, federal public investment plans and programs have multiple-objective functions, one major element of which is environmental quality.[11]

8. U. S. Water Resources Council, *Summary Analysis,* Washington, July 1972.

9. President's Reorganization Plan No. 3 of 1970, July 9, 1970, effective December 2, 1970.

10. *Environmental Quality,* Chapters 4 and 7.

11. Only time will tell whether political commitment to environmental quality will be as strong and consistent as has been the case, for example, for economic growth and full employment during the twenty-five years following enactment of the Employment Act of 1946. But the continuing strength of public support for environmental quality values, and concomitant actions by the courts, and federal and state legislatures, support the view that environmental quality is now firmly established as a national goal.

The United Kingdom

Concern for air and water pollution is of long standing in Britain, predating that in the United States. But, as in this country, the environmental quality issue as such has emerged only recently. In October 1969, the Labour Government took the first official step concerning overall environmental quality policy by giving the newly established Secretary of State for Local Government and Regional Planning coordinating responsibilities for all governmental action on control of environmental pollution.[12] These responsibilities had previously been scattered among ten ministries. This action was followed by setting up in February, 1970, a standing Royal Commission on Environmental Pollution,[13] and in May, 1970, by issuance of a White Paper on protection of the environment.[14] The White Paper was very low key in tone and did little more than sketch the nature of problems, the current situation, and modest proposals for governmental action. Thus, when the Labour Government was turned out in the summer of 1970, only the first preliminary steps had been taken to deal with environmental quality as a major policy issue.

The environmental issue was given considerable attention by the Conservative Government in its White Paper on central government reorganization of October 15, 1970, which announced establishment of a super-Department of the Environment, unifying the Ministries of Housing and Local Government, Public Buildings and Works, and Transport under a Secretary of State for the Environment.[15] In the U. S. context this amounts to creating a super-department combining the present Departments of Housing and Urban Development, and Transportation plus the Environmental Protection Agency and the regional development functions of the Economic Development Agency and the Appalachian Regional Commission. Although the mandate of this new Department was much broader than environmental quality, the agency was charged with such major environmental matters as "the preservation of amenity, the protection of the coast and countryside, the preservation of historic towns and monuments, and the control of air, water, and noise pollution." [16]

12. *The Protection of the Environment: The Fight Against Pollution,* Cmnd 4373, H.M.S.O., May, 1970, p. 6.

13. Royal Commission on Environmental Pollution, *First Report,* Cmnd 4585, H.M.S.O., February, 1971, pp. iii, iv.

14. *The Protection of the Environment,* p. 6.

15. *The Reorganization of Central Government,* Cmnd 4506, H.M.S.O., October 15, 1970.

16. *Reorganization of Central Government,* Cmnd 4506.

Of special significance was the bringing together of land use, transport, regional development, and environmental quality programs and policy into a single organization with representation at the highest level—via a Secretary of State.

The U. K. stance with respect to environmental quality can be inferred in part from the first report of the Royal Commission on Environmental Pollution filed in February 1971.[17] An admirably well-balanced document, this report states the problem as one of how to strike a balance between "benefits gained from economic and technological achievements" and losses from "deterioration of the environment." In principle, resources devoted to reducing pollution, as compared with other claims, should depend on the relevant costs and benefits. Because of great difficulties in making such calculations, the report noted that decisions may still have to be made in the absence of satisfactory information on costs and benefits, and "in the end . . . must still reflect subjective value judgments."

In contrast to the United States action on NEPA, the U. K. organizational and policy initiatives did not, by themselves, raise environmental quality to the level of a national objective; there was no policy statement to this effect in the British enabling legislation analogous to the statement in NEPA. Yet U. K. initiative appears to give increasing importance to the objective.

In some respects, the change in U. K. policy toward environmental quality has been less striking than in the United States. British legislation and on-going programs in air and water pollution, noise and oil pollution have a much longer history than those in the United States. For example, national noise abatement legislation dates from 1960 in the U. K., and has only recently (1972) been enacted in the United States.

Thus, the change in Britain has been less in new legislation than in more active use of existing laws. Public awareness of environmental problems has increased, and various public alarms over toxic wastes—lead and cyanide—and DDT have generated a more sensitive attitude on the part of government. But, no political party in the U. K. has raised the environmental question to the level of a major issue. And, while the issue has attained considerable importance with the younger generation, it has failed to gain widespread public concern on a par with issues of economic growth, inflation, unemployment, and housing.[18]

17. Royal Commission on Environmental Pollution, *First Report*.
18. Based, in part, on personal communication from David Pearce, Department of Economics, The University of Southampton, November 30, 1972.

Although it is hazardous to generalize on this issue, it appears that the general policy stance of the British is to avoid all out commitment of national resources to grand programs of environmental improvement and to favor a careful balancing of economic and environmental costs and benefits for specific projects and restricted programs. From available evidence one could infer that environmental quality has a status that is subordinate to economic objectives. But, at least some U. K. public investments take into account the objective of environmental quality.[19]

Air Quality Programs

The United States

In the 1970 amendments to the Clean Air Act, the United States made a significant shift toward national minimum standards for both ambient air quality and emissions from stationary sources and automobiles.[20] Primary ambient air standards, based on health criteria, and secondary standards, based on "aesthetics, vegetation and materials," were established on a uniform national basis for six pollutants: particulate matter, sulfur oxides, carbon monoxide, hydrocarbons, oxides of nitrogen, and photochemical oxidants. Implementation was left to the states which must meet primary standards within three years and secondary standards within a "reasonable" time.[21] National emission standards for 1975 model automobiles require that emissions of carbon monoxide and hydrocarbons be 90 per cent below the 1970 standards and, for 1976 model cars, emissions of oxides of nitrogen be 90 per cent below 1970 levels. Emission standards for major new stationary sources, including fossil fuel steam generators, portland cement plants, sulfuric and nitric acid plants, and large incinerators, have also been established on a national basis.

The Act leaves considerable flexibility to states and local "air quality control regions," typically metropolitan areas, in determining how to

19. For example the extensive studies of the Commission on the Third London Airport (the Roskill Commission) included exhaustive examination of environmental quality aspects as well as economic costs and benefits. See Commission on the Third London Airport, *Report,* London, H.M.S.O., 1971. However, U. K. nationalized industries are not required to consider environmental effects. Also, to date, the Department of the Environment does not require that appraisals of road investments incorporate an evaluation of environmental effects. (David Pearce, November 30, 1972.)

20. Public Law 91–604 (December 31, 1970).

21. States can, if they wish, establish higher standards than those established by the EPA under provisions of the Act.

achieve the primary and secondary standards. But it has already become obvious that some large urban areas will have extreme difficulty in meeting the primary standards by 1973. EPA has already granted 2-year extensions to 18 states with such problem areas.[22] Other regions enjoy levels of air quality above the secondary standards, and a recent Federal District Court ruling enjoins the EPA from approving state implementation plans that would allow significant deterioration of existing air quality in such regions.[23]

Metropolitan areas with serious problems of meeting primary standards would have, in terms of our approach, an objective function of the following form:

$$\min \sum_{t=1}^{T} \frac{M_t(x_t) + K_t(x_t)}{(1+r)^t},$$

subject to a set of ambient air constraints on pollutants, of the form $P_i \leq P_i^*$, where P_i is the concentration per specified time period of pollutant $_i$ and P_i^* is the national standard for maximum concentration of pollutant i.

This approach to air quality management requires detailed knowledge of the cost and production functions of the various means available to state and local governments for reducing pollution at its sources and controlling its spatial and temporal distribution. An essential element is some means or model of relating changes in quantities and rates of pollution at the sources to changes in concentration in the general environment of the airshed. Although such air quality diffusion models are now in common use in air quality management, their predictive reliability in complex airsheds with many emission sources is not high. Also, the costs of various means of control, including fuel and process changes, emission control devices, and timing of emissions are not well understood. Thus, the task of attaining a least-cost means of achieving the national standards is a formidable one.

A more fundamental issue is raised, however, by the primary national standards for ambient air quality. They are based on health criteria which are admittedly incomplete and "somewhat controversial." [24] Inso-

22. This extension was declared invalid by the Court of Appeals of the District of Columbia on January 31, 1973. The ruling requires the States to submit implementation plans by April 15, 1973, as required by the Clean Air Act. "Natural Resources Defense Council v. EPA," *Law Week* Vol. 41: p. 2,400.

23. Sierra Club v. Ruckelshaus, 4 ERC 1205, 2 ELR 20262 (D.D.C. 1972).

24. *Environmental Quality,* Third Annual Report of Council on Environmental Quality, p. 7.

far as can be determined no specific information on the cost of achieving these standards in the most difficult problem areas was used in establishing the standards.[25] It seems reasonable to assume, however, that the implicit weight α or opportunity cost associated with meeting national standards will vary widely among communities. Within some reasonable range, ambient air quality standards should reflect both the health effects and the opportunity costs of achieving the standards; and because opportunity costs are likely to vary widely, no single set of national standards can meet this test.

In summary, the United States has adopted a strategy combining the national minimum ambient air standards approach, with strict national emission controls on major *new* sources, both stationary and mobile. In attempting to achieve national emission standards on 1975 and 1976 model automobiles and on major new industrial plants, the EPA will be forced to balance costs and technological feasibility against the levels of residual emissions to be allowed and the timing of required actions. It is assumed that both industry and the federal government will strive for least-cost means of meeting emission standards for new sources, and the government has attempted to explore the cost implications of national emission standards. But, differential local cost implications of meeting national ambient air standards were not taken into account in setting these standards; in effect the federal government is imposing implicit α's of unknown, but probably high, values on problem communities. In any event, the α's, are likely to have a wide range of values as among communities.

The United Kingdom

The British have a long experience in dealing with air pollution. National control over major industrial sources has been in effect for over 100 years.[26] Following the London smog episode of 1954, the Clean Air Act of 1956 (subsequently expanded in 1968) was directed at controlling pollution by domestic smoke; more recently there has been increasing concern about automobile pollution.

25. The Council on Environmental Quality, EPA, and the Department of Commerce sponsored a series of studies of economic impact of meeting air and water quality standards on the general economy and selected key industries most likely to be severely affected. But microeconomic studies of costs of meeting such standards in problem urban areas were not included in these analyses. See *The Economic Impact of Pollution Control,* U.S.G.P.O., March, 1972.

26. The Alkali Act of 1863; current industry legislation is embodied in the Alkali, etc. Works Regulation Act, 1906 and the Alkali, etc. Works Order, 1966.

British policy is almost exclusively directed at control of emission sources. No national or local ambient air quality standards have been adopted; nor is there any strong pressure to do so.

The historic British approach, developed in control of industrial sources by the Alkali Inspector, is labeled "best practicable means." Under this approach, the Government requires industries to install control equipment, based on the best available technology, to the extent that costs are within reach of the industrial organizations concerned.[27] "Best practicable means" is defined in the context of changing technology, the prevailing political and economic climate, and public attitudes toward pollution; in the rare case of dispute over a specific case, the meaning is interpreted by the courts. The system is quite flexible, and although there are no formal emission standards, the Alkali Inspectorate is able to adopt working emission standards or "presumptive limits" as a result of experience with different types of plants.[28] There is some evidence that working standards are being upgraded, under the influence of increased public concern for the environment.

The means to be used may include changes of fuel or production process, control at points of emission from the plant, and decisions on height of chimneys. Some consideration is given to existing pollution in the vicinity, and topography, micrometeorology, and presence of tall buildings in establishing height of chimneys.

Control of pollution from domestic smoke, whose source has been the traditional British open grate fire, is in the hands of local authorities. The Clean Air Acts of 1956 and 1968 give local authorities power to establish "smoke control areas" within which all smoke emissions are illegal. In such areas, householders and other establishments have been required to switch to use of smokeless fuels. Local authority grants to householders of 70 per cent of the cost of conversion are available.[29] Spectacular reductions in smoke emissions have been achieved in some urban areas, notably

27. As of 1970, 2,200 industrial plants, mainly electricity generation, cement, ceramics, chemical, petroleum and petro-chemical, and iron and steel, were regulated by the national Alkali Inspectorate; 30,000 other industrial premises were under control of some 1,100 local governments. *The Protection of the Environment,* p. 6.

28. World Health Organization, Regional Office for Europe, *Long Term Programme in Environmental Pollution Control in Europe: Trends and Developments in Air Pollution Control in Europe,* Copenhagen, 1971, pp. 8–9. See also U. K. Ministry of Housing and Local Government, *106th Annual Report on Alkali, etc. Works, 1969,* H.M.S.O., 1970, and U. K. Programmes Analysis Unit, *An Economic and Technical Appraisal of Air Pollution in the United Kingdom,* Chilton, Didcot, Berks, 1972.

29. Of the total cost of conversion, the central government bears 40 per cent, local government 30 per cent, and the householder, 30 per cent.

London, but some areas in the north of England have made little progress, and in 1970 the Government was considering using its powers under the 1968 Clean Air Act to compel laggard local authorities in badly polluted areas to make smoke control orders.[30]

As of 1971, Britain had not developed a national program of automobile emission control remotely comparable to that in the United States. The British attitude at that time is reflected in two quotations: "In Europe, due to the difference in climatic conditions, air pollution from petrol-engined vehicles presents a different and less acute problem (than in problem sections of the United States), and the development of a completely pollution-free car might not be the most sensible use of resources" [31] and "There is no firm evidence in Britain that the present level of pollutants (emitted by road vehicles) is a hazard to health, even in busy city streets, although smoke from diesel engines can be very offensive." [32] Britain is carefully watching experiences in the United States and countries in Western Europe and is conducting research along a number of fronts including the health effects of long-term exposure to automobile emissions.[33]

In contrast to the United States, Britain has not attempted to impose national ambient air quality standards; in terms of our model, it has not elevated the air quality objective to the level of a constraint that must be met. Expert opinion is skeptical that such minimum standards on specific pollutants can, in fact, be directly related to human health. Also, in contrast to the United States, Britain has not emphasized modelling and research on relating emissions to ambient air quality on an airshed basis. There is little or no information on the levels and national distribution of pollutants other than smoke and sulfur dioxide.[34] In terms of our model, the production functions are not used to provide information for formulating specific air pollution control measures. Major industrial-source emissions are controlled as uniformly as possible throughout the country and, except under special circumstances, the control requirements that are imposed do not vary with the amount of air pollution already existing in the area. On the other hand, the British are extremely sensitive to cost of control. The "best practicable means" approach emphasizes least-cost means of reducing harmful emissions, and stops short of levels of control at which marginal costs become exorbitant.

30. *The Protection of the Environment,* p. 9.
31. *The Protection of the Environment,* p. 11.
32. Royal Commission on Environmental Pollution, *First Report,* p. 12.
33. U. K. Programmes Analysis Unit, *Appraisal of Air Pollution.*
34. U. K. Programmes Analysis Unit, *Appraisal of Air Pollution.*

In summary, the British system emphasizes the cost functions, and places little emphasis on defining local emission—ambient air quality relationships or air quality—benefit functions. It is a national system which, through rejection of national minimum ambient air standards, avoids high α's in regional airsheds, but which makes no pretense to achievement of least-cost means of attaining specific levels of ambient air quality. The implicit national weight given to air quality, as related to economic development as conventionally measured, appears to be lower than the weight inferred from United States policy, if not performance. The British have consistently worked toward achieving air quality goals since 1956, and the weight given to the goals appears to be increasing relative to other national goals. But the British have attempted to strike a reasonable and consistent balance between reduction of emissions and the costs of doing so.

Water Quality Programs

The United States

The trend of national policy as revealed in the Federal Water Pollution Control Act of 1965 and the recently enacted 1972 Amendments is toward more uniform national standards on all streams, coupled with strict national effluent standards on point sources, based upon "best practicable" and "best available" technology.[35] The 1972 Amendments establish as a national goal that discharge of pollutants into navigable waters be eliminated by 1985. This and the generous financing provisions of the Amendments (total authorizations of $24.6 billion over the next three fiscal years) go far beyond the Administration's original recommendations.[36]

Passing over the details of the complex and changing water quality programs and policy of the United States, we can summarize the current status and trends as follows:

1. The requirements that states develop federally-acceptable stream standards and implementation plans, presumably related to existing

35. Public Law 92–500.

36. The President has recently reduced the allotment of waste treatment construction grants from $11.2 billion provided by the Act for fiscal years 1973 and 1974 to $5 billion.

and projected uses of streams, are consistent with a regional water quality management approach in which standards and means for achieving them could be developed in a benefit-cost or least-cost framework. Yet, as the Delaware Estuary example has shown, even where the least-cost means of achieving various quality levels have been developed through detailed systems analysis, the least-cost solution was rejected in favor of an "equitable apportionment" means which was significantly more costly.[37]

2. With few exceptions, there has been no serious, consistent attempt to relate point-source effluent requirements to ambient water quality. Secondary treatment is now generally regarded as a uniform requirement, and in certain problem areas, tertiary treatment of point sources is being strongly advocated. Only now is the federal government becoming aware of the possibility that complete point-source control of pollution in some basins may have only limited effects on stream quality of some important reaches, because of the dominance of pollutants arising from uncontrolled runoff.[38]

3. The trend in national legislation toward strong point-source effluent control, through federal permits, strict effluent standards, and major subsidies for construction (but not for operation) of local public treatment plants, is almost sure to lead to extremely high private and public costs, and to inability to even approach least-cost solutions. Prohibitive costs may well be avoided as the terms "best practicable means" and "best available means" are interpreted in the bargaining process between industry and regulatory agencies. But failure to relate these regulatory and subsidy measures to a regional water quality management framework, which ties all major sources of pollutants (point and nonpoint) to ambient water quality, will lead to grossly inefficient uses of national sources.

In terms of our model, current U.S. policy, at least in principle, seeks to:

$$\min \sum_{t=1}^{T} \frac{M_t x_t + K_t x_t}{(1 + r)^t},$$

37. See Edwin Johnson, "A Study in the Economics of Water Quality Management," *Water Resources Research* Vol. 3, No. 2 (1967): 303 and Ralph Porges, "Regional Water Quality Standards," *Journal of the Sanitary Engineering Division,* Proceedings of the ASCE (June 1969): 427.

38. *Environmental Quality,* p. 16.

subject to meeting a set of ambient water quality standards

$$Q_i^s \geq Q_i^{s*}, \text{ and}$$

subject to meeting specific effluent standards often expressed as levels of treatment (secondary, tertiary),

$$Q_i^E \geq Q_i^{E*}.$$

The Q_i^{s*} have been set often with little regard to the cost of meeting them. The Q_i^{E*} have been set without any regard to the differential contribution to stream quality that may be involved in uniform application throughout the nation, a state, or a water quality region.

Insufficient emphasis is given to defining the production function [$y = f(x)$] between resource inputs to pollution control measures (x) and "outputs" in terms of improvements in water quality [(y) which is almost always a vector].

The 1972 Amendments do include some extremely worthwhile provisions for water quality research, planning, and management that may lead to significant improvements in the long run. For example, the EPA Administrator is authorized to conduct, and provide grants for, research on almost all aspects of pollution; problem sources, physical, economic, and social effects, and methods of management and control. Substantial grants are provided to states for comprehensive planning and development of long-range programs. Area-wide waste treatment management plans, focused primarily on large metropolitan areas, are required to be established, providing opportunities for achieving economies of scale. While these research, planning, and management provisions are likely to lead to long-term payoffs, they will have little effect on improving the efficiency of expenditure of the $18 billion of federal construction grants authorized for the three fiscal years 1973–1975.

Until recently, the United States has invested very little in water quality management in relation to the seriousness of the pollution problem. Limited resources have forced at least a modest amount of engineering type least-cost discipline, even though more systematic economic approaches to regional water quality management have been dismissed as too complex, impractical, or even as giving a "license to pollute." Now that major resources are being committed to the cleanup task, however, the need for systematic approaches is much greater; as a very minimum we need a strategy of avoiding prohibitively costly projects, and insuring that major resources are not committed to projects that have only minor effect on water quality.

The United Kingdom

Since 1951, new discharges of industrial or sewage effluent require consent of the applicable River Authorities, and since 1961 all such discharges, including those in existence before October 1, 1951, require such consent. Although consent may be refused, the law stipulates that it "is not to be unreasonably withheld." No uniform effluent standards are imposed by law, but in practice most River Authorities have adopted the standards developed by a Royal Commission in 1912. These so-called Royal Commission standards have become the norm in British pollution control.[39]

Ambient or in-stream water quality standards are not used, and there is only a tenuous link between effluent standards and in-stream water quality objectives.

As of 1970 there was still widespread noncompliance with the consent conditions established by the River Authorities. Only about one-half of the reporting municipalities were in compliance with Royal Commission standards. Surprisingly, industrial compliance was reported to be better than that of municipalities. In the absence of specific national grants for sewage treatment, British communities (as, until recently, in the United States) placed low priority on investments in sewage treatment. Although national loans are available for these investments, the long drawn out technical review of such projects by the national ministry concerned provides another excuse for local communities to delay. In summary, the British have had only partial success after 20 years of effort in achieving compliance with effluent standards whose norm (the Royal Commission standards) are not overly stringent—perhaps equivalent to or slightly better than a good-practice secondary treatment plant in the United States.

Even with this mixed record, ambient river quality has shown some improvement over the past 15 years. A 1970 survey of conditions in England and Wales showed that, compared with 1958, the percentage of mileage of grossly polluted nontidal rivers (4.38 per cent) and of tidal rivers (11.7 per cent) had declined, while, at the other extreme, percentage of

39. *Taken for Granted: Report of the Working Party on Sewage Disposal,* U. K. Ministry of Housing and Local Government/The Welsh Office, H.M.S.O., 1970, p. 7.

mileage of unpolluted rivers (76.2 per cent) had increased. While overall gains were modest, the line had been held against deterioration.[40]

Under leadership of the National Water Resources Board, a systematic approach to planning for water resource management was undertaken for the River Trent, which in its upper and middle reaches is now grossly polluted. The objectives of the study, which was in the form of a multidisciplinary research program, were as follows:

1. "To determine the different ways in which the River Trent, its tributaries and other waters in, or which can be brought into, the Trent area may be used to satisfy the expected demands in that area, or elsewhere, for water for domestic, industrial, agricultural, and amenity use, of quality suitable for each."
2. "To evaluate the costs and benefits of each of those ways as a guide to determining the most efficient solution." [41]

The River Trent study represents the first major attempt in Britain to apply benefit-cost analysis and systems analysis techniques, including computer modeling to problems of water quantity and quality of an entire river basin. The need for more systematic analysis than that undertaken by staffs of the River Authorities was well recognized by professionals in the Water Resources Board, the Department of the Environment, and the national water research institutes. In the River Trent research studies the British were adapting United States water resource systems theory and practice, including such examples as the Delaware River Estuary Study, to their major problems of water quantity and quality, which had historically been dealt with by separate local agencies and even by separate professional and research groups.

The need for merging the planning and management of water quantity and quality led the Government in late 1971 to propose creation of ten new, multipurpose regional water authorities to replace the 29 existing river authorities.[42] The new authorities will assume all of the water conservation, water quality control, recreation, and navigation responsibilities of the existing authorities as well as the water supply and sewage disposal responsibilities previously performed by local governments, and re-

40. *Report of a River Pollution Survey in England and Wales, 1970,* Department of the Environment/The Welsh Office, H.M.S.O., London, 1971.

41. Water Resources Board, *Trent Research Program,* Interim Report, Reading, 1970.

42. "Reorganization of Water and Sewage Services, Government Proposals and Arrangements for Consultation," Department of the Environment, December 2, 1971.

sponsibilities for canals and river navigation previously discharged by the central government. In effect, these ten agencies will be comprehensive river basin management organizations with authority to develop new resources and to regulate water abstractions and emissions.

In terms of our model, British water quality management has been characterized by a concern for costs. Although effluent standards (the Royal Commission norms) have constituted the principal management strategy, in practice cost has been an important element in the design and timing of investments to meet the standards. Until recently, as in the Thames and Trent River studies, production function relationships between resource inputs in treatment plants and ambient stream quality have not been investigated in detail. In the past 20 years, only two national surveys of in-stream quality were conducted, one in 1958 and the second in 1970. As stated above, no national in-stream quality standards are imposed.

The Trent studies and the reorganization of river authorities point to probable adoption by Britain of some elements of the Kneese water quality management approach, with concern for developing relationships between effluents and in-stream quality, systematic search for least-cost means for achieving higher levels of stream quality, and possible use of charges to control effluents and finance needed investments. Emphasis on the Royal Commission effluent standards will decrease, as other management alternatives such as river purification lakes and timing of releases of effluents will receive increasing attention. Benefit measurement, especially of recreation and fisheries, will play a modest role. Most important, each river authority will have rather complete control of planning and management of water quantity and quality, subject only to broad national policy and financial constraints.

The objective function of a typical river authority will probably be of the form

$$\min \sum_{t=1}^{T} \frac{M_t x_t + K_t x_t}{(1 + r)^t},$$

subject to a set of local ambient water quality standards

$$Q_i^s \geq Q_i^{s*}$$

The set of Q_i^{s*} would not be fixed a priori, but would be selected parametrically; that is, the cost implications of moving from one level of Q_i^{s*} to another would be investigated.

Perhaps a set of effluent standards, based on the Royal Commission model,

$$Q_i^E \geq Q_i^{E*},$$

would also be present at least for a time.

Conclusion

In the United States, practice in air and water quality management has lagged far behind the developed theory of regional management on an airshed or watershed basis. Although some attention is given to regional management for achieving least-cost, or greatest net benefit solutions, U.S. policy has been moving toward adoption of uniform minimum national standards for both ambient air and water quality and for control of effluents, with inadequate regard for the cost-effectiveness of these approaches. In contrast, Britain has operated its air and water quality management programs with little concern for the theory of regional quality management, but with concentration on levels of effluent control that can be obtained at reasonable cost. Britain now appears ready to move toward the regional water quality management approach, and to supplement its reasonable cost of treatment strategy with one of finding least-cost means of meeting various levels of ambient stream standards.

One can summarize by noting that the United States is long on the theory of optimization and short on its practice; while the British, somewhat skeptical of optimization approaches that involve large investments of research and information, are more likely in the short-run to develop a reasonable application of the theory, at least in the water resource field.

COMMENT

A. Myrick Freeman, III, Bowdoin College

In this paper Professor Hufschmidt has examined the environmental management policies of the United States and the United Kingdom from the point of view of a public investment analyst and has attempted to infer something about the nature of the implied objective functions from the laws and other public documents. The paper includes some very useful descriptive material on the present shape of air and water pollution control policies in both countries. I found these descriptions and the comparisons of basic approaches to be very informative. But I am uneasy about those parts of the paper that include statements about the form of the objective function and the implied weights or relative values placed on economic efficiency and environmental quality. My skepticism is grounded in a doubt that the political system articulates consistent weights or values.

First I will discuss the concept of an objective function and two kinds of uses to which it can be put. Then I will outline a model of the political system and use it to interpret the political response to the environmental problem. This model and the interpretation of U.S. pollution control policy are offered as alternatives to Hufschmidt's analysis of the situation. Finally I will turn to some specific observations concerning Hufschmidt's interpretation of the National Environmental Policy Act of 1969.

Basically the role of an objective function is to reduce unlike or noncommensurable objects or states to some common measure such as dollars, utility, or social welfare. This is done through the use of weights or valuations to transform noncommensurate values into commensurate or additive values. The resulting summation is an index of the degree to which a given project or program achieves the goals embodied in or expressed by the weights.

When they are available and when certain conditions are met, we can use market prices as the weights. An objective function using market prices or market-related shadow prices is said to be based on an efficiency criterion. But for some kinds of outputs or states, such as environmental quality and income distribution, there may not be any prices or weights revealed by the market for making these outputs commensurate with other elements in the objective function. We then have what is sometimes called the "multiple objective problem." The multiple objectives referred to are those other objectives or criteria in addition to economic efficiency. The problem arises from the absence of market determined weights or values for converting the different objectives into a common rubric. The multiple objective problem is a problem of valuation.[1] Hufschmidt's paper is about the value or weight attached to environmental quality.

It is appropriate to distinguish between two kinds of analysis which make use of the concept of an objective function. The first I would characterize as *normative analysis* or policy design. One engaged in this form

1. See A. Myrick Freeman, III, "Project Design and Evaluation with Multiple Objectives," *The Analysis and Evaluation of Public Expenditures: The PPB System—A Compendium of Papers* (Washington, D.C., U. S. Government Printing Office, 1969), pp. 565–576.

of analysis takes an objective function as given or postulates it, and follows the function through to its logical conclusions. The aim is to rank or compare different projects or programs.

Of course a basic question here is where the analyst gets the objective function and what elements and weights should go into it? This has sparked a debate whether economists should use an objective function limited to economic effciency (i.e., those outputs to which prices can be attached) or whether it should include income distribution and now environmental quality as arguments.

I will not pursue this question. This is not Hufschmidt's topic. But I want to observe that this debate has induced some economists to undertake a different kind of analysis, something which I would characterize as the *positive analysis of objective functions*. By this I mean the formulation and testing of hypotheses about the elements of and weights of the objective functions actually used by policy makers.[2] Or to put it differently, this work represents an attempt to infer a revealed objective function from the observed choices or actions taken by policy makers.

The logic behind the positive analysis of objective functions is clear. If the terms at which one output (e.g., income redistribution) can be exchanged for another (e.g., efficiency) are known, that is if the marginal rate of transformation among outputs is known; if the actual position on the MRT surface chosen by policy makers is known; and if, in rough terms, rationality or consistency in public choice is assumed, then the implied marginal rate of substitution or valuation of one output in terms of the other can be readily calculated.[3] It is the assumption of rationality which is most troublesome.

I see Hufschmidt's paper as a somewhat informal kind of positive analysis of the relative weights given to environmental values and economic efficiency values as revealed by recent policy decisions. It is this aspect of the paper which is most troublesome to me. And the basis of my apprehension is the kind of political model which I believe underlies Hufschmidt's analysis.

2. See, for example, Robert H. Haveman, *Water Resources Investment and the Public Interest* (Nashville: Vanderbilt University Press, 1965); Burton Weisbrod, "Income Redistribution Effects and Benefit Cost Analysis," in Samuel B. Chase, editor, *Problems in Public Expenditure Analysis* (Washington, D.C.: The Brookings Institution, 1968); A. Myrick Freeman, III, "Income Distribution and Social Choice: A Pragmatic Approach," *Public Choice* VII (Fall 1969): 3–21; and K. Mera, "Experimental Determination of Relative Marginal Utilities," *Quarterly Journal of Economics* 83, No. 3 (August 1969).

3. See Freeman, "Income Distribution and Social Choice."

One can only impute some normative significance for policy analysis to revealed choices if the assumptions described above are satisfied. And if one wishes to pursue this line of inquiry, it is appropriate to ask, "What is your model of or theory of the political process which encourages you to assume these conditions are met?" At issue is whether Congress and the agencies have made or ever can make choices about trade-offs which are internally consistent and in accord with generally accepted value judgments or criteria.

One of the strongest advocates of this type of analysis is Arthur Maass.[4] Maass' discussions emphasize "community-oriented choices" which reflect the general interests of society as a whole. Maass argues that decision makers give different answers to questions depending on the way in which the question is framed and the environment in which the decision maker is called upon to make his choices. Also he argues that we can and should create government institutions for the "purpose of framing the question so as to elicit the *right,* or in our case, community-oriented, response." [5]

Maass' rather informal model is quite different from and at odds with the political models of Downs, McKean, Wildavsky, and Margolis. The latter are micromodels of the political process. They assume that individuals acting in the political process are utility maximizers who assess the gains and costs to themselves of the choices they make. Some of these models emphasize the fact that the gains and costs as perceived by decision-makers may be quite different from the social benefits and costs of these decisions. These models cast doubt on the likelihood that public choices will be either internally consistent over time and across programs or in conformity with generally acceptable value judgments. For example, what kind of objective function can be inferred from public policy decisions concerning the Aid to Dependent Children program, depletion allowances, and the Farm Price Support Program.

I don't propose that the question of which kind of model applies (if either) has been settled, or that much progress has been made. We need a far better understanding of the political decision-making process before we offer interpretations about the results of this process in terms of social welfare or objective functions. We also should be cautious about giving officials more advice about how to incorporate environmental quality or

4. See Arthur Maass, "Benefit-Cost Analysis: Its Relevance to Public Investment Decisions," *Quarterly Journal of Economics* 79, No. 2 (May 1966): 208–226; and Arthur Maass, "Public Investment Planning in the United States: Analysis and Critique," *Public Policy* 18, No. 2 (Winter 1970): 211–243.

5. Maass, "Public Investment Planning," pp. 236–238.

income distribution or regional development into their multiple-objective decision-making processes. Our analytical techniques are way ahead of our understanding of how these techniques are used and perhaps perverted in the political process, or even whether they will be used.

Hufschmidt does not provide us with a model of the political process to help in interpreting the policy action he studies. But I would argue that the implicit model underlying Hufschmidt's paper is more like that of Maass than those of Downs, McKean, et. al. In this section I will briefly sketch out an alternative micromodel of the political process and use it to interpret recent U.S. environmental policies with respect to air and water pollution control. This model focuses attention on both those policy makers who are responsible for setting and carrying out pollution control policies and the places in the political structure where decisions on policy are made.[6]

One of the functions of a political system is to reconcile or resolve conflicting interests. Many of these interests are economic in nature. With respect to the pollution issue, the conflict is over how much cleaning up of pollution is actually going to be done, and who is going to pay for it. An important element in understanding the politics of pollution is to see whether a given policy step resolves this conflict or attempts to obscure the conflict or postpone its effective resolution.

The political actors include the elected legislators and administrators as well as the appointed bureaucrats who are charged with carrying out or enforcing pollution control policies. Each of the political actors has a constituency, i.e., a group whose diverse views the actor must take into account when making decisions, because the constituency controls or at least affects his political future. The constituencies of the elected officials include all eligible voters; but in some cases, for example congressional committee chairmen, they include other elements as well. The constituencies of the bureaucrats include members of the legislative body, especially committee chairmen, senior bureaucrats and elected officials who control appointments and programs, and the specific groups affected by the policies or programs under the supervision of the bureaucrats.

Political actors may be thought of as maximizing a utility function which, in the case of elected officials, would depend upon electoral support. For bureaucrats, arguments in the utility function could include prestige, probability of promotion, budget and program size, and probability of moving to lucrative private employment at a later date.

6. See also A. Myrick Freeman, III and Robert H. Haveman, "Clean Rhetoric and Dirty Water," *The Public Interest* No. 28 (Summer 1972): 63–65.

If an elected official chooses a policy which takes something away from a group, he may lose the support of that group. He would be more likely to choose such a policy if the losers were small in number and/or ability to influence other elements of his support. In general political actors have incentives to find policies which hide the costs or shift the costs to less influential elements of their constituencies, to postpone decisions (every decision has a cost), and to avoid the costs of the decision by shifting the responsibility for making it to another place in the political system.

An important element of the model is the locus of the actual decision, i.e., where in the political structure the final decision which resolves the conflict and distributes the benefits and costs is actually made. In examining the locus of decision making, two criteria are relevant: accountability and accessibility. Ideally political actors should be accountable to all groups affected by their decisions; and all affected groups should have access to the locus of decision. However, such perfection does not characterize our political system. Institutional arrangements such as the congressional seniority system and loose campaign financial regulations that enable economic power to be transformed into political power undermine both accountability and accessibility. When such structural imperfections persist, political actors gain leeway to avoid those decisions likely to alienate major sources of support.

At least in part on the basis of this model, I am led to a quite different interpretation of the U.S.'s political response to the environmental problem. The principal feature of federal legislation dealing with air and water pollution is that responsibility for implementing pollution control plans is assigned to the states. Each state is required to set air and water quality standards and to develop and implement programs for attaining these standards. For the most part states have chosen the enforcement-regulation approach based on the issuance of permits to individual dischargers and backed up by case-by-case legal enforcement of detected violations. I want to emphasize three aspects of the federal policy.

1. Federal legislators have consistently shifted the burden for making difficult decisions to the states (for example the setting of standards, or the issuing of discharge permits), or to decision makers within the federal bureaucracy (EPA).[7] Furthermore the guidelines accompanying

7. The shift away from reliance on states under a federal framework and toward federally imposed standards which are the principal features of the Clean Air Act of 1970 and the Clean Water Amendments of 1972 is not inconsistent with the model

these delegations of authority have been vague and poorly defined leaving considerable discretion to the ultimate decision makers and expanding the scope for negotiation and bargaining on a case-by-case basis. As a result, those who make the real decisions about who cleans up how much are typically both less accountable and less accessible than the legislators who pass the basic legislation. Federal legislation typically passes with large majorities, sometimes unanimously. One can only conclude—if everybody is for it they must have found some way to duck the real issue.

2. Setting environmental quality standards is a meaningless exercise unless effective mechanisms are developed for achieving the standards. Almost without exception, states have placed primary reliance on some form of licensing accompanied by judicial enforcement of the license terms. The political aspects of this procedure are quite unfavorable to effective pollution control.[8] The adoption of a licensing system does not resolve the political conflicts inherent in pollution policy. Instead, there is a continuing political conflict over license applications, terms and limits on discharges, and enforcement. Furthermore, these battles are fought on terms that are advantageous to dischargers for several reasons. The political issues are removed from the legislative arena and placed in a bureaucratic one where accountability and accessibility are less. The choices are more likely to be framed in technical terms that make it difficult to articulate the public interest. Finally, each discharger has large incentives to devote resources and energy toward swinging the decision his way; the public interest is diffuse and because few people have the incentive or the command of resources, likely to be poorly represented.

3. Policy-makers have been willing to subsidize industrial dischargers wherever this could be hidden in tax-depreciation formulae or municipal cost-sharing programs.

In sum, the political system has responded to the emerging environmental problem by shifting the real decisions from the federal to the state

---

sketched out here. This shift can be viewed as a rational political response to public pressure to do something, or at least to give the appearance of doing something. And furthermore, it is not at all clear that this greater federal involvement represents a movement toward more effective policy. In fact developments as of early February 1973 suggest an imminent breakdown of the implementation of air quality programs in major urban areas such as New York City and Los Angeles.

8. See A. Myrick Freeman, III and Robert H. Haveman, "Residuals Charges for Pollution Control: A Policy Evaluation," *Science* 177, No. 4046 (July 1972): esp. 328–329.

level and from legislatures to bureaucratic agencies. It has tended to make decisions in arenas where there is less accountability and accessibility, and to avoid final resolutions of the political conflicts in favor of piecemeal, fragmented decisions. All of these tendencies work against the public interest in pollution control.

Now let me offer some comments on the inferences drawn by Hufschmidt concerning the National Environmental Policy Act (NEPA) of 1969. Hufschmidt interprets NEPA as raising environmental quality to the level of a national objective in a multiple-objective function. In a narrow formal sense this is true, because it is the stated purpose of the Act itself. But when the Act and the subsequent actions of governmental agencies are examined from the perspective of the political model previously outlined, it is not clear that there was any change in federal policy in an operational sense.

First, although Congress said that environmental values must also be weighted in the decision process, it placed the responsibility for weighing these factors and making the actual decisions in the hands of the executive branch.

Second, Congress gave no guidance as to what kind of analytical framework should be used or what weight should be placed on environmental considerations relative to other values in the decision process. The exact language reads, "Will insure that presently unquantified environmental amenities and values may be given appropriate consideration in decision making along with economic and technical considerations." But what meaning can be given to "appropriate" in this context?

Finally the agencies to whom the Act is directed have responded in most part only in a pro forma way. In general the agencies have taken steps to meet the technical requirements of the Act, i.e., the drafting of impact statements, but even this has required the prodding of the courts. But it is difficult to find examples of agencies voluntarily making decisions which are different than they would have been in the absence of the Act. Rather there is some justification for the charge that impact statements have become vehicles for justifying or rationalizing politically predetermined decisions (the Alaskan pipeline is a case in point). In fact one could argue that the agencies, having been handed a hot potato by Congress, have deftly passed it on to the federal judiciary. Whatever impact NEPA has had on decision making and environmental policy has come about through court orders and injunctions rather than through prompt and bold agency response to a clear statement of national policy objectives.

Index

Abatement. *See* Residuals management and control
AFL-CIO (American Federation of Labor-Congress of Industrial Organizations), 374, 382, 383
Air pollution. *See* Air quality
Air pollution, acute episodes of: in Donora, Pa., 325, 327; in London, 325, 327, 444; in Los Angeles County, 329; in Meuse Valley (Belgium), 325; morbidity during, 326–327; mortality during, 325, 326–328; in New York City, 326–327, 328. *See also* Atmospheric inversions; Health effects of air pollution
Airport and Airway Development Act of 1970, 367
Airports: in Chicago, 361; factors affecting location of, 360–394; in London, 390–392, 394 (*see also* Roskill Commission); in Minneapolis-St. Paul area, 352–394 (*see also* Joint Committee; Metropolitan Airports Commission; Metropolitan Council; Wold-Chamberlain Field); in New York, 390
Airport Zoning Act, 370
Air quality: atmospheric factors affecting, 181–183, 213, 328, 330, 333–343 passim, 348; benefits of improving, 137–141; control of, 3, 118–119, 123, 240–242, 244–245, 248, 256, 266, 442–447; effect of, on health, 4, 100, 108n, 139–141, 325–350, 442, 443, 445, 446; effect of, on land values, 141; programs and policies to protect, 442–447; regional differences in, 131–132, 160, 171–172, 442–443, 444; residuals affecting, 123, 181–183, 202–212, 262–263, 325–345 passim, 442. *See also* Atmospheric dispersion models; Atmospheric inversion; Environmental quality; Health effects of air pollution; Residual materials; Residuals management and control
Air Quality Implementation Planning Program (U.S.), 182
Air Transportation Association, 359
Alkali Act (U.K.), 445
Almon technique, 341, 342
Aluminum Association, 124
American Iron and Steel Institute, 124
Anderson, R., 141
Andrews, R. N. L., 438n
Appropriation. *See* Ownership
Army Corps of Engineers, 122
Arrow, K., 29, 44, 66, 72, 94
Assimilation of wastes, 31, 99, 100, 101
Atmospheric dispersion models, 181–183, 214–215, 230, 235, 443
Atmospheric inversions, 128, 328, 330
Availability: vs. appropriation, 77–79; forced, 81–82, 86, 87; and production, 79–82
Aynsley, E., 312n
Ayres, R. U., 243n, 278n

Baumol, W., 3, 93, 101n, 102, 128–131 passim, 161, 166, 397n, 402n, 417n, 421n, 429n, 430n, 431
Becker, G., 137, 164
Becker, W. H., 327
Benefits: diminishing marginal, 114; of environmental control policies, 113–114, 118–119, 134–135, 136–141, 146–150, 156–157; of increase in stock of social capital, 20; methods of estimation of, 133–134, 156, 166; social compared with private, 71; sociographic distribution of, 172. *See also* Cost-benefit analysis
Benefit-cost analysis. *See* Cost-benefit analysis
Benstock, M., 121n
Black, R. J., 208n
Bohm, P., 133
Borgelt, M., 374, 375, 379, 383
Bowen-Samuelson requirement, 25

Bower, B. T., 4, 102n, 105, 110, 116, 117, 121, 122, 171n, 175n, 199n, 234n, 276n, 277n, 278n, 279n, 303n, 319, 322, 323n, 437n
Boyd, J. H., 171n
Bradford, D., 402n
Bradley, W., 326, 327
Bramhill, D., 105
Brooks, Mrs. G., 366, 381
Brown, G., 133n
Buchanon, J., 133

Cahill, W. J., Jr., 312n
Caldwell, L. K., 435n
Carlos Avery Wildlife Management Area (Minneapolis-St. Paul), 360, 361, 362, 368, 370, 378, 379
Capital, private: allocation of, 17, 48; compared with social overhead capital, 9–10, 11; demand for, derived from environmental services, 48, 103, 245, 248; investment in, 17–18, 45, 46; relationship of, to output, 11–13; stock of, 14, 15, 17–22, 31–32, 46
Capital, social overhead: allocation of, 16, 17; compared with private capital, 9–10, 11; compared with pure public goods, 10–11; cost of services of, 14; definition of, 9, 16–17; depreciation of, 20; investment in, 17–22; management of, 9–10, 11; neutrality of, 12, 13, 14, 15; relationship of, to output, 11–13; social product of, 15; stock of, 14, 15, 17–22
Centralized and decentralized economies, 43, 76
Chicago, Illinois: airport in, 361; air quality in, 131–132, 158, 333, 334, 336–343, 344, 345, 349
Choguill, C., 239n
Christy, F. T., Jr., 398
Chrysler Corporation, 125
Ciriacy-Wantrup, S. U., 133
Citizens League (Minneapolis), 366
Clarke, E., 133, 159
Clean Air Act: U.K. (1956, 1968), 444, 445, 446; U.S. (1970), 442
Climatological conditions: effect on air quality of, 99, 128–129, 165–166, 182; and health effects of air pollution, 326–344 passim, 348. *See also* Exogenous environmental factors
Coase, R. H., 2, 78, 79, 96–97, 132, 160, 400, 411, 418
Coca-Cola Midwest, Inc., 124
Common property, natural resources of, 398, 399. *See also* Public goods
Companion, M., 374, 383–384
Compensation for external effects, 157, 161; and adjustment of trade patterns, 402–404; agencies for, 387–388, 390, 394; and asymmetry in gains and losses, 406–407; forms of, 76, 387, 390, 408, 412, 428, 433; transnational systems for, 399, 400, 421. *See also* Negotiation of payments for externalities
Computer costs and capability: as limitation on model size, 174, 176, 178, 193, 213, 230–233; and nonlinear models, 190–191, 231–232
Conflicting interests, resolution of, 384–389, 390–391, 457, 459–460
Congestion: in use of social overhead capital, 11, 12, 13, 24; increasing costs of, 98–99
Constraints on functional relationships: air quality, 446; budget, 23, 51, 74, 137, 408, 409; capital stock, 31; in decomposed models, 179–180; in environmental models, 183–187, 215–216; environmental quality, 32, 192, 236–237, 277, 393, 437, 443; waste treatment, 29–30; water quality, 448–449, 453–454
Consumers, gross income of, 51, 52. *See also* Households
Consumption: as constrained by ownership, 75; definition of, 74; vs. ownership, 74–75; effect of, on utility levels, 13, 23, 74
Continuous Air Monitoring Program, 331
Control Data Corporation, 362
Cootner, P. H., 207n, 276
Cost-benefit analysis: application of, to airport location problems, 389, 391–393, 394; application of, to comparison of environmental policies, 142–157, 164, 436–437; application of, to households, 135–136, 137–141; application of, to

political process, 456; need for, 3, 93, 133–134; use of, in formulation of government policies, 441–448 passim, 451, 452–453. *See also* Cost-effectiveness models
Cost conditions: availability of data on, 284; effect of effluent charges on, 119; effect of emission regulation on, 144–145, 164, 306–308, 312–316, 317, 318; effect of externalities on, 134–135, 136–137, 164, 404–405; effect of residuals management practices on, 277, 284, 285, 286–288, 300–301, 302–305, 306–308, 312–316, 319; effect of, on size of industry, 60; of firms vs. industry, 319, 322–324; increasing, of emission regulation, 107, 154, 164, 262, 263, 268, 297, 299; increasing, of environmental damage, 98–99; of new vs. existing plant, 318, 319, 322–323, 324
Cost-effectiveness models, 241, 243, 266, 267, 272, 274, 315
Costs, administrative: of direct regulation, 105–106; of distributing social services, 16, 103; of effluent charges, 103; of mixed policy programs, 108, 245; of varying emission levels, 160. *See also* Costs, transaction
Costs, social: divergence of, from private costs, 71, 78; effect of random factors on, 115; effect of substitution and firm scale on, 153–154; internalization of, 67, 88, 95, 96, 160–161; of pollution control activities, 101, 102, 113–114, 141–142, 144–147, 151–155, 240–243, 244–248, 264; sociographic distribution of, 172; of use of social capital, 14, 15, 16, 17, 20, 22. *See also* Costs, transaction; External effects
Costs, transaction: of government policies, 3, 122; and institutional changes, 72–73, 88; of private negotiation, 96, 414, 416, 432–433; as reason for market failure, 3, 94; and supply of public goods, 131; and transfrontier externalities, 417–419
Council of Economic Advisers, 438
Council of Environmental Quality, 317, 438
Credit instruments, market for, 50, 51, 52
Crocker, T., 141

Dales, J. H., 76, 133
Damage functions: definition of, 99–100; flow vs. stock, 98, 101–102; forms of, 98–99
Damages: legal, 71; transnational, 399–400, 406–430
d'Arge, R., 4, 427–434
Davis, O., 1
Delaware River Basin Commission, 209, 212n, 213
Delaware River Estuary, 119, 200, 448, 451; characteristics of, 212, 213; residuals inputs of, 209–212
Delaware Valley region: atmospheric quality of, 213; geographical description of, 174, 199–202; sources of residuals in, 178, 200, 202–212
Delson, J. K., 207n, 277n
Demsetz, H., 67, 72, 94
Denver, Colorado, 331, 333, 334, 336, 349
Department of the Environment (U.K.), 391, 440–441
Depletability: definition of, 82–83; lack of, 86; of private goods, 87, 91; of public goods, 92
Direct regulation: combined with effluent charges, 113–115, 128; compared with discharge fees or taxes, 3, 102, 103, 106, 119–120, 125, 130–131, 238; compared with output restriction, 142, 150–151, 164–165; compared with subsidies, 105n, 121; costs of, 97n, 105–106, 114–115, 119, 122, 123, 125, 144–154; enforcement of, 106, 122; flexibility of, 108, 126, 129, 159–160; forms of, 61, 97–98, 121–123, 134, 144–150; optimal use of, 114–115; success of, 122–123, 125
Discharge fees: adjustment to, 111–112, 197–208; combined with environmental quality standards, 107–108, 113–115; compared with direct controls, 3, 102, 103, 106, 119–120, 125, 130–131, 238; compared with other taxes, 27, 54, 128; compared with subsidies, 105; to control air quality, 118–119; to control water quality, 116–118, 119–120; costs and

benefits of, 102, 103, 104, 113; effectiveness of, 116–121; effect of changes in, 48–49, 53–54, 62–63, 297, 299; effect of, on allocation of resources, 27, 48, 237; effect of, on price levels, 53, 62–63; elasticity of emissions with respect to, 117–118; and environmental quality standards, 36, 49, 54, 61–62; as imputed prices, 35, 49, 266; optimal rate of, 114–115, 166; as policy tool, 2, 3, 27, 36, 95; revenues from, 61, 121; shifting of, 104, 129; use of, in regional models, 174–175; variations in, 104, 128–129, 159, 165–166
Discount rate, on social investment, 22
Disposability, 76, 80
Dolbear, F. T., Jr., 408
Dominance principle, 68, 71, 83, 84–85, 88, 91
Donora, Pennsylvania, 325
Dorr, J. L., 372, 379
Dreze, J. H., 89
Dunne, D., 372, 379, 383, 384

Eckhardt, R. B., 328
Effluents. *See* Residual materials
Effluent charges. *See* Discharge fees
Effluent standards. *See* Environmental quality standards
Electric energy: fuel used in production of, 253–254; reduction in consumption of, 118; residuals in production of, 118–119, 262. *See also* Industry studies: coal-electric energy, electric utility
Elliot, R., 117, 118n
Ellsaesser, H. W., 330
Emissions. *See* Residual materials
Emission controls. *See* Direct regulation; Residuals management and control
Emission rights, sale of, 159, 165
Emission standards. *See* Environmental quality standards
Emission taxes. *See* Discharge fees
Employment: effect of environmental policies on, 44, 54; in imperfect labor markets, 52–53. *See also* Labor; Wage rates
Environmental impact statements, 367, 438, 439, 460
Environmental policy, in U.K.: on air quality, 444–447; broadening of, 440–442, 453; compared with U.S., 440–441, 447, 450; of Labour and Conservative governments, 440; on water quality, 450–453
Environmental policy in U.S.: on air quality, 442–444; broadening of, 437–439, 453, 460; compared with U.K., 440–441, 447, 450; state implementation of, 458–460; on water quality, 447–450
Environmental Protection Agency, 119, 154, 178n, 181, 207, 208n, 209, 219n, 317, 331, 438, 439, 443, 444, 449
Environmental quality: and assimilation of wastes, 31; imputed prices for, 34–35, 454; measurement of, 3, 30–31, 57; national policies concerning, 435, 438–460; regional differences in, 131–132, 158–159, 160–162, 163, 444, 445–446, 448; shifts in, 128–129; and utility levels, 32, 63; variation in, over time, 160, 163. *See also* Environmental quality standards
Environmental Quality Act of 1969, 367
Environmental Quality Standards: for air, 239, 242, 269; combined with discharge fees, 36, 49, 54, 61–62, 108, 119–120, 238; costs of, 144–146, 312–316, 317–318; effect of changes in, 55, 146, 147, 304–306, 319, 322, 323; effect of exogenous factors on, 113; maximum vs. minimum, 129; methods of achieving, 102, 241, 459; as policy tool, 28, 54–55, 101–102, 106–107, 129, 237, 318, 437, 450, 452–453; variation in, 129–130, 134, 154–155, 159–160, 163, 165, 306, 308n; for water, 116–117, 119–120, 447–448
Ethridge, D. E., 117, 118n
Euler-Hamilton equations, 37
Euler-LaGrange conditions, 20, 31
Exclusion principle: applied to social institutions, 67, 72, 82; and liability, 92, 94; and ownership, 75, 77, 79, 85–86, 88
Exogenous environmental factors: effect of, on atmospheric conditions, 112, 113, 182; effect of, on social costs, 115; and flexibility in controls, 108, 128–129, 134, 158–160, 163, 165–166; health effects of, 326–344 passim, 348; modification of, 235; uncertainty caused by, 99, 100, 103,

129. *See also* Climatological conditions
External effects: caused by pollution, 2–3, 23–24, 398; definition of, 23, 25, 65, 71, 73, 91, 92, 93, 94, 397–398; and forced ownership, 78–79, 92; as function of private consumption, 23–25, 94, 401, 407–408, 431; as function of production, 24–25, 94, 401; institutional correction of, 87, 91, 92, 93, 398; internalization of, 67, 88, 95, 96, 160–161, 398, 399; market regulation of, 96, 398; national vs. transnational, 428–429; and nonrejectability, 77, 81, 92, 94; nonuniformity in, 159–160; and perfect selectivity, 80–81, 92; real vs. monetary, 94; and size of market, 92–94; substitution between, 157; taxes or bribes to correct, 76, 408, 412, 428, 433; transfrontier, 397–434. *See also* Compensation for external effects; Costs, social

Factor mobility, international, 402n, 419–425, 426–427, 431
Federal Aviation Administration, 359
Federal Water Pollution Control Administration, 171n, 200n
Feedback effects, of residuals management activities, 254–255, 259, 262, 266–267, 271, 272, 273–274, 310, 316–317
Fiacco, A. V., 193n, 195n
Fisher, C. E., 117
Foley, D., 89
Ford, D., 267n
Frankel, R., 207n, 277n
Freeman, A. M., 454n, 455n, 457n, 459n
Free rider problem, 67, 96, 143, 159, 426, 433
Fuels: household use of, 245, 249, 445–446; reduction in use of, 144, 146, 148–150, 153, 164–165; substitution in type of, 118, 123, 153, 155, 245, 256, 294, 445–446

Games: application of, to transnational negotiation, 412–415, 416, 432; dominance principle in, 68, 71, 83, 84–85, 88, 91; as form of economic model, 91, 94; institutional structure and rules of, 67–69, 71, 88; Nash type, 69, 412, 413
Gaussian plume formation, 182
Gauss-Seidel method, 188
Gaver, K. M., 341
Geisel, M. S., 325n
General Accounting Office, 120
General Motors, 124–125
Glasser, M., 327, 328, 330, 334, 340
Glass Manufacturers Institute, 124
Global environment, problems of, 4, 401
Glumack, R. G., 373n
Goldsmith, J. R., 329–330
Government. *See* Institutional arrangements; Legal institutions
Greenburg, L., 326, 330, 333, 340
Griffin, J. M., 118–119, 324n

Haavelmo, T. A., 44
Hadley, G. H., 179n
Haefele, E. T., 172, 191n, 351n
Hall, L. M., 372n
Hamilton equations, 32, 33
Hamm (Theodore) Brewing Co., 124
Harned, J., 119n, 120
Harper, D. V., 354n, 355n, 359n, 360n, 361n, 362n, 376n, 377n
Harrod neutrality, 43
Haveman, R., 89n, 455n, 457n, 459n
Health effects of air pollution, 4, 100, 108n, 139–141, 442, 443–444, 445, 446. *See also* Morbidity associated with air pollution; Mortality associated with air pollution
Hearon, W. M., 277n, 303n
Herfindahl, O., 435n
Hetland, J. L., Jr., 372n, 385, 388
Hexter, A. C., 329–330
Highland Park, Minnesota, Community Council, 382
Historical preservation, 392, 440
Hodgson, T. A., Jr., 328–329, 331, 344
Hoffman, Councilman, 375, 379, 383
Households: costs and benefits to, of environmental quality controls, 136–141, 164; labor input of, 245, 259; pollution caused by, 143, 207–208; waste management options for, 214, 236
Howe, C. W., 176n, 317n
Hufschmidt, M., 5, 436n, 453, 454, 455, 457, 460

Illinois Power Co., 254n

Imputed prices. *See* Shadow prices
Inada conditions, 28, 30
Income: changes in disposable, 53, 55; gross consumer, 51, 52
Income, distribution of: effect of environmental factors on, 9; effect of effluent charges on, 62; effect of transnational externalities on, 399; equalization of, 16; optimization of, by social agencies, 10
Income taxes. *See* Taxes, income
Institute of Industrial Research, 254n
Industrial models: compared to firm models, 319, 322, 324; condensation of, 177–178; inclusion of, in regional models, 177–178, 236, 242; size of, 176; types of, 175–176, 282, 323–324; uses of, 317–318, 319
Industry: aggregation of sectors of, 244–245, 248–249, 267–268; data on, 282–289, 319, 322; production residuals of, 27, 28, 46, 53, 57, 104n, 142, 202, 209–212, 248–249, 252–254, 276–285, 289–318, 319; size of, 60, 105; structure of, 66–67, 104, 106, 107n, 130–131. *See also* Cost conditions
Industry studies: canning, 285, 286; coal-electric energy, 282, 308–316, 322, 324n; electric utility, 118–119, 269–272; paper manufacturing, 282, 283, 285, 289–290, 291–293, 302–308; petroleum, 282, 283, 297–302; steel, 269–272, 290, 294–296; sugar beet, 117, 282, 319
Institutional arrangements: and elected vs. appointed officials, 376, 381–384, 388–389, 393; importance of, 23, 26, 384–385, 456; international, 399, 400–401, 421, 428, 430, 432; legal (*see* Legal institutions); legislative, 172, 232, 353, 358, 380, 389–394, 459, 460; need for changes in, 72, 73, 87, 92, 93, 94, 237–238, 458; and optimality, 66, 67, 71, 92; and property rights, 67, 72, 75; regional, 172, 232, 237, 276; single function vs. multifunction, 376–381, 384, 385, 386–387, 390, 393; structure of, 98, 176, 384, 385–386, 387–388, 390–391
Instruments of public policy. *See* Public policy tools
Interest rates, 35, 42, 43
International trade: to adjust transnational externalities, 431, 433, 434; effects of domestic environmental policies on, 4, 402–403
Investment: demand for, 44–45; increasing costs of, 18; optimization of, 19–22; in private and social capital, 17–22
Isard, W., 239
Izraeli, O., 141

Jackson, J., 4, 389, 393, 395
Johnson, E., 119, 448n
Johnson, R. N., 375
Joint Committee, on Minneapolis-St. Paul airport: composition of, 373, 381; functions of, 373; vote of, on airport site, 374–375, 383, 384
Jorvig, R., 367n
Judge, J. F., 377n

Kamien, M., 133
Kelly, R. A., 3, 181n, 234, 318n
Kissin, J., 239n
Kittle, R. W., 288
Kneese, A. V., 1, 102n, 105, 110, 116, 117, 119n, 120, 121, 123, 133, 171n, 199n, 206n, 234n, 243n, 276, 278n, 279n, 319, 397n, 435n, 437n, 452
Kohn, R., 3, 119n, 240n, 242n, 243n, 245n, 248n, 249n, 254n, 255n, 259n, 272, 273, 274
Krier, J., 121
Kuhn-Tucker theorem, 30, 37
Kurz, M., 29, 44

Labor: cost of, in abatement activities, 245; equilibrium in market for, 52–53; imperfections in market for, 27, 52–53; as input from household sector, 245, 259; short-run demand for, 45–46, 50, 54; supply of, 32, 42–43, 51, 63. *See also* Employment; Wage rates
LaGrange, Euler-: conditions, 20, 32, 58; function, 37, 137, 143; multipliers, 16, 20, 58–59
Large numbers, and negotiation of externalities, 96–97
Lave, L., 4, 100, 108n, 326, 347–350 passim
Ledbetter, J., 312n

Ledyard, J. O., 89n
Legal institutions: assessment of damages by, 76; international, 399, 400–401, 421; remedial actions of, 71–72, 121, 122–123; and rights of ownership, 75; role of, in environmental policy, 438, 439, 443, 445, 459, 460. *See also* Institutional arrangements; Liability
Leontief, W., 239, 240, 248, 262, 267n, 268, 272, 273, 274
LeVander, Governor (Minn.), 365
Levine, L., 382
Lewis, J. P., 392n
Liability: based on ownership, 75, 83–87, 91; effect on international transfers of wealth, 407, 411, 415; effect of, on international output adjustments, 420, 424–427; institutional arrangements to enforce, 71–72, 92, 121, 122–123, 409–410n; and international negotiations, 413, 418–419, 432; in international relations, 406, 407, 409–411, 415–416, 428, 433, 434; knowledge of, 86–87; lack of, 92, 399; types of, 87, 94, 409–411
Lichfield, N., 392n
Lindahl prices, 66, 89
Liu, B., 239n, 240n, 244, 248, 254n, 268, 273
Location: of airports, 351–394; of industry, 160–162, 284, 419–425, 426–427, 431
Löf, G. O. G., 117, 206n, 207n, 276, 277n, 303n, 319
London, England: air pollution in, 155, 325; airport location in, 390–392, 394 (*see also* Roskill Commission)
Los Angeles, California, air shed, 123, 329, 336n
Loucks, D. P., 176n

Maass, A., 456, 457
Maler, K. G., 2, 56, 58–63 passim, 122n
Mangasarian, O., 33
Marcus, M., 119
McCarroll, J., 326, 327
Market failure: attributes of, 91, 93; caused by transaction costs, 3, 94; correction of, 87, 91, 92. *See also* External effects
Marks, D. H., 176n
Martins, G. W., 375
Martin-Tikvart dispersion model, 182
Marx, W., 121n
Matheson, W. K., 366n
Maun, J. A., 372
May, R. M., 188
Medical services, cost of attributed to air pollution, 139–140
Mera, K., 455n
Metropolitan Airports Commission (Minneapolis-St. Paul): financial support of, 354–355, 365; functions and jurisdiction of, 353–354, 364, 381, 387; and Ham Lake airport site, 352, 360, 362–363, 364, 365, 368–369, 377–378; members of, 354, 356–357, 381–382, 388; and the Metropolitan Council, 359, 373, 381, 385; and Northwest Airlines, 364–365; staff of, 353, 354, 357–358, 376–377, 381, 386; support of second airport by, 365–366, 375–376, 377
Metropolitan Council (Minneapolis-St. Paul): Development Guide Committee of, 355, 373, 375; functions and jurisdiction of, 353, 355–356, 373; and Ham Lake airport site, 363–364; legislative nature of, 353, 358, 380, 381, 383, 384, 387, 388; members of, 353, 356, 358–359, 371, 383, 384, 388, 394; and the Metropolitan Airports Commission, 359, 373, 381, 385; and Northwest Airlines, 383; staff of, 386; support of single airport by, 367, 369, 372, 380
Meuse Valley, Belgium, 325
Meyer, R., 2–3, 90–94 passim
Miernyk, W. H., 239, 248
Mills, E. S., 1, 105, 239n
Minneapolis-St. Paul metropolitan area: alternative airport sites in, 353, 359–360, 362, 363, 367, 372, 373–374, 379–380; demand for airspace in, 359, 366; intercity rivalry in, 353, 379–380, 382; and relative representation on Metropolitan Council, 354, 371; Sewer Board of, 353; and tax support of airport, 365, 369, 372, 377, 382; Transit Authority of, 353, 380; union jurisdiction in, 374–375, 379, 382
Minnesota state legislature, 353, 373, 376

Minnesota, University of, 368
Mishan, E. J., 133, 392n
Money: changes in supply of, 50–51, 52; equilibrium in market for, 52
Monopoly: and direct regulation of polluting activities, 106, 130–131; and effectiveness of effluent charges, 104, 107n, 130–131; elimination of, by consumer contracts, 66–67; as public goods problem, 67
Moral suasion: limitations of, 124–125; use of, in emergency situations, 108–109, 123–124, 126
Morbidity associated with air pollution, 326, 327, 329n
Mortality associated with air pollution: adjusted for population size, 327, 330, 336; age groups affected by, 325, 326, 327, 328; effect of climatological factors on, 326, 327, 328, 330, 331, 333, 334, 336, 337, 340, 341, 342, 343, 344, 348; cross section vs. time series data on, 341, 344, 349–350; cycles in, 331–333, 334; daily rate of, 329, 330, 331, 334, 344; effect of specific pollutants on, 329, 330, 331, 334, 336, 337, 340, 343, 345, 347; and previous health problems, 325, 326–327, 328, 329, 343–344, 349; effect of interaction of pollutants on, 330, 337; other factors affecting, 328, 330, 332, 337, 340, 341, 348; time lag in, 326, 328–329, 330–331, 333, 340, 341, 344, 345, 347. *See also* Air pollution, acute episodes of; Health effects of air pollution; Morbidity associated with air pollution
Mosher, J. C., 329–330
Multiplier, abatement, 240–241, 242–243, 256–264, 266, 272

National Environmental Policy Act (1969), 438, 441, 454, 460
National Safety Council, 328
National Science Foundation, 1
National sovereignty: compared with individual rights, 428; definition of, 398–399; and payment of environmental damages, 400, 427
National Water Commission, 176n
National Water Resources Board, 451
Negotiation, of payments for transnational externalities, 400, 404, 406, 410–411, 415, 428, 429, 430; effect of liability on, 413–414, 415, 416, 418–419; effect of transaction costs on, 418–419; effect of uncertainty on, 415, 419; efficiency effect of, 430–431, 432, 433; incentives to undertake, 412, 413, 414–415, 416; involvement of intranational groups in, 433
New Jersey, Delaware Valley region of, 174, 200
New York City, 123–124, 326–327, 328, 330, 336n, 349, 390
New York region, industrial residuals in, 276
New York Times, 123–124
Noll, R., 107n
Nonintervention, policy of, 131–132, 143–144
Nonuniformity of environmental conditions: and sale of emission rights, 159, 165; and uniform standards, 129–130, 134, 154–155, 158–160, 163, 306, 308n; and zoning system, 159, 165
Norsworthy, J. R., 239n, 248n
Northwest Airlines: and airport financing, 365, 372–373; as major employer, 353, 361, 364, 381, 383; and opposition to additional airport, 365, 366, 372, 380; and opposition to Ham Lake airport site, 361, 364, 365, 366, 380
Nourse, H. O., 239n
Nyrob, D., 361n, 365

Oates, W., 3, 93, 101n, 102, 128–131, 166
O'Connor, D. J., 187n
Odum, H. T., 185n, 186
Olson, S., 366n
Optimization: global, 420–424, 427, 430; of growth, 32–36; of investment rates for private and social capital, 17–22; Paretian, 59, 61, 66, 71, 431; of production changes and waste treatment, 27, 233–234, 279; of public and private goods, 25; in regional models, 174–175, 179, 193, 195–198, 215–216, 225–229, 232; technological, 57, 58; and use of environmental submodels, 183, 187, 232
Orchard-Hayes, W., 177n

Organization for Economic Cooperation and Development (OECD), 407, 409–410n
Output, industrial: available vs. appropriated quantities of, 78–79; changes in type of, 318; control of availability of, 79, 82; equilibrium in market for, 51; restriction of, to control emissions, 27, 150–151, 153, 236; short-run supply of, 46, 47
Ownership: and appropriation, 74, 75; as basis of damage payments, 77; of common natural resources, 398, 400, 427; as constraint on consumption, 75; distinguished from availability, 78–79; distinguished from consumption, 74–75, 76–77; and exclusion principle, 75, 77, 85–86; forced, 78–79, 80–81, 82, 86, 87, 91, 92, 428; legal-social determination of, 77; rights stemming from, 75

Pareto optimality, 25, 59, 61, 66, 71, 411, 412
Pasquill point source, 182
Patinkin monetary model, 49, 63
Pearce, D., 441n, 442n
Penalty functions, 175; parameters of, 217, 225; use of, to eliminate constraints, 192, 193, 194–195, 214–216, 236
Pennock, G., 374–375, 383
Pennsylvania, Delaware Valley region of, 174, 199–200
Penrose, E. T., 18–19
Phelps, E. B., 31, 184, 192, 213
Philadelphia, Pennsylvania, 178n, 207, 208, 331, 333
Phoenix, Arizona, 124
Pigou, A. C., 2, 160
Pigovian taxes, 421–422, 423
Pittsburgh, Pennsylvania, 123, 131
Plott, C., 2–3, 90–94
Plourde, C. G., 98n
Policy tools: changes in, 144; and exogenous factors, 100–101; mixture of, 95, 98, 104–105, 107, 111–115, 125, 126, 128; relative costs of, 119–120, 142; relative effectiveness of, 102, 115–121; and type of damage, 98; varieties of, 93, 95. *See also* Direct regulation; Discharge fees; Environmental quality standards; Moral suasion; Price incentives; Prohibition; Public production; Subsidies; Taxes
Political processes: decision-making function of, 456–460; and environmental issues, 4, 131, 351–352, 384–385, 454–457; utility concepts applied to, 456, 457–458
Pollutants. *See* Residual materials
Pollution control. *See* Direct regulation; Discharge fees; Moral suasion; Prohibition; Public production; Residuals management and control; Subsidies; Taxes
Pontyagin's maximum principle, 32, 33, 56, 58, 63
Price incentives, as tool of environmental policy, 95, 97, 102, 125. *See also* Discharge fees; Subsidies; Taxes
Prices: changes in, 45; determination of, in barter economy, 48–49, 62–63; determination of, in monetary economy, 27–28; effect on, of change in effluent charge, 53, 62–63; international, 402–404, 415, 419–420, 424–427, 431
Prices, imputed or shadow. *See* Shadow prices
Private goods: characteristics of, 23, 87–88, 91; and externalities, 81, 94
Production: adjustments in, 419–424; distinguished from distribution, 77–78; effect of externalities on, 401–405; as function of inputs to waste treatment, 29–30; as function of private and social capital, 11–13; residuals of, 27, 28, 29, 57, 142–143, 249, 276–277, 278–279. *See also* Output; Production methods
Production methods: cost of changing, 144–145; effect of, on production of residuals, 57, 142, 171–172, 234, 235, 278–279, 289–297, 307–308, 308–310
Product, social, 15, 25–26
Profits: and allocation of international resources, 419–427; effect of price incentives on, 105; effect of waste treatment inputs on, 60–61; maximization of, 35–36, 44–45, 59
Prohibition, of damaging activities, 101
Property rights, specification of, 68–69. *See also* Exclusion principle; Ownership
Public goods: consumption of, 65–66;

definition of, 10–11, 12, 23, 25, 26, 65–67, 81–82, 87, 91–92; optimal level of, 23
Public policy instruments. *See* Policy tools
Public production of waste treatment services, 97, 110

Quality of environment. *See* Environmental quality; Environmental quality standards

Ramsey theory of optimum growth, 17, 27
Random variables. *See* Exogenous environmental factors
Raphael, D. L., 239n, 255n
Recovery of residual materials: as additional rather than substitute production, 258, 273, 302; constraints on, 279; costs of, 278–279, 286–289, 295, 304, 306–308; feedback effects of, 259n, 263, 267; public response to, 124; technology of, 278–279; 280, 297, 304; value of material salvaged by, 245, 302–304, 306
Recycling. *See* Recovery of residual materials
Redheffer, R. M., 190n
Reed, Councilman, 375
Regional environmental models, 181–191; computer program for, 214–232; decomposition of, 178–180; vs. disaggregated research, 238; for residuals management, 191–236; size of, 174, 176, 180; types of, 171–173, 174–175; use of industry submodels in, 177–178, 318
Regional Plan Association (N.Y.), 276
Regional planning and development: economic and environmental objectives of, 439; as objective of water resources policy, 437, 448, 449–450; in the United Kingdom, 392, 440–441, 450–451
Reid, J., 159
Rejectability: definition of, 76–77; and externalities, 77, 81; lack of, 79, 82, 86, 87–88, 92
Residual materials: control of (*see* Residuals management and control); generated in consumption, 207–208; generated in control activities, 240–241, 242–243, 254–264, 266–267, 271; generated in production, 27, 28, 46, 53, 57, 104n, 142, 202, 209–212, 248–249, 252–254, 276–285, 289–318, 319; generation and discharge models for, 174–175, 202, 234, 276; relationships among, 171, 207–208, 233, 234, 235, 276–277, 297, 299, 304, 310–312; simultaneous consideration of, 171, 207–208, 233, 234; variability in rates of discharge of, 182–183, 285–286, 290; variables affecting, 279–280, 289–296, 316, 317. *See also* Air quality; Recovery of residual materials; Water quality
Residuals management and control: of auto emissions, 124–125, 132n, 155, 160, 442, 444, 446; capacity constraints on, 268; costs of, 48, 61, 152–154, 157, 240–242, 245, 262–264, 268, 272–273, 299–307, 443; feedback effects of, 254–255, 259, 262, 266–267, 271, 272–273, 310, 316–317; by local authorities, 117–118, 445–446, 449–451; market for, 46–47, 49, 62; methods of, 62, 63, 107, 118–119, 144–150, 157, 204–206, 208, 214, 133–134, 235, 240–242, 244–245, 248, 269–270, 399, 443, 445; and mortality rates, 344, 345; multiplier effect of, 240–241, 242–243, 256–264, 266, 272, 273–279; policies concerning, 435–460; standards for, 442–444, 445. *See also* Discharge fees; Recovery of residual materials; Subsidies; Taxes
Resources for the Future, Inc., 1, 4, 171, 175, 181, 234, 275, 276, 280, 282, 288, 319, 322, 323, 324n
Ridker, R., 137
Rivers and Harbors Act of 1899, 122
River Authorities (U.K.), 450, 451, 452
Roberts, D. J., 89
Roberts, M., 322
Roe, D., 374
Rogers, P. P., 196n
Rohlich, G. A., 212n
Rolfe, S., 119n, 120
Rosen, J. B., 195–196n
Roskill Commission, and location of third London airport, 390–394
Royal Commission on Environmental Pollution, 440–441
Rubin, D., 375n

Ruhr Valley (W. Germany), 116
Russell, S., 3, 172, 176n, 191n, 195n, 206–207n, 216n, 234, 235, 258n, 275n, 277n, 280, 281, 282n, 290n, 297n, 302, 318

St. Louis, Mo., airshed: air pollution control in, 3, 240, 243–245; air pollution sources affecting, 242, 248–249, 252–254, 262, 268; daily pollution levels in, 331, 333, 334, 336, 349
Samuelson, P., 10–11, 23, 25, 53, 89
Sawyer, J. W., 282n
Scale: changes due to output restriction, 153–154; constant returns to, 12, 18; economies in residuals management, 110, 132
Seagraves, J., 117, 118n
Secretary of Health, Education and Welfare, 367
Secretary of the Interior, 367
Secretary of State for Local Government and Regional Planning (U.K.), 440
Secretary of Transportation, 367
Selectivity of supply: definition of, 80–81; of private goods, 80–81, 83, 87–88, 91; of public goods, 81, 82, 87, 91–92
Seskin, E., 4, 100, 109n, 326, 347–350
Sewer charges. *See* Discharge fees, to control water quality
Shadow prices: of capital, 34, 60; of consumption goods, 59; emission fees as, 266; of environmental quality, 34–35, 454; of labor, 60; of pollution control, 264–266, 274; of social capital, 10, 15–17, 19–22; of waste disposal services, 35
Sharman, F. A., 392n
Small numbers: and lack of markets, 72–93; and private negotiations, 96
Smith, R., 207n
Smith, S. C., 279n
Smith, V. L., 89n
Sokolnikoff, I. S., 190n
Speas, R. Dixon, Associates, 367, 368n, 369
Spofford, W. O., 3, 172, 191n, 195n, 216n, 234, 235, 238n, 318n
Sporn, P., 322
Starrett, D., 402n
Stauffer, T. R., 322
Steady states: conditions for, in growth models, 27, 36, 37–41; of environmental quality, 31, 180, 187–189
Streeter, H. W., 184, 192, 213
Streeter-Phelps equation, 31
Strotz, R., 59n
Subsidies: compared with direct controls, 105n; compared with effluent charges, 105, 120; effectiveness of, 120–121, 125, 459; funds for payment of, 61; to reduce polluting activities, 105, 120–121; tax credits as, 120–121; for waste treatment, 35–36, 60, 61, 97, 120–121. *See also* Price incentives
Substitution: among factors of production, 12, 118, 119, 154, 164, 276, 280; among fuels, 118, 123, 153; between public and private goods, 25; in consumption, 118, 119; timing of, 155

Taxes: to adjust transfrontier damages, 421, 423, 425, 426–427; credits on, 120–121; on effluents or emissions (*see* Discharge fees); on fuels, 150; as function of amount of pollution, 59–60; head taxes, 59, 61; as incentive to reduce pollution, 97; income, 2, 50, 51, 54
Taylor series, 38, 196
Technical change: effect on water consumption, 276; in growth models, 43; to increase rejectability, 94; increasing returns to, 61; induced by external effects, 405–406; in residuals management, 256, 306–308, 317, 318
Technical specifications, to control waste emissions, 123n
Teller, A. A., 239n, 248n
Tennessee Valley Authority, 311n
Thomann, R. V., 184n, 187n
Threshold effects: in environmental assimilation, 99, 100, 101; and environmental quality controls, 108
Tideman, N., 133, 159
Tikvart, Martin-, dispersion model, 182
Time lags: in health effects of air pollution, 326, 328–329, 330–331, 333, 340, 341, 344, 345, 347; in implementing antipollution measures, 103, 106, 107, 109, 126
Time preference, of governments, 41–42

Tolley, G., 3, 163–167
Transaction costs. *See* Costs, transaction
Transformation rates, between public and private goods, 25
Trent River, 451
Tucker, Kuhn-, theorem, 30, 37
Turvey, R., 133

Uncertainty of environmental conditions. *See* Exogenous environmental factors
Union Electric Co., 254n
United Nations Stockholm Conference on the Human Environment, 436
United States Atomic Energy Commission, 438
United States Bureau of the Census, 276
United States Department of Commerce, 231
United States Congress, 317
United States Geological Survey, 209, 213n
United States Senate Select Committee on Water Resources, 275
United States Water Resources Council, 438–439
Universities-National Bureau Committee for Economic Research, 1
Upton, C., 102n, 133
Utility levels: effect of externalities on, 401, 408–409; as function of consumption, 74; as function of environmental quality, 27, 31–32, 57, 60, 136–137; as function of private consumption and social capital, 13–14; as function of production of public goods, 65–66; maximization of, 15–17, 19, 23, 32–33, 37, 59, 74, 136–137; over time, 17; and political processes, 456, 457–458
Uttormark, P. D., 212n
Uzawa, H., 2, 11, 16, 18, 22, 23, 26

Vandermeulen, D. C., 57
Vaughn, W. J., 176n, 282n, 290n
Veldhuizen, H., 312n
Vickrey, W., 104n

Wage rates: effect of air quality on, 141; in imperfect markets, 27, 35, 52–53
Walras's law, 48, 52
Washington, D.C., air quality in, 331, 333
Waste disposal and treatment. *See* Residuals management and control
Waste products. *See* Residual materials
Water consumption, 117–118, 175, 275–276, 277, 279
Water Pollution Control Act, 317–318, 447, 449
Water quality: and control policies, 93, 116–120, 447–453; exogenous factors affecting, 185, 212–213; models of changes in, 183–187, 214, 235; regional problems of, 171, 172, 447–448, 449, 450–452; residuals affecting, 117–118, 183–186, 202–212, 222–223; standards for, 184, 219–220, 447–448, 450; treatment options to improve, 174, 204–206
Weisbrod, B., 455n
Western Europe, environmental policy in, 427–438. *See also* Environmental policy, in U.K.
Westphal, L., 397n
White Papers (U.K.): on central government reorganization, 440; on protection of the environment, 440
Williams, R. B., 184–185n
Wilmington, Delaware, wastewater treatment facility, 207
Wilson, J. Q., 351n
Wold-Chamberlain field (Minneapolis-St. Paul), 350–381 passim
Wolozin, H., 133
Wright, C., 133

Young, R. A., 176n

Zangwill, W. I., 192–193n
Zerbe, R., 133, 160
Zwick, D., 121n